Lecture Notes in Computer Science

Lecture Notes in Computer Science

Vol. 270: E. Börger (Ed.), Computation Theory and Logic. IX, 442 pages. 1987.

Vol. 271: D. Snyers, A. Thayse, From Logic Design to Logic Programming. IV, 125 pages. 1987.

Vol. 272: P. Treleaven, M. Vanneschi (Eds.), Future Parallel Computers. Proceedings, 1986. V, 492 pages. 1987.

Vol. 273: J.S. Royer, A Connotational Theory of Program Structure. V, 186 pages. 1987.

Vol. 274: G. Kahn (Ed.), Functional Programming Languages and Computer Architecture. Proceedings. VI, 470 pages. 1987.

Vol. 275: A.N. Habermann, U. Montanari (Eds.), System Development and Ada. Proceedings, 1986. V, 305 pages. 1987.

Vol. 276: J. Bézivin, J.-M. Hullot, P. Cointe, H. Lieberman (Eds.), ECOOP '87. European Conference on Object-Oriented Programming. Proceedings. VI, 273 pages. 1987.

Vol. 277: B. Benninghofen, S. Kemmerich, M.M. Richter, Systems of Reductions. X, 265 pages. 1987.

Vol. 278: L. Budach, R.G. Bukharajev, O.B. Lupanov (Eds.), Fundamentals of Computation Theory. Proceedings, 1987. XIV, 505 pages. 1987.

Vol. 279: J.H. Fasel, R.M. Keller (Eds.), Graph Reduction. Proceedings, 1986. XVI, 450 pages. 1987.

Vol. 280: M. Venturini Zilli (Ed.), Mathematical Models for the Semantics of Parallelism. Proceedings, 1986. V, 231 pages. 1987.

Vol. 281: A. Kelemenová, J. Kelemen (Eds.), Trends, Techniques, and Problems in Theoretical Computer Science. Proceedings, 1986. VI, 213 pages. 1987.

Vol. 282: P. Gorny, M.J. Tauber (Eds.), Visualization in Programming. Proceedings, 1986. VII, 210 pages. 1987.

Vol. 283: D.H. Pitt, A. Poigné, D.E. Rydeheard (Eds.), Category Theory and Computer Science. Proceedings, 1987. V, 300 pages. 1987.

Vol. 284: A. Kündig, R.E. Bührer, J. Dähler (Eds.), Embedded Systems. Proceedings, 1986. V, 207 pages. 1987.

Vol. 285: C. Delgado Kloos, Semantics of Digital Circuits. IX, 124 pages. 1987.

Vol. 286: B. Bouchon, R.R. Yager (Eds.), Uncertainty in Knowledge-Based Systems. Proceedings, 1986. VII, 405 pages. 1987.

Vol. 287: K.V. Nori (Ed.), Foundations of Software Technology and Theoretical Computer Science. Proceedings, 1987. IX, 540 pages. 1987.

Vol. 288: A. Blikle, MetaSoft Primer. XIII, 140 pages. 1987.

Vol. 289: H.K. Nichols, D. Simpson (Eds.), ESEC '87. 1st European Software Engineering Conference. Proceedings, 1987. XII, 404 pages. 1987.

Vol. 290: T.X. Bui, Co-oP A Group Decision Support System for Cooperative Multiple Criteria Group Decision Making. XIII, 250 pages. 1987.

Vol. 291: H. Ehrig, M. Nagl, G. Rozenberg, A. Rosenfeld (Eds.), Graph-Grammars and Their Application to Computer Science. VIII, 609 pages. 1987.

Vol. 292: The Munich Project CIP. Volume II: The Program Transformation System CIP-S. By the CIP System Group. VIII, 522 pages. 1987.

Vol. 293: C. Pomerance (Ed.), Advances in Cryptology – CRYPTO '87. Proceedings. X, 463 pages. 1988.

Vol. 294: R. Cori, M. Wirsing (Eds.), STACS 88. Proceedings, 1988. IX, 404 pages. 1988.

Vol. 295: R. Dierstein, D. Müller-Wichards, H.-M. Wacker (Eds.), Parallel Computing in Science and Engineering. Proceedings, 1987. V, 185 pages. 1988.

Vol. 296: R. Janßen (Ed.), Trends in Computer Algebra. Proceedings, 1987. V, 197 pages. 1988.

Vol. 297: E.N. Houstis, T.S. Papatheodorou, C.D. Polychronopoulos (Eds.), Supercomputing. Proceedings, 1987. X, 1093 pages. 1988.

Vol. 298: M. Main, A. Melton, M. Mislove, D. Schmidt (Eds.), Mathematical Foundations of Programming Language Semantics. Proceedings, 1987. VIII, 637 pages. 1988.

Vol. 299: M. Dauchet, M. Nivat (Eds.), CAAP '88. Proceedings, 1988. VI, 304 pages. 1988.

Vol. 300: H. Ganzinger (Ed.), ESOP '88. Proceedings, 1988. VI, 381 pages. 1988.

Vol. 301: J. Kittler (Ed.), Pattern Recognition. Proceedings, 1988. VII, 668 pages. 1988.

Vol. 302: D.M. Yellin, Attribute Grammar Inversion and Source-to-source Translation. VIII, 176 pages. 1988.

Vol. 303: J.W. Schmidt, S. Ceri, M. Missikoff (Eds.), Advances in Database Technology – EDBT '88. X, 620 pages. 1988.

Vol. 304: W.L. Price, D. Chaum (Eds.), Advances in Cryptology – EUROCRYPT '87. Proceedings, 1987. VII, 314 pages. 1988.

Vol. 305: J. Biskup, J. Demetrovics, J. Paredaens, B. Thalheim (Eds.), MFDBS 87. Proceedings, 1987. V, 247 pages. 1988.

Vol. 306: M. Boscarol, L. Carlucci Aiello, G. Levi (Eds.), Foundations of Logic and Functional Programming. Proceedings, 1986. V, 218 pages. 1988.

Vol. 307: Th. Beth, M. Clausen (Eds.), Applicable Algebra, Error-Correcting Codes, Combinatorics and Computer Algebra. Proceedings, 1986. VI, 215 pages. 1988.

Vol. 308: S. Kaplan, J.-P. Jouannaud (Eds.), Conditional Term Rewriting Systems. Proceedings, 1987. VI, 278 pages. 1988.

Vol. 309: J. Nehmer (Ed.), Experiences with Distributed Systems. Proceedings, 1987. VI, 292 pages. 1988.

Vol. 310: E. Lusk, R. Overbeek (Eds.), 9th International Conference on Automated Deduction. Proceedings, 1988. X, 775 pages. 1988.

Vol. 311: G. Cohen, P. Godlewski (Eds.), Coding Theory and Applications 1986. Proceedings, 1986. XIV, 196 pages. 1988.

Vol. 312: J. van Leeuwen (Ed.), Distributed Algorithms 1987. Proceedings, 1987. VII, 430 pages. 1988.

Vol. 313: B. Bouchon, L. Saitta, R.R. Yager (Eds.), Uncertainty and Intelligent Systems. IPMU '88. Proceedings, 1988. VIII, 408 pages. 1988.

Vol. 314: H. Göttler, H.J. Schneider (Eds.), Graph-Theoretic Concepts in Computer Science. Proceedings, 1987. VI, 254 pages. 1988.

Vol. 315: K. Furukawa, H. Tanaka, T. Fujisaki (Eds.), Logic Programming '87. Proceedings, 1987. VI, 327 pages. 1988.

Vol. 316: C. Choffrut (Ed.), Automata Networks. Proceedings, 1986. VII, 125 pages. 1988.

Vol. 317: T. Lepistö, A. Salomaa (Eds.), Automata, Languages and Programming. Proceedings, 1988. XI, 741 pages. 1988.

Vol. 318: R. Karlsson, A. Lingas (Eds.), SWAT 88. Proceedings, 1988. VI, 262 pages. 1988.

Vol. 319: J.H. Reif (Ed.), VLSI Algorithms and Architectures – AWOC 88. Proceedings, 1988. X, 476 pages. 1988.

Vol. 320: A. Blaser (Ed.), Natural Language at the Computer. Proceedings, 1988. III, 176 pages. 1988.

Vol. 321: J. Zwiers, Compositionality, Concurrency and Partial Correctness. VI, 272 pages. 1989.

Vol. 322: S. Gjessing, K. Nygaard (Eds.), ECOOP '88. European Conference on Object-Oriented Programming. Proceedings, 1988. VI, 410 pages. 1988.

Vol. 323: P. Deransart, M. Jourdan, B. Lorho, Attribute Grammars. IX, 232 pages. 1988.

Lecture Notes in Computer Science

Edited by G. Goos and J. Hartmanis

362

Björn Lisper

Synthesizing Synchronous Systems by Static Scheduling in Space-Time

Springer-Verlag

Berlin Heidelberg New York London Paris Tokyo Hong Kong

Author

Björn Lisper
Department of Telecommunication and Computer Systems
The Royal Institute of Technology
S-100 44 Stockholm, Sweden

CR Subject Classification (1987): C.5.4, F.1.2

ISBN 3-540-51156-3 Springer-Verlag Berlin Heidelberg New York
ISBN 0-387-51156-3 Springer-Verlag New York Berlin Heidelberg

Printed in Germany

Printing and binding: Druckhaus Beltz, Hemsbach/Bergstr.
2145/3140-543210 – Printed on acid-free paper

Preface

This LNCS edition is, in some sense, the second edition of this thesis. The first one, defended for the doctoral degree in February 1987, was released as a research report at the Department of Numerical Analysis and Computer Science, the Royal Institute of Technology, Stockholm, Sweden. When it was clear that Springer-Verlag would publish the thesis in its Lecture Notes in Computer Science series, I seized the opportunity to improve the original manuscript. Thus, the following changes have been made:

- Some of the notation has been altered, in order to adhere to generally accepted standards. Thus, the term "multiple-assignment" has been replaced with "concurrent single-assignment" which I feel is more descriptive. Furthermore, ordinary function notation is now used for application of substitutions. This was previously denoted by the symbol "/".
- The definition of input and output functions of substitution fields has been changed to allow negative time delays. This makes no difference whatsoever in the interesting cases, but it removes some slight inconsistencies that were present between the definitions in chapter 8. Some proofs in this chapter had to be minimally altered in order to cover the introduced new cases.
- A few lemmas in the appendices were found to follow more or less directly from results already proven in the literature. For these lemmas the given proofs have been replaced by appropriate references. Some other proofs were unnecessarily complicated and have been simplified.
- A chapter with conclusions has been added.
- The reference list has been updated, in order to include relevant papers that have occurred since the first edition.
- Various misprints and typing errors have been fixed. Linguistic errors have been corrected. The style of writing has, hopefully, been somewhat improved.

Stockholm, February 1989 Björn Lisper

Abstract

A framework for synthesis of synchronous concurrent systems with local memory is developed. Given an output specification of the system, a *cell action structure*, which is a fixed set of atomic events with data dependencies, can be derived. The events are computation steps where output values are computed from inputs. This set can be explicitly mapped into a *space-time*, so that the data dependencies are embedded in a *communication ordering*. The space-time is a model of the events in the target hardware and the communication ordering represents constraints on the communication possible between events.

Cell action structure, mapping and space-time together constitute a description of the resulting execution pattern. From this description, a description of the minimal fixed hardware required to support the execution pattern can be derived. The hardware description is a directed labelled graph, where the nodes represent processing elements (cells) and edges communication lines with discrete delays. Various performance criteria are also defined, by which different execution patterns for the same cell action structure can be compared.

Transformation of space-time is also considered. By such transformations new hardware architectures for a certain cell action structure can be derived from old ones.

The range of synchronous systems possible to synthesize in the way described in this thesis spans from simple systolic arrays to more complicated configurations like n-cube and perfect shuffle networks.

Key words: Synchronous systems, parallel processing, systolic arrays, computational events, cell actions, recurrences, space-time schedules, index vectors

Table of Contents

1. Introduction

This thesis deals with the issue of parallel systems: how to synthesize a parallel architecture for a certain problem to be solved, and how to utilize a given parallel architecture for the same thing. We will consider *synchronous*, tightly coupled distributed systems only. These consist of processing elements, or *cells*, connected to other cells by clocked communication lines. Examples of such systems are *systolic arrays*, [KLe80], pipelined array processors (where each pipeline stage is a "cell"), synchronous boolean n-cubes and others.

Much of the thesis is devoted to the development of a mathematical framework where it is possible to express facts about such systems and about how computations are carried on them. Thus, we must also formulate what we mean by computing and how the task of carrying out a certain computation is tied to a certain architecture.

Computing is an issue of classical automata theory, but this theory is not suitable for our purposes. The reason is that automata theory is strictly serial in its nature. A classical automaton is a mathematical model of the von Neumann machine which consists of a serial processor and a global, limited memory. The global memory is modelled by a finite set of states and the serial processor is modelled by considering strict sequences of state transitions only (or transitions from sets of states to sets of states, if nondeterministic automata are considered). This seriality makes classical automata highly inconvenient as a means of describing the parallel execution of algorithms.

It is desirable then to define what we mean by computing, in a way less related to a specific architecture and more related to the actual problem to be solved. The obvious choice is to consider functional semantics; a computation is to evaluate an output function (or several functions) for some given inputs. It is well-known that this makes parallelity possible. In a purely applicative language, for instance, all arguments of a function can be evaluated in parallel since they are independent of each other. We are, however, *not* going to consider applicative parallelity. The reason is that applicative evaluation is based on a demand-driven approach. The evaluation proceeds in a top-down fashion from the top function application down through the computational tree to the leaves: the input variables and constants which have no arguments to evaluate. Instead we are interested in a data-driven approach, where the evaluation starts at the bottom by making constants and input variables available, and then proceeds upwards by applying functions to their evaluated arguments.

The data-driven approach is preferred since we want to describe and synthesize systems which are clearly data-driven or can support data-driven executions. An execution on such a system starts by making input variables and constants available to different cells. It then proceeds through cells computing intermediary values, which are sent over the communication lines to other cells. These cells in turn use the values as arguments for further function applications. In this way the computation goes on until the final output values are calculated; these values will be the values of functions built up recursively from the functions the cells have applied to the intermediary values.

Some efforts have been done to find suitable mathematical tools for dealing with systems of this type. One approach is to model the clocked, distributed hardware by a labelled directed multigraph, where the nodes represent cells and the edges represent communication lines, [MR82,Me83]. By labelling each edge with a sequence of variables and each node with a sequence of function applications it is possible to derive a system of equations, which can be solved for unknown sequences of variables. If the hardware system is expected to deliver the values of the output functions at certain lines at certain times, the correctness of the system can be verified by checking the solution for the corresponding variables in the sequences for these lines.

Another approach is to describe the action of the system by a *space-time algorithm*, [C83]. A space-time algorithm is a system of recursion equations where each equation is labelled by a space-time coordinate. Each point in space corresponds to a cell. An equation having a space-time coordinate means that the cell at the corresponding space point performs the assignment defined by the recursion equation at the time given by the coordinate. The behaviour of the space-time algorithm is defined as the least fixed-point solution of the system of recursion equations; the result is a function from space-time to values, that tells for each space-time coordinate which function is being evaluated there.

Both models above have in common that they are *descriptive*, they can be used to model existing systems and how functions are evaluated by such systems. Thus, they can be used to verify that the system evaluates the functions correctly. It is, however, not possible to model the process of *synthesis*, that is: to *derive* a system that evaluates given functions, by these mathematical tools. Synthesis is to a limited extent possible using a third, *transformational* approach, [JWCD81,KLi83,WD81,W81]. Here, there is again an underlying graph model where the nodes represent functional units and the edges represent communication lines. As before there is a time sequence of variables associated with each edge. The new feature introduced is a *delay operator*. This can be seen as a functional unit that delivers its input sequence delayed one time step as output sequence. By adding and removing delay elements, according to certain transformation rules, a given system can be transformed into a new one while leaving the output functions invariant.

Instead of having delay units as nodes, each edge in the graph can be labelled with an integer-valued delay. Using this formulation a general theorem has been proved, which states that any "semisystolic" system with some zero delay lines can be transformed into a "systolic" system where all edges have a delay greater than zero (the systolic conversion theorem, [LS81]). This is an interesting result since zero delay lines usually should be avoided; in a clocked hardware system they correspond to lines unbroken by latches, and long such lines need more time and power to get charged or discharged. Therefore, the clock frequency can often be increased if such lines are avoided.

The transformational methods above provide a limited possibility of synthesis. They do not, however, solve the problem of how to find a starting point for the transformations, that is: an *initial* system that evaluates the output functions correctly. The basic problem is that the formulation is still essentially a description of the hardware, and what can be done is to find out what functions a given system is really computing. For the purpose of synthesis one needs a model that starts off *with the functions to be computed.* From this specification there should be a way to find methods, or algorithms, that tell *how* to compute the functions. *The algorithms should be formulated independently of any intended hardware*! First when an algorithm has been formulated, the connection between algorithm and hardware should be made. The result is a description of the resulting hardware and how the algorithm is carried out on it. Alternatively, if the intention is to use an existing architecture, a hardware description is given and the task is then to find a way to execute the algorithm on the architecture specified by the description. Finally, there must be ways to *evaluate* the resulting solution with respect to performance criteria, so that different solutions can be compared and the in some sense "best" solution can be selected. Thus, it must be possible to do the following in a mathematical framework suitable for synthesis:

- Give a system specification as the function(s) the system is to compute.
- Specify algorithms in a hardware-independent way, and check if an algorithm meets a given system specification (computes the desired functions).
- Specify the target hardware, either totally, which means that an existing architecture is going to be used, or partially, which means that constraints are given that the resulting hardware must fulfil.
- Connect algorithm and hardware in a correct way, giving a way to execute the algorithm on the given architecture. In the case of partially specified hardware there must also be a way to derive the actual resulting architecture.

- Formulate various performance criteria, by which different solutions to the same synthesis problem can be compared. Examples of such criteria are: total execution time, number of cells needed, average cell utilization, cell complexity, complexity of connection topology, "possibility of hardwiring" etc.
- Formulate constraints on the system as a whole, not only on the resulting hardware. For instance, "result x must be present at cell c at time t", or "total execution time should not be greater than t".

In this thesis a mathematical model of computation is developed which is aimed to be suitable for hardware synthesis. The approach is *event-oriented*, an algorithm is expressed as a set of events where each event is a single step in the algorithm. The events do not take place independently of each other, there are *precedence constraints* which means that certain events must precede others. To give readers unfamiliar with this kind of approach a flavour of the idea, a simple example from everyday life is presented. Those readers who are already familiar with event-oriented descriptions can safely skip it.

Example 1.1: Consider the task of switching a car wheel. This consists of several subtasks: The car must be lifted by a jack, the hub cap, the bolts and the old wheel removed, the new wheel put in place, the bolts fastened, the hub cap put back in place, and finally the car can be taken off the jack.

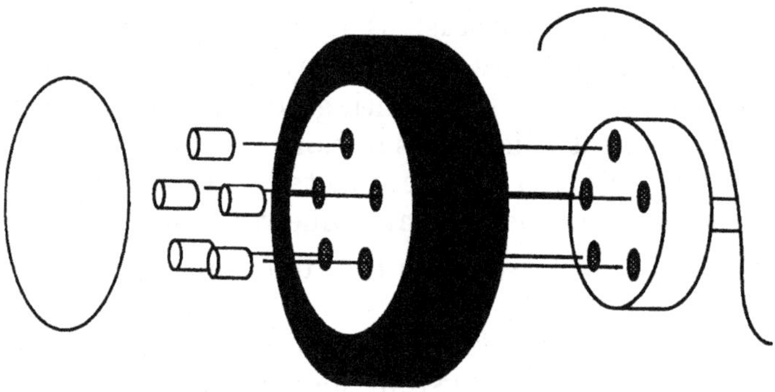

Figure 1.1: To switch a wheel

The carrying out of each such subtask can now be seen as an atomic event. Some events cannot take place before others: before the new wheel is put in place the old one must have been removed, for instance, and before the old wheel is removed the bolts must be removed. These precedence constraints can be presented in a diagram, where an arrow from event a to event b means that a must precede b. (Note that only the "direct" precedences are indicated. Indirect precedences of type "a must precede c since a must precede b and b must precede c" are not shown.)

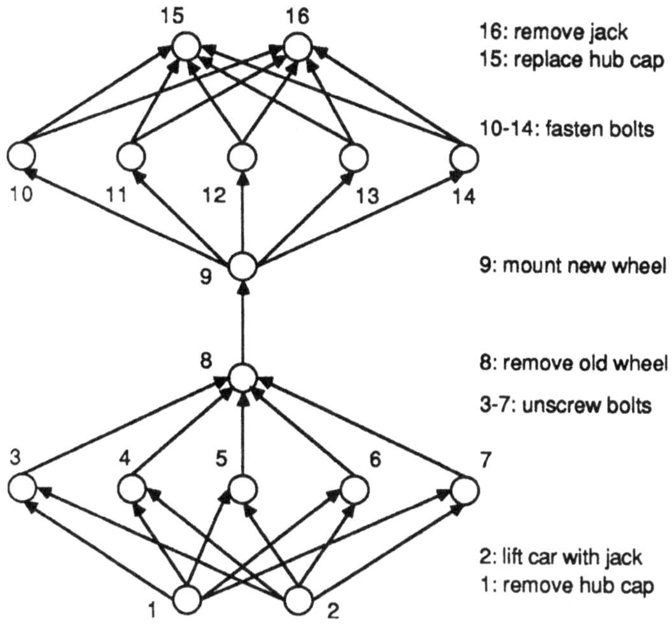

Figure 1.2: Precedence diagram for wheel switching

For convenience, the subtasks have been numbered from 1 to 16. Note that there are no cycles in the graph. If there were cycles, then there would be events that could never take place since they would have to be preceded by themselves. A diagram of the type above is essentially a PERT flowchart describing task dependencies, [Kn69].

Let us now make some assumptions, some of which are admittedly not entirely realistic. Assume first that each subtask, like removing a bolt, putting back a wheel etc. takes one time unit to perform. Assume further that we have access to an unlimited number of mechanics, numbered from 1 and upwards. Let us finally assume that the mechanics can switch instantly from performing one subtask to performing another one, and that they will never interfere with each other while performing subtasks simultaneously. Thus, it should for instance be possible for two mechanics to remove bolts at the same time without interfering with each other. Under these assumptions we can make a *schedule* of the subtasks, that tells each mechanic what to do for each time.

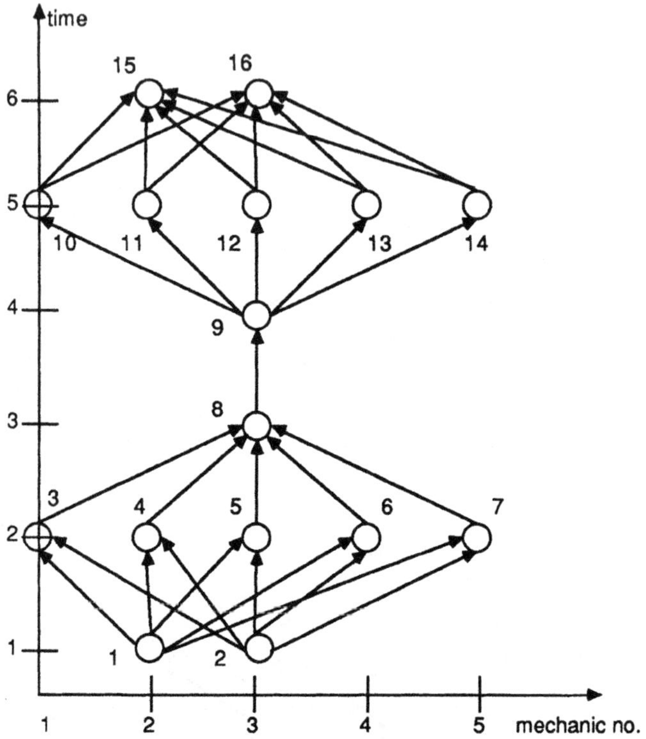

Figure 1.3: Schedule of wheel switching

Before the tasks are scheduled we introduce a *time*, that is common for all mechanics. Thus, we have a set of pairs of the form $\langle \text{time}, \text{mechanic} \rangle$, where $\langle t, m \rangle$ represents mechanic m at time t. The schedule is now given by a function f, that maps each subtask s to be done to a time-mechanic pair $\langle t, m \rangle$. $f(s) = \langle t, m \rangle$ means that subtask s is performed by mechanic m at time t. If there is no s such that $f(s) = \langle t, m \rangle$, then mechanic m is idle at time t. Note that the precedence constraint on the subtasks must be preserved into time. If s precedes s', then s must be mapped to a time less than s'. Note also that once f is determined, we know which mechanics that will ever have anything to do. In the given example these are the mechanics 1 to 5. The other mechanics (from 6 and upwards) can

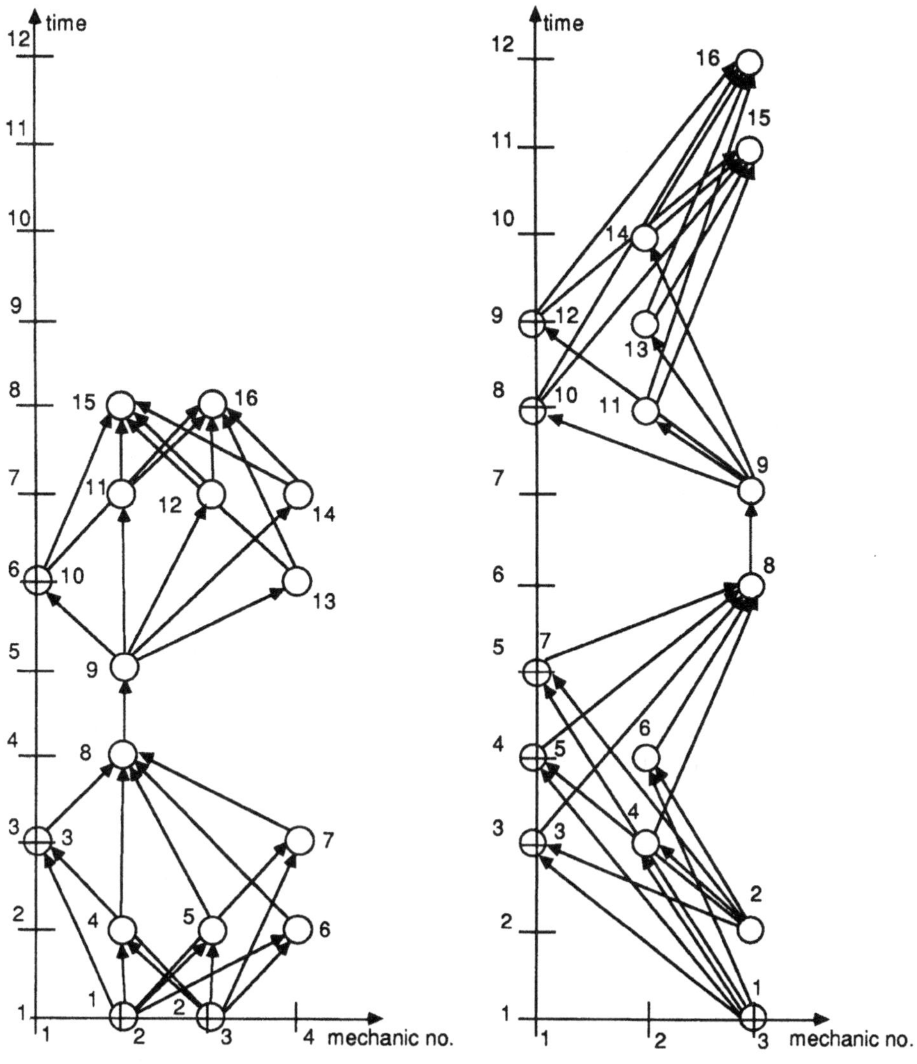

Figure 1.4: Another wheel switching schedule. A wheel switching schedule with limited mechanic capability.

be fired as well, without affecting our ability to switch wheels in the way given by f.

In general there are several possible schedules. Figure 1.4 gives an example of two more schedules for switching wheels. It is possible to put further constraints on the schedule, in order to reflect limitations of the resources that are available. Let us for instance consider a case where we have three mechanics only; moreover mechanics 1 and 2 can only fasten and remove bolts and mechanic 3 is only capable of handling the jack and the wheels. The second schedule in figure 1.4 meets these constraints.

The scheduling methodology described so far relies on one fundamental assumption: *it is assumed that we have total knowledge in advance about the subtasks and the precedence constraints between them.* Thus, the scheduling function can also be determined in advance. It is, however, quite possible to imagine cases when this assumption does not hold. Consider, for instance, the following scenario.

Assume that we may switch wheels on cars which previously had their wheels switched by very careless mechanics, who sometimes forgot to fasten some bolts. Assume further that the hub caps cover the bolts completely, so that it cannot be seen if a bolt is present or not without the hub cap being removed. *Now we cannot know exactly what set of subtasks that is to be scheduled until the hub cap is removed*, since we do not have to remove a bolt that is not there. So to achieve total information about what is to be done some subtasks must be performed first.

In the case of incomplete à priori information there are two possible approaches. The first one is to schedule all possible subtasks in advance, regardless of if they always are going to be performed or not. In the example above this approach would lead to schedules, where the mechanics always try to remove all the bolts even if some are not present. The second approach is to consider, instead of a single schedule, a *scheduling strategy* where different schedules are chosen depending on the information that is successively gathered while the subtasks are carried out. ■

Let us now assume that we are to carry out an algorithm instead of switching a wheel. Exactly as in the wheel-switching example the algorithm can be broken down into subtasks, or *cell actions*. These cell actions can be scheduled in the same way as the wheel-switching subtasks. The difference is that we do not have mechanics to carry out the cell actions; instead we will use processors, or cells. The set of possible locations for cells will be called a *space* and the set of pairs of cell locations and times will be called a *space-time*. The schedule is then expressed as an explicit mapping from cell actions to space-time.

Several attempts have been made to describe the parallel schedule of algorithms as a mapping of cell actions (or corresponding objects) into a space-time. The earliest work in the area known to the author is the work concerned with parallellizing FORTRAN DO-loops for execution on multiprocessor systems, [KMC72,La74,Mu71]. Later it was discovered that the methods developed there could be applied to synthesis of regular hardware structures, [MiWi84,Mo82]. In these works the event structure is somewhat obscured by the fact that the semantics is given by sequential constructs in an imperative language, like FORTRAN or Pascal. Other papers give a more event-oriented description of the algorithm to be performed, as systems of recurrence equations or concurrent assignments [CS83, C85A, C85B, C86A, C86B, C86C, DI86, JRK87, Li83, Q83, QG85, Raj86, RK88], or using a more graph-theoretic approach [Ku86,Le83,Le84,RFS82A,RFS86].

In this thesis the development will be made according to the outline sketched here. We will thus let the specification of a system be a set of functions from inputs to outputs. We will study how these functions can be computed using a set of predefined "primitive" functions, assumed to have known implementations. The steps in the computations can be arranged into the previously mentioned cell actions. These are computations suitable to perform by one cell in one time step. There will be precedence constraints; some cell actions produce output values that are used as inputs by other cell actions, and then the "producer" must execute before the "consumer". Cell actions and precedence constraints together form our notion of an "algorithm".

The desired structure of the hardware will be described by a space-time augmented by a *communication ordering*, that describes the communication that can take place between different events in space-time. The implementation of an algorithm is now expressed as an explicit mapping from cell actions to space-time, such that the communication requirements given by the precedence constraints for the cell actions do not violate the communication constraints given by the communication ordering.

The result of this procedure is a schedule of cell actions. This schedule tells for each cell what it will do at each time. If the communication ordering specifies the communication possibilities of the system completely, then we have obtained an implementation of the algorithm on existing hardware. If not, it is possible to find a minimal communication ordering that supports the communication requirements. This corresponds to synthesis.

The organization of the thesis is as follows: chapter two contains a compilation of the notation and the definitions of basic mathematical concepts that are used throughout the thesis. In chapters three to six, a computational model is developed that is suitable for our purposes. Chapter seven contains a definition of a hardware model and it also treats how to link algorithms with hardware. In chapter eight some performance criteria for schedules of algorithms on hardware structures are defined. In order to formulate some of these criteria a more detailed hardware model has to be developed. This model turns out to be interesting by itself. Chapter nine treats how to utilize a regular algorithm structure to reduce the complexity of the scheduling task by the aid of *index vectors* and *linear mappings*. The techniques developed in this chapter turn out to be useful in practical situations. Chapter ten contains an attempt to extend the computational model to cover data-dependent executions, where the schedule may be altered depending on information that is not available in advance. In chapter eleven, finally, some examples of applications are given. These examples range from synthesis of systolic arrays for algorithms in linear algebra to schedules of recursive-partitioning type algorithms, like the Fast Fourier Transform, on such topologies as the n-cube and the perfect shuffle network.

The appendices are of two different types. Appendices A to D contain lemmas about set closures, binary relations, composition of functions of several variables and substitutions. These results are of a general nature and their proofs are included here for completeness. Appendices E to L contains proofs of lemmas, propositions and theorems in the text.

2. Preliminaries

2.1 Basic notation

The following notation will be used throughout this thesis.

predicate logic:

$[\ldots]$	quantifier range
$\forall a \in A[p(a)]$	$\forall a[a \in A \implies p(a)]$
$\exists a \in A[p(a)]$	$\exists a[a \in A \wedge p(a)]$

sets:

N	the natural numbers (non-negative integers)
Z	the set of integers
Z^+	the positive integers
R	the real numbers
$\lvert A\rvert$	the cardinality of the set A
$P(A)$	the power set (set of all subsets) of A
i_A	the identity function on A
$A \setminus B$	$\{ a \mid a \in A \wedge a \notin B \}$, the set difference of A and B

(partial) functions:

$f(A)$	$\{ f(a) \mid a \in A \}$, the image of A under f.
$f\vert_{A'}$	$\{ \langle a,b \rangle \mid a \in A' \wedge \langle a,b \rangle \in f \}$, f restricted to A'.

families, Cartesian product, tuples:

Let I be a countable set and let A be any set. A *family* $(a_i \mid i \in I)$ of elements in A is a mapping $\varphi: I \to A$ where $a_i = \varphi(i)$ for all $i \in I$. I is called the *index set* of the family $(a_i \mid i \in I)$. The image of I under φ is denoted by $\{ a_i \mid i \in I \}$. If $(A_i \mid i \in I)$ is a family of sets, then the union and intersection of the elements of $\{ A_i \mid i \in I \}$ will be denoted by

$$\bigcup(A_i \mid i \in I)$$

and

$$\bigcap(A_i \mid i \in I)$$

respectively. We define $\bigcup(A_i \mid i \in \emptyset) = \emptyset$. If I is nonempty, then the *Cartesian product*

$$\prod(A_i \mid i \in I)$$

is defined as the set of all functions $f: I \to \bigcup(A_i \mid i \in I)$, for which $f(i) \in A_i$ for all $i \in I$. The *tuple* $\langle a_i \mid i \in I \rangle$ is the element f in $\prod(A_i \mid i \in I)$, for which $f(i) = a_i$ for all $i \in I$. For finite index sets $I = \{i_1, \ldots, i_n\}$ we might sometimes use the conventional tuple notation $\langle a_{i_1}, \ldots, a_{i_n} \rangle$. $\prod(A_i \mid i \in \emptyset)$ is considered to contain the *empty tuple* $\langle\rangle$ as its single element. Since a tuple is a function, it holds for all $I' \subseteq I$ that $\langle a_i \mid i \in I \rangle|_{I'} = \langle a_i \mid i \in I' \rangle$. Clearly $\langle a_i \mid i \in I \rangle|_{I'}|_{I''} = \langle a_i \mid i \in I \rangle|_{I''}$ when $I'' \subseteq I' \subseteq I$.

For functions $g: \prod(A_i \mid i \in I) \to A$, the notation $g(a_i \mid i \in I)$ will be used for the value in A, that g assigns to $\langle a_i \mid i \in I \rangle$. $g(A_i \mid i \in I)$ denotes the image of $\prod(A_i \mid i \in I)$ under g.

For all $i \in I$, we define the mapping $e_i^I: \prod(A_k \mid k \in I) \to A_i$ by

$$e_i^I(a_k \mid k \in I) = a_i.$$

e_i^I is called a *projection*. Cf. [G79].

The following notation will be used for *function composition*; let g be a function from $\prod(A_k \mid k \in I)$ to A, and for all $k \in I$ let g_k be a function from $\prod(A_i \mid i \in I_k)$ to A_k. Then, if $I' = \bigcup(I_k \mid k \in I)$, for all $a \in \prod(A_i \mid i \in I')$ we define

$$g \circ \langle g_k \mid k \in I \rangle (a) = g(g_k(a|_{I_k}) \mid k \in I).$$

Thus, $g \circ \langle g_k \mid k \in I \rangle$ denotes g composed with the functions g_k, $k \in I$, which is a function from $\prod(A_i \mid i \in I')$ to A. If I has a single element k, then we will write $g \circ g_k$ for $g \circ \langle g_k \rangle$. We also define $g \circ \langle g_k \mid k \in I' \rangle$ when $I' \subseteq I$: for all $k \in I$, set

$$g_k^I = \begin{cases} g_k, & k \in I' \\ e_k^I, & k \notin I'. \end{cases}$$

Then $g \circ \langle g_k \mid k \in I' \rangle = g \circ \langle g_k^I \mid k \in I \rangle$.

Equality of tuples: tuples, or elements of a Cartesian product, are as mentioned special functions from an index set to some set; equality between tuples therefore means equality between functions. In tuple notation, $\langle a_i \mid i \in I \rangle = \langle b_i \mid i \in I' \rangle$ if and only if $I = I'$ and $a_i = b_i$ for all $i \in I$.

binary relations (Let R, R' be binary relations over a set A, let $A' \subseteq A$):

$RR' = \{ \langle a, a' \rangle \mid \exists a'' \in A[aRa'' \wedge a''R'a'] \}$	relation composition
$R^0 = \omega_A = \{ \langle a, a \rangle \mid a \in A \}$	
$R^n = RR^{n-1}$, all $n \in Z^+$	n-th power of R
$R^+ = \bigcup(R^n \mid n \in Z^+)$	transitive closure
$R^* = R^+ \cup R^0$	transitive and reflexive closure
$R\|_{A'} = \{ \langle a, a' \rangle \mid a, a' \in A' \wedge \langle a, a' \rangle \in R \}$	restriction of R to A'

We will write $sp(a, R)$ for $\{ a' \mid a'Ra \}$. R is called *finite-downward* iff R is a strict partial order (s.p.o) and if $sp(a, R)$ is finite for all $a \in A$.

equivalence relations and partitions:

A/ε	the partition given by the equivalence relation ε
$[a]\varepsilon$	the block in A/ε containing a
ε_f	the equivalence relation induced by f
$\varphi_{A/\varepsilon}: A \to A/\varepsilon$	the natural mapping $A \to A/\varepsilon$, $\varphi_{A/\varepsilon}(a) = [a]\varepsilon$ when $a \in A$

2.2 Non-strict functions

We now give a definition of non-strict functions. Cf. [PA85]. Note that we do not introduce the concept of undefined data elements in our definition. In the following we consider functions $f: \prod(A_i \mid i \in I) \to A_f$.

Definition 2.1: Let $I' \subseteq I$ and let $b \in \prod(A_i \mid i \in I')$. Then $f|_b: \prod(A_i \mid i \in I \setminus I') \to A_f$ is given by:

$$\forall a \in \prod(A_i \mid i \in I)[a|_{I'} = b \implies f|_b(a|_{I\setminus I'}) = f(a)]$$

$f|_b$ is simply the restricted function that remains when the constant elements in b are substituted for the corresponding arguments of f.

Definition 2.2: Let $I' \subseteq I$. f is *independent of* I' iff $f|_b = f|_{b'}$ for all $b, b' \in \prod(A_i \mid i \in I')$.

Definition 2.3: f is *non-strict* iff there exists an $I' \subset I$, a $b \in \prod(A_i \mid i \in I')$ and a nonempty $I'' \subseteq I \setminus I'$ such that $f|_b$ is independent of I''.

This definition captures what was stated less formally above; if we substitute the (non-empty) constant tuple b for the corresponding arguments to f, then the remaining function $f|_b$ becomes independent of its arguments indexed by I''. Of course, a *strict* function is a not non-strict function. One non-strict function is of particular interest to us:

Definition 2.4: For every value domain D, $if_D: \{true, false\} \times D \times D \to D$ is defined by:

$$\forall x, y \in D[if_D(true, x, y) = x \wedge if_D(false, x, y) = y].$$

This function becomes independent of its second or third argument, respectively, when its first argument is *false* or *true*.

2.3 Basic definitions and properties concerning heterogeneous algebras

In this thesis, the computational model that is used will be formulated using concepts of universal, heterogeneous algebra. Therefore, we give some basic definitions concerning such algebras. Cf. [BL70,G79].

Definition 2.5: An *algebra* $\mathcal{A}$ is a pair $\langle \mathcal{S}; F \rangle$ where:

1. $\mathcal{S}$ is a nonempty set of sets. Each element in $\mathcal{S}$ is called a *phylum*, or *sort*, of $\mathcal{A}$.
2. F is a set of finitary operators. For each $f \in F$, there is a finite index set I_f such that $f \notin I_f$. Every $f \in F$ is a mapping

$$f: \prod(S_{if} \mid i \in I_f) \to S_f$$

where, for all $i \in I_f$, holds that $S_{if} \in \mathcal{S}$ and $S_f \in \mathcal{S}$.

$n(f) = |I_f|$ is the *arity* of f. Note that $I_f = \emptyset$ is allowed, in which case f is a mapping $\{\langle\rangle\} \to S_f$, i. e., a constant of type S_f. This kind of mapping is also called a *nullary operator*. S_f is called the *sort* of f. If $\mathcal{S}$ is a singleton, then the algebra is said to be *homogeneous*, otherwise it is *heterogeneous*.

Sometimes we will add the unary identity functions, on the sorts of an algebra, as operators to the algebra. This yields a new algebra, given by the following definition:

Definition 2.6: Let $\mathcal{A} = \langle \mathcal{S}; F \rangle$ be an algebra. Then the *identity closure* of $\mathcal{A}$ is $i(\mathcal{A}) = \langle \mathcal{S}; F \cup \{ i_S \mid S \in \mathcal{S} \} \rangle$.

In the following we will be interested in functions built up recursively using the operators of some algebra. Such functions are called *polynomials* (cf. [G79]):

Definition 2.7: Let $\mathcal{A} = \langle \mathcal{S}; F \rangle$ be an algebra. Let $(A_x \mid x \in X)$ be a family of sets, such that $X \cap F = \emptyset$ and $\{ A_x \mid x \in X \} \subseteq \mathcal{S}$. $\mathcal{P}^{(X)}(\mathcal{A})$ is the set of *polynomials in X over $\mathcal{A}$*, and $S: \mathcal{P}^{(X)}(\mathcal{A}) \to \mathcal{S}$ is the *sort function* of $\mathcal{P}^{(X)}(\mathcal{A})$. They are defined by:

1. The projections e_x^X, for all $x \in X$, belong to $\mathcal{P}^{(X)}(\mathcal{A})$, and $S(e_x^X) = A_x$.
2. For all $f \in F$, if $p_i \in \mathcal{P}^{(X)}(\mathcal{A})$ and $S(p_i) = S_{if}$ for all $i \in I_f$, then $f \circ \langle p_i \mid i \in I_f \rangle$ is an element of $\mathcal{P}^{(X)}(\mathcal{A})$ and $S(f \circ \langle p_i \mid i \in I_f \rangle) = S_f$.
3. The elements of $\mathcal{P}^{(X)}(\mathcal{A})$ are only those obtained from 1 or from applying 2 finitely many times.

The elements of the index set X can be considered names of unbound variables. e_x^X is the function that corresponds to the unbound variable x. Note that any polynomial in $\mathcal{P}^{(X)}(\mathcal{A})$ is considered a function of all $x \in X$, even if it is independent of some variables in X. For every nullary operator f in $\mathcal{A}$, $f \circ \langle\rangle$ is a polynomial according to 2 above. These polynomials are constant-valued functions, and they belong to $\mathcal{P}^{(X)}(\mathcal{A})$ for any X.

Next we consider *formal expressions.* These are objects built up exactly as polynomials. Cf. *polynomial symbols* [G79], *word algebras* [BG81], *expressions* [Be72], and the *theory of expressions* [MaWa84].

Definition 2.8: Let $\mathcal{A} = \langle \mathcal{S}; F \rangle$ be an algebra. Let $(A_x \mid x \in X)$ be a family of sets, such that $X \cap F = \emptyset$ and $\{A_x \mid x \in X\} \subseteq \mathcal{S}$. $\mathcal{E}^{(X)}(\mathcal{A})$ is the set of *expressions in* X *over* $\mathcal{A}$, and $S: \mathcal{E}^{(X)}(\mathcal{A}) \to \mathcal{S}$ is the *sort function* of $\mathcal{E}^{(X)}(\mathcal{A})$. They are defined by:

1. All $x \in X$ belong to $\mathcal{E}^{(X)}(\mathcal{A})$, and $S(x) = A_x$.
2. For all $f \in F$, if $\mathbf{p}_i \in \mathcal{E}^{(X)}(\mathcal{A})$ and $S(\mathbf{p}_i) = S_{if}$ for all $i \in I_f$, then $f \bullet \langle \mathbf{p}_i \mid i \in I_f \rangle$ is an element of $\mathcal{E}^{(X)}(\mathcal{A})$ and $S(f \bullet \langle \mathbf{p}_i \mid i \in I_f \rangle) = S_f$.
3. The elements of $\mathcal{E}^{(X)}(\mathcal{A})$ are only those obtained from 1 or from applying 2 finitely many times.

Symbols denoting expressions will be written in boldface, "$\mathbf{p}$". "$\bullet$" denotes composition of operator with a tuple of expressions. The expressions given by 2 above are called *compound expressions.* Expressions given by 1 are called *variables.* Expressions of form $f \bullet \langle\,\rangle$ are compound expressions according to 2, but they are also called *constants.* If f is unary, we will write $f \bullet \mathbf{p}$ for $f \bullet \langle \mathbf{p} \rangle$. For any compound expression $\mathbf{p} = f \bullet \langle \mathbf{p}_i \mid i \in I_f \rangle$ we will refer to f as the *top operator* of $\mathbf{p}$. When dealing with operators with a familiar notation we will not enforce the notation introduced here; thus, we will for instance write "$a + b$" instead of "$+ \bullet \langle a, b \rangle$".

The basic difference between formal expressions and polynomials is the notion of equality. Since polynomials are functions, equality between polynomials is the same as equality between functions. Equality between expressions, on the other hand, is defined by:

Definition 2.9:

E1. For all $x, y \in X$, $x = y$ regarded as expressions if and only if they are equal in X.

E2. For all variables x and compound expressions $\mathbf{p}$, $x \neq \mathbf{p}$.

E3. If two compound expressions $f \bullet \langle \mathbf{p}_i \mid i \in I_f \rangle$ and $f' \bullet \langle \mathbf{p}'_i \mid i \in I_{f'} \rangle$ are equal, then $f = f'$ and $\mathbf{p}_i = \mathbf{p}'_i$ for all $i \in I_f$.

Two expressions are therefore equal if and only if they are constructed in the same way. Since $\mathcal{E}^{(X)}(\mathcal{A})$ is defined recursively, we can prove properties of its elements by induction on its structure. To do this we use the following *induction principle* for expressions (cf. [G79,MaWa84]):

Definition 2.10: (induction principle for $\mathcal{E}^{(X)}(\mathcal{A})$):
For every predicate q on $\mathcal{E}^{(X)}(\mathcal{A})$ holds that if:
1a. For all $x \in X$, $q(x)$, and:
1b. For all constants $f \bullet \langle\rangle$, $q(f \bullet \langle\rangle)$, and:
2. For all compound expressions $f \bullet \langle \mathbf{p}_i \mid i \in I_f \rangle$,
$\forall i \in I_f[q(\mathbf{p}_i)] \implies q(f \bullet \langle \mathbf{p}_i \mid i \in I_f \rangle)$:
Then, for all $\mathbf{p} \in \mathcal{E}^{(X)}(\mathcal{A})$, $q(\mathbf{p})$.

Note that case 1b could have been included under case 2 in the induction principle, but it is kept separate since it is intuitively more appealing to think of the treatment of the constants as a base case. This induction principle will be extensively used when proving properties about formal expressions. A corresponding induction principle can be stated for polynomials, [G79].

It is sometimes interesting to know which variables are present in an expression. We will therefore, for every $\mathbf{p} \in \mathcal{E}^{(X)}(\mathcal{A})$, define a set *varset*($\mathbf{p}$) that contains exactly the variables in $\mathbf{p}$, cf. [MaWa84]. We also extend the definition to sets of expressions.

Definition 2.11: *varset* is defined by:
vs1. For every expression $\mathbf{p} \in \mathcal{E}^{(X)}(\mathcal{A})$:
If $\mathbf{p} \in X$, then $varset(\mathbf{p}) = \{\mathbf{p}\}$.
If $\mathbf{p} = f \bullet \langle \mathbf{p}_i \mid i \in I_f \rangle$, then $varset(\mathbf{p}) = \bigcup(\, varset(\mathbf{p}_i) \mid i \in I_f \,)$.
vs2. For any $E \subseteq \mathcal{E}^{(X)}(\mathcal{A})$, $varset(E) = \bigcup(\, varset(\mathbf{p}) \mid \mathbf{p} \in E \,)$.

Substitutions are transformations on expressions where expressions are substituted for variables. An inductive definition can be found in [MaWa84]. Our definition here is based on the view of substitutions as functions. It is similar to the standard definition as found in, for instance, [H80]:

Definition 2.12: A *substitution* σ in $\mathcal{E}^{(X)}(\mathcal{A})$ is a partial function $X \to \mathcal{E}^{(X)}(\mathcal{A})$, where $S(x) = S(\mathbf{p})$ for all $\langle x, \mathbf{p} \rangle \in \sigma$. σ is extended to a homomorphism $\mathcal{E}^{(X)}(\mathcal{A}) \to \mathcal{E}^{(X)}(\mathcal{A})$ in the following way:
$\mathbf{p} \in X$: $\sigma(\mathbf{p}) = \mathbf{p}$ when $\langle \mathbf{p}, \sigma(\mathbf{p}) \rangle \notin \sigma$.
$\mathbf{p} = f \bullet \langle \mathbf{p}_i \mid i \in I_f \rangle$: $\sigma(\mathbf{p}) = f \bullet \langle \sigma(\mathbf{p}_i) \mid i \in I_f \rangle$.
Equality between substitutions is equality between sets.

Whenever there is no risk of confusion, we will use the same notation for a substitution and its extension. The set of all substitutions in $\mathcal{E}^{(X)}(\mathcal{A})$ is denoted $\mathcal{S}^{(X)}(\mathcal{A})$.

Note that we define a substitution σ as a *partial* function. This is in contrast to [H76,H80], where a substitution is defined as a total function for which $\sigma(x) = x$ almost everywhere. The reason is that we will use substitutions in a somewhat different context than usual, where it sometimes is natural to treat them as sets of pairs. Set equality between substitutions is also non-standard: usually two substitutions σ_1, σ_2 are considered equal iff $\sigma_1(\mathbf{p}) = \sigma_2(\mathbf{p})$ for all expressions $\mathbf{p}$. These notions of equality are not equivalent. Set equality implies application equality, but application equality only implies set equality modulo pairs of the form $\langle x, x \rangle$. Note, further, that the finiteness restriction is dropped. Apart from this, the difference is very minor. Standard results about substitutions, not relying on finiteness or involving equality between substitutions, will carry over to substitutions as defined here. Since a substitution $\sigma \in \mathcal{S}^{(X)}(\mathcal{A})$ is a partial function $X \rightarrow \mathcal{E}^{(X)}(\mathcal{A})$, the restricted function $\sigma|_{X'}$ is also a substitution in $\mathcal{E}^{(X)}(\mathcal{A})$ for all $X' \subseteq X$. A special substitution worth mentioning is the *empty substitution*, $\emptyset$. $\emptyset(\mathbf{p}) = \mathbf{p}$ for all expressions $\mathbf{p}$.

Definition 2.13: For any substitution $\sigma \in \mathcal{S}^{(X)}(\mathcal{A})$, the *domain*, *dom*$(\sigma)$ is $\{\, x \mid \langle x, \mathbf{p} \rangle \in \sigma \,\}$ and *range*$(\sigma) = \bigcup(\, \mathit{varset}(\mathbf{p}) \mid \langle x, \mathbf{p} \rangle \in \sigma \,)$.

Cf. [MaWa84]. A pair $\langle x, \sigma(x) \rangle$ in a substitution σ will sometimes be denoted $x \leftarrow \sigma(x)$.

Definition 2.14: For all substitutions $\sigma_1, \sigma_2 \in \mathcal{S}^{(X)}(\mathcal{A})$ we define

$$\sigma_1\sigma_2 = \{\, x \leftarrow \sigma_1(\sigma_2(x)) \mid x \in \mathit{dom}(\sigma_2) \,\} \cup \sigma_1|_{\mathit{dom}(\sigma_1) \setminus \mathit{dom}(\sigma_2)}.$$

Theorem 2.1: *For all substitutions* σ_1, σ_2 *and expressions* $\mathbf{p}$, $\sigma_1\sigma_2(\mathbf{p}) = \sigma_1(\sigma_2(\mathbf{p}))$.

Proof. See appendix D. ■

Thus, definition 2.14 is consistent with function composition of the extended homomorphisms. Note that substitution composition, with substitutions defined as partial functions, will not be properly defined by just demanding $\sigma_1\sigma_2(\mathbf{p}) = \sigma_1(\sigma_2(\mathbf{p}))$ for all $\mathbf{p}$. This condition does not, as mentioned previously, imply set equality.

$\mathcal{P}^{(X)}(\mathcal{A})$ and $\mathcal{E}^{(X)}(\mathcal{A})$ are closed under the operations in F. Thus, there are algebras $\langle \mathcal{P}^{(X)}(\mathcal{A}); F \rangle$ and $\langle \mathcal{E}^{(X)}(\mathcal{A}); F \rangle$, and we can speak about homomorphisms between them.

Definition 2.15: ϕ, the *natural homomorphism* $\mathcal{E}^{(X)}(\mathcal{A}) \rightarrow \mathcal{P}^{(X)}(\mathcal{A})$, is defined by:

1. $\phi(x) = e_x^X$, all $x \in X$.
2. for all compound expressions $f \bullet \langle \mathbf{p}_i \mid i \in I_f \rangle$ in $\mathcal{E}^{(X)}(\mathcal{A})$, $\phi(f \bullet \langle \mathbf{p}_i \mid i \in I_f \rangle) = f \circ \langle \phi(\mathbf{p}_i) \mid i \in I_f \rangle$.

The following important theorem is well-known from similar developments (see, for instance, [G79]). We therefore give it without proof:

Theorem 2.2: *The function* ϕ *is onto.*

ϕ may be used to define a set of identities on $\mathcal{E}^{(X)}(\mathcal{A})$:

Definition 2.16: The relation $=_\phi$ on $\mathcal{E}^{(X)}(\mathcal{A})$ is defined by: for any $\mathbf{p}, \mathbf{p}' \in \mathcal{E}^{(X)}(\mathcal{A})$

$$\mathbf{p} =_\phi \mathbf{p}' \iff \phi(\mathbf{p}) = \phi(\mathbf{p}').$$

The following theorem is not hard to prove in our framework developed here. It holds in the case of unsorted universal algebras (see [G79], where it follows from theorem 26.2). Since the extension to many-sorted algebras does not affect the proof significantly we omit the proof here:

Theorem 2.3: $=_\phi$ *is closed, that is:*

1. $\mathbf{p} =_\phi \mathbf{p}$
2. $\mathbf{p} =_\phi \mathbf{p}' \implies \mathbf{p}' =_\phi \mathbf{p}$
3. $\mathbf{p} =_\phi \mathbf{p}' \wedge \mathbf{p}' =_\phi \mathbf{p}'' \implies \mathbf{p} =_\phi \mathbf{p}''$
4. $\mathbf{p}_i =_\phi \mathbf{p}_i'$ *for all* $i \in I_f \implies f \bullet \langle \mathbf{p}_i \mid i \in I_f \rangle =_\phi f \bullet \langle \mathbf{p}_i' \mid i \in I_f \rangle$
5. $\mathbf{p} =_\phi \mathbf{p}' \implies \sigma(\mathbf{p}) =_\phi \sigma(\mathbf{p}')$

3. Output specification

As mentioned in the introduction, we want to specify the action of a system entirely by functions from inputs to outputs. This leads to the following definition:

Definition 3.1: Let X be a countable index set. An *output specification in* X is a nonempty, countable set of functions O. Each $f \in O$ is a function $f: \prod(A_x \mid x \in X) \rightarrow A_f$ and it is dependent on only a finite number of inputs. $\prod(A_x \mid x \in X)$ is the *domain* of O.

A system given by an output specification will usually be implemented using more primitive functions than the ones in the specification. This naturally leads to the study of algebras where these primitive functions are operators. If we deal with scientific computations, then we are probably interested in functions built up by arithmetical operations on real numbers. If we study bit-level systems, then an algebra with logical operators is likely to be more appropriate. For an algebra $\mathcal{A}$, the polynomials in $\mathcal{P}^{(X)}(\mathcal{A})$ are exactly the functions with inputs X that we can realize using the operators of $\mathcal{A}$. Therefore, it is interesting to know if we can use polynomials in $\mathcal{P}^{(X)}(\mathcal{A})$ to somehow meet a given output specification. (In the following A will denote $\prod(A_x \mid x \in X)$).

Definition 3.2: Let $f: A \rightarrow A_f$. Let $\mathcal{A}$ be an algebra. *f is realizable in $\mathcal{A}$* iff there exists a function $\eta_f: A \rightarrow \mathcal{P}^{(X)}(\mathcal{A})$ such that $f(a) = \eta_f(a)(a)$ for all $a \in A$. η_f is a *representation function of f with respect to $\mathcal{A}$* and $\eta_f(A)$ is a *polynomial representation of f in $\mathcal{A}$*.

If O is an output specification with domain A, then *O is satisfiable in $\mathcal{A}$* iff for all $f \in O$ there exists an $\eta_f: A \rightarrow \mathcal{P}^{(X)}(\mathcal{A})$ such that f is realizable in $\mathcal{P}^{(X)}(\mathcal{A})$. In this case $\eta: O \times A \rightarrow \mathcal{P}^{(X)}(\mathcal{A})$ is defined by $\eta(f, a) = \eta_f(a)$ for all $f \in O$ and $a \in A$. η is called a *representation function* of O with respect to $\mathcal{A}$ and $\eta(O, A)$ is a *polynomial representation of O in $\mathcal{A}$*.

Definition 3.2 gives the condition for when a function f is realizable by a set of polynomials. Which polynomial that corresponds to f might, however, vary with the inputs. That is, the structure of the way of computing f using the operators in $\mathcal{A}$ may be *data-dependent*. An interesting case is when f always corresponds to the same polynomial:

Definition 3.3: Let $\mathcal{A}$ be an algebra and let f be a function realizable in $\mathcal{A}$. *f is structurally data-independent in $\mathcal{A}$* iff there exists a representation function η_f of f with respect to $\mathcal{A}$ such that $\eta_f(a) = \eta_f(a')$ for all $a, a' \in A$. The output specification O is structurally data-independent in $\mathcal{A}$ iff every $f \in O$ is structurally data-independent in $\mathcal{A}$.

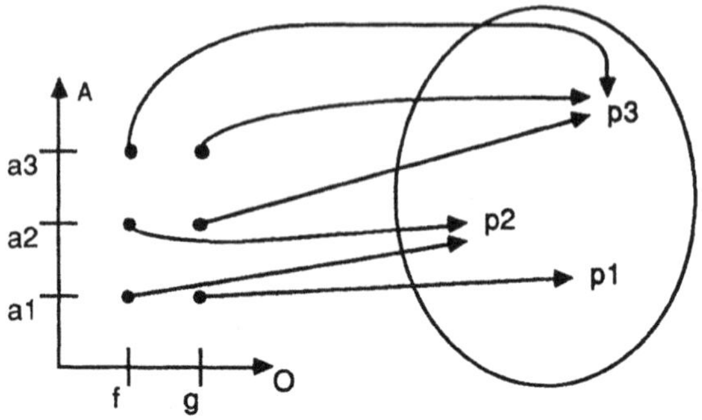

Figure 3.1: Representation function.

Proposition 3.1: *f is structurally data-independent in $\mathcal{A}$ iff f is a polynomial over $\mathcal{A}$.*

Proof. A structurally data-independent function f is equal to the function $\eta_f(a) \in \mathcal{P}^{(X)}(\mathcal{A})$ (which is the same polynomial regardless of a). ■

Corollary 3.1: *O is structurally data-independent in $\mathcal{A}$ iff $O \subseteq \mathcal{P}^{(X)}(\mathcal{A})$.*

Thus, structural data-independence for a function in an algebra $\mathcal{A}$ is the same as the function being a polynomial over $\mathcal{A}$. We will later see that structural data-independence of a function f makes possible an à priori scheduling of the evaluation of f.

Example 3.1: In order to explain the difference between structural data-dependence and -independence, the following example of a structurally data-dependent output specification is shown. Let us define a function g that gives the value computed by Newton-Raphsons method for solving $f(x) = 0$. Consider the following recursive definition of polynomials p_n in $\{x\}$ over $\mathcal{A} = \langle\{R\}; \{-, /, f, f'\}\rangle$, where f' is the derivative of f:

$$p_0 = e_x^{\{x\}}$$

$$n > 0: \qquad p_n = p_{n-1} - f(p_{n-1})/f'(p_{n-1})$$

Define a representation function η_g by: for all $a \in R$, $\eta_g(a) = p_n$ for the least $n > 0$ such that $|p_n(a) - p_{n-1}(a)| \leq \epsilon$.

This defines the output function g, the approximation of a solution to $f(x) = 0$, in the input variable x that gives the start value. Now it is easy to see that although g, provided the algorithm terminates, is realizable in $\mathcal{A}$, g is not structurally data-independent in general. This is because different start values may lead to different numbers of iteration corresponding to different p_n. ■

There is a relationship between structural data dependence and non-strict operators. Given a function f that is structurally data-dependent in $\langle \mathcal{S}; F \rangle$, it is sometimes possible to use the non-strict *if*-operator (definition 2.4) to construct a single polynomial that is always equal to f. Then f is structurally data-independent in $\langle \{B\} \cup \mathcal{S}; \{if_{A_f}\} \cup F \rangle$, where $B = \{true, false\}$. Sufficient conditions for this construction are established in the next theorem.

Theorem 3.1: *(if-conversion) Let $f: A \to A_f$ be realizable in $\mathcal{A} = \langle \mathcal{S}; F \rangle$. Assume that there exists a representation function $\eta_f: A \to \mathcal{P}^{(X)}(\mathcal{A})$ such that $\eta_f(A) = \{p_1, \ldots, p_n\}$ is finite. Assume further that each condition $\eta_f(a) = p_i$ is expressible as a polynomial c_i in $\mathcal{P}^{(X)}(\langle \{B\} \cup \mathcal{S}; \{if_{A_f}\} \cup F \rangle)$. Then f is structurally data-independent in $\langle \{B\} \cup \mathcal{S}; \{if_{A_f}\} \cup F \rangle$.*

Proof. See Appendix E. ■

Note that the function g defined by Newton-Raphsons method in the previous example is not expressible as a polynomial by *if*-conversion, unless there is an upper bound for the number of iterations.

Let us finally mention an alternative approach. Instead of considering representation functions and polynomials, the functions in an output specification could have been defined recursively using least fixpoint semantics, [Ma74]. The reason why we do not choose this approach is that such definitions actually hide *to much.* For instance, structural data-independence becomes harder to define, and we are especially interested in structurally data-independent functions here.

Example 3.2: Reconsider the function g in the previous example. It can be defined as the least fixpoint solution to the following recursive equation, where $\xi(x) = x - f(x)/f'(x)$:

$$F(x) = if(|\xi(x) - x| \leq \epsilon, \xi(x), F(\xi(x)))$$

■

4. Formal expressions as schemes for evaluation

As mentioned in the preliminaries, the difference between polynomials and formal expressions is in the different notions of equality. The following example may clarify this. Consider two expressions $x_1 \cdot (x_2 + x_3)$ and $x_1 \cdot x_2 + x_1 \cdot x_3$ in an algebra where "·" distributes over "+". The polynomials $\phi(x_1 \cdot (x_2 + x_3))$ and $\phi(x_1 \cdot x_2 + x_1 \cdot x_3)$ are equal, since they will both yield the same output for the same inputs. The expressions, however, are *not* equal according to definition 2.9. (E3 is for instance violated since the top operators "·", "+" are not equal.)

Why do we consider formal expressions? The reason is that they can be given a particular interpretation: *constants can be seen as making a constant value available, variables as making an input value available and compound expressions as representing the application of the outermost operator to the intermediate data produced by the immediate subexpressions,*. Thus, an expression **p** together with its subexpressions constitutes a scheme for computing $\phi(\mathbf{p})$. When we speak about an *evaluation* of an expression, we mean the action for it given by the interpretation above. In this section we will state some basic properties of expressions. We will also make some definitions that will be useful under our interpretation of expressions.

Definition 4.1: Let O be an output specification in X with domain A, that is satisfiable in $\mathcal{A}$ with representation function η. Let $E \subseteq \mathcal{E}^{(X)}(\mathcal{A})$.

1. E is an *evaluable set of expressions*, if and only if for all $\mathbf{p} \in E$
   $\mathbf{p} = f \bullet \langle \mathbf{p}_i \mid i \in I_f \rangle \implies \mathbf{p}_i \in E$ for all $i \in I_f$.
2. The *evaluable closure* $ec(E)$ of E is the smallest evaluable set of expressions such that $E \subseteq ec(E)$.
3. If $\eta(O, A) \subseteq \phi(E)$, then E is an *expression representation* of O and $ec(E)$ is a *computational scheme* for O.

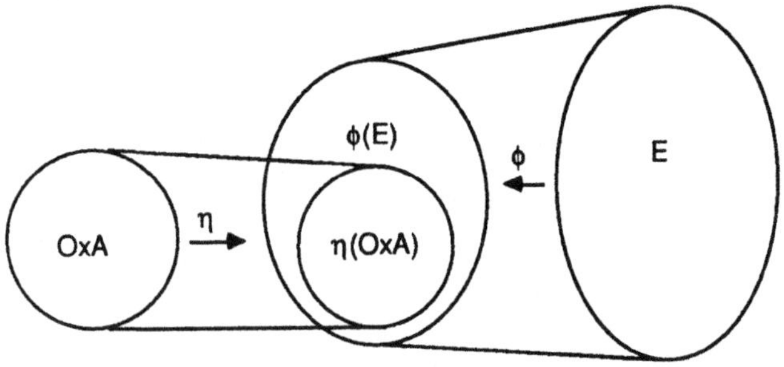

Figure 4.1: An expression representation.

"Set of expressions" will sometimes be abbreviated by "SoE". For all expressions in an evaluable set of expressions E, all their subexpressions are also in E. E therefore contains every step, expressed as operators in the algebra $\mathcal{A}$ applied to subexpressions, necessary to compute all polynomials in $\phi(E)$ according to the expressions chosen to represent them.

By theorem 2.2, an expression representation $E \subseteq \mathcal{E}^{(X)}(\mathcal{A})$ always exists for any output specification O that is satisfiable in $\mathcal{A}$. E contains expressions corresponding to all polynomials that may ever be used to give a value for a function in O. The computational scheme $ec(E)$ contains every step, that might ever become necessary to perform in order to compute the functions in O according to the expressions in E. The evaluation of the expressions in E will produce, among others, the values of the functions in O for the given inputs. If O is structurally data-dependent in $\mathcal{A}$, then we cannot tell until we know all inputs *which* expression in E that will give the value of a certain function in O. If, on the other hand, O is structurally data-independent, then we will always know *in advance* which expression(s) that corresponds to a certain function.

The evaluable closure can be defined in an alternative way, inductively over expressions (cf. [Be72]):

Theorem 4.1: *The evaluable closure can be alternatively defined by:*

1. *For every expression* $\mathbf{p} \in \mathcal{E}^{(X)}(\mathcal{A})$:
   1a. *If* $\mathbf{p} \in X$, *then* $ec(\{\mathbf{p}\}) = \{\mathbf{p}\}$.
   1b. *If* $\mathbf{p} = f\bullet\langle \mathbf{p}_i \mid i \in I_f \rangle$, *then* $ec(\{\mathbf{p}\}) = \bigcup\langle ec(\{\mathbf{p}_i\}) \mid i \in I_f \rangle \cup \{\mathbf{p}\}$.
2. *For any* $E \subseteq \mathcal{E}^{(X)}(\mathcal{A})$, $ec(E) = \bigcup\langle ec(\{\mathbf{p}\}) \mid \mathbf{p} \in E \rangle$.

Proof. See appendix F. ■

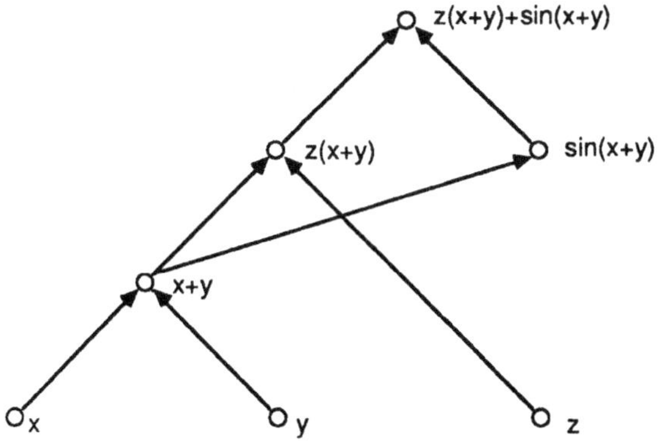

Figure 4.2: Dependency graph

Definition 4.2: The binary relation $\prec_e$ on $\mathcal{E}^{(X)}(\mathcal{A})$ is given by the following: Let $\mathbf{p}', \mathbf{p} \in \mathcal{E}^{(X)}(\mathcal{A})$. Then $\mathbf{p}' \prec_e \mathbf{p}$ iff $\mathbf{p} = f \bullet \langle \mathbf{p}_i \mid i \in I_f \rangle$ and, for some $i \in I_f$, $\mathbf{p}' = \mathbf{p}_i$.

Under our interpretation of expressions as representing events where operators are applied to intermediate data, the relation $\prec_e$ defined above obtains a special meaning: $\mathbf{p}' \prec_e \mathbf{p}$ *means that the result of evaluating* $\mathbf{p}'$ *is used as input when evaluating* $\mathbf{p}$. Thus, $\prec_e$ can be considered a *data dependence relation* on $\mathcal{E}^{(X)}(\mathcal{A})$.

Example 4.1: The graph in figure 4.2 illustrates the data dependencies on the evaluable closure of the set of expressions $\{z \cdot (x+y) + \sin(x+y)\}$ in $\mathcal{E}^{(\{x,y,z\})}(\langle\{R\};\{+,\cdot,\sin\}\rangle)$. ■

Theorem 4.2: $\prec_e^+$ *is finite-downward.*

Proof. See appendix F. ■

The transitive closure $\prec_e^+$ determines the amount of parallelity that is possible when evaluating a set of expressions. If $\mathbf{p}' \prec_e^+ \mathbf{p}$, the evaluation of $\mathbf{p}'$ has to take place before the one of $\mathbf{p}$. If $\mathbf{p}' \not\prec_e^+ \mathbf{p}$ and $\mathbf{p} \not\prec_e^+ \mathbf{p}'$, then they may take place concurrently. Theorem 4.2 assures that any evaluation of an expression has to be preceded by only finitely many other evaluations of expressions. Thus, the evaluation of any expression can be scheduled in finite time. We can also note that $\langle \mathcal{E}^{(X)}(\mathcal{A}), \prec_e^* \rangle$ is an *elementary event structure*, [NPW81].

Usually we will consider subsets E of $\mathcal{E}^{(X)}(\mathcal{A})$, which means that the restricted relation $\prec_e|_E$ and its transitive closure are of interest. When E is evaluable we have the following result:

Lemma 4.1: *For any evaluable* $E \subseteq \mathcal{E}^{(X)}(\mathcal{A})$, $\prec_e|_E^+ = \prec_e^+|_E$.

Proof. See appendix F. ■

5. Computational events and computation networks

In this chapter we will introduce *computational events* (c.e.'s). They constitute the most basic events that we will consider. A computational event is labelled with an expression and it then represents an evaluation of that expression. This means that it is possible to describe computations where the same expression is evaluated several times, since more that one computational event may be labelled with the same expression. Had we chosen the expressions themselves to be the basic events of computation, there would not have been any way of representing multiple evaluations of expressions.

We will also consider *computation networks*, which is the usual way to define the DAG or tree representations of computation considered in the literature. See for instance [HJ81]. Finally, we will investigate under which circumstances a computation network is equivalent to a set of computational events.

5.1 Computational events

Definition 5.1: Let D be a countable set and let E be an evaluable set of expressions. If there exists a function $\psi: D \to E$ that is onto, then we say that D is a *set of computational events (SoCE) of* E.

Computational events will be denoted in roman: "p" in order to distinguish them from expressions, "**p**", and polynomials, "p". Every computational event p is labelled with an expression $\psi(\mathrm{p})$ and p represents an evaluation of $\psi(\mathrm{p})$ in E. This interpretation of a computational event can be formalized as follows:

Definition 5.2: Let $\psi: D \to E$ be onto. For every computational event $\mathrm{p} \in D$, its *output function with respect to* ψ, $o_\psi(\mathrm{p})$, is $\phi \circ \psi(\mathrm{p})$.

Thus, the meaning of a computational event is a function; namely the polynomial induced by the expression that labels the computational event.

Given a set of computational events D of E, we are interested in finding data dependence relations on D that are consistent with $\prec_e$ on E. That is, since $\prec_e$ describes the transmission of data necessary to evaluate all expressions in E, the relation on D should describe the transmission of data necessary to compute the output functions of all computational events in D. There are in general several possibilities to accomplish this. Therefore, we define not only one relation, but a whole class of relations on D satisfying the above. The definition is based on the following observation: given a computational event p of a compound expression $\psi(\mathrm{p})$, then, for each $\mathbf{p}'$ such that $\mathbf{p}' \prec_e \psi(\mathrm{p})$, there may be *several* computational events p' such that $\psi(\mathrm{p}') = \mathbf{p}'$. However, for each argument of $\psi(\mathrm{p})$, *only one* of the computational events of the corresponding subexpression can provide the input to p. Each possible choice of computational event providing the input gives a distinct data dependence relation.

Definition 5.3: Let $\psi: D \to E$ be onto. Then $comp(D, \psi)$ is the set of computational events p in D for which $\psi(\mathrm{p})$ is a compound expression $f \bullet \langle \mathbf{p}_i \mid i \in I_f \rangle$ and $n(f) > 0$.

$(\, s_{\mathrm{p}} \mid \mathrm{p} \in comp(D, \psi)\,)$ is a *family of selector functions with respect to D and ψ* iff: for every p in $comp(D, \psi)$, where $\psi(\mathrm{p}) = f \bullet \langle \mathbf{p}_i \mid i \in I_f \rangle$, s_{p} is a function $I_f \to D$ such that $\psi(s_{\mathrm{p}}(i)) = \mathbf{p}_i$ for all $i \in I_f$.

Definition 5.4: Let $\psi: D \to E$ be onto. For every family of selector functions $SF = (\, s_{\mathrm{p}} \mid \mathrm{p} \in comp(D, \psi)\,)$, $\prec_{SF}$ is defined by:

1. For all p in $comp(D, \psi)$ and for all $i \in I_f$, $s_{\mathrm{p}}(i) \prec_{SF} \mathrm{p}$.
2. For no other p', p in D, $\mathrm{p}' \prec_{SF} \mathrm{p}$.

The *class of data dependence relations on D with respect to ψ and E*, $\mathcal{R}(D, \psi, E)$, is the set

$$\{ \prec_{SF} \mid SF \text{ is a family of selector functions with respect to } D \text{ and } \psi \}.$$

For any $\prec \in \mathcal{R}(D, \psi, E)$, $\langle D, \prec \rangle$ is a *computational event structure of E under ψ*.

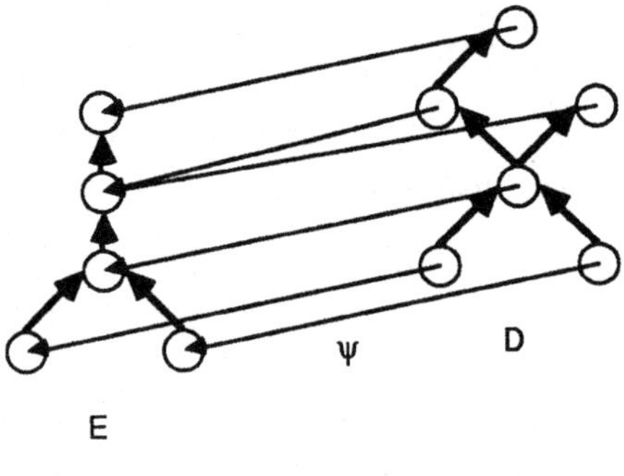

Figure 5.1: Set of computational events, set of expressions, labelling function ψ, data dependence relation.

Consider a set of computational events with a labelling function ψ that is one-to-one. Then there is exactly one computational event for each expression. Each expression will be evaluated only once and any input is therefore produced exactly once. It is thus not surprising, that a *unique* data dependence relation on the set of computational events in this case is determined by ψ and $\prec_e$.

Proposition 5.1: *Assume that* $\psi: D \rightarrow E$ *is onto and* $1-1$. *Then,* $\mathcal{R}(D, \psi, E)$ *has exactly one element* $\prec$ *given by*

$$\mathrm{p} \prec \mathrm{p}' \iff \psi(\mathrm{p}) \prec_e \psi(\mathrm{p}')$$

for every two $\mathrm{p}, \mathrm{p}' \in D$.

Proof. See appendix G. ∎

Corollary 5.1: $\mathcal{R}(\mathcal{E}^{(X)}(\mathcal{A}), I, \mathcal{E}^{(X)}(\mathcal{A})) = \{\prec_e\}$, *where* I *is the identity function on* $\mathcal{E}^{(X)}(\mathcal{A})$.

So if we want to consider every expression to be the computational event of evaluating itself, then no complications arise. The data dependence relation on expressions considered as computational events is the same as the original one.

Theorem 5.1: *For every* $\prec \in \mathcal{R}(D, \psi, E)$, $\prec^+$ *is finite-downward.*

Proof. See appendix G. ∎

This result is the counterpart to theorem 4.2 for data dependence relations on sets of expressions. It is important since it implies that the output function of any computational event is computable, provided that the operators in the underlying algebra are computable. Furthermore, we can note that $\langle D, \prec^* \rangle$ is an elementary event structure for every $\prec$ in $\mathcal{R}(D, \psi, E)$.

Example 5.1: Reconsider the set of expressions in example 4.1. Choose the set of computational events $D = \{1, \ldots, 8\}$ and define ψ by:

$\psi(1) = x$
$\psi(2) = y$
$\psi(3) = z$
$\psi(4) = (x + y)$
$\psi(5) = (x + y)$
$\psi(6) = \sin(x + y)$
$\psi(7) = z \cdot (x + y)$
$\psi(8) = z \cdot (x + y) + \sin(x + y)$

We have two computational events labelled with addition of x and y, an operation that consequently will be performed twice. Figure 5.2 shows two possible data dependence relations on D. ∎

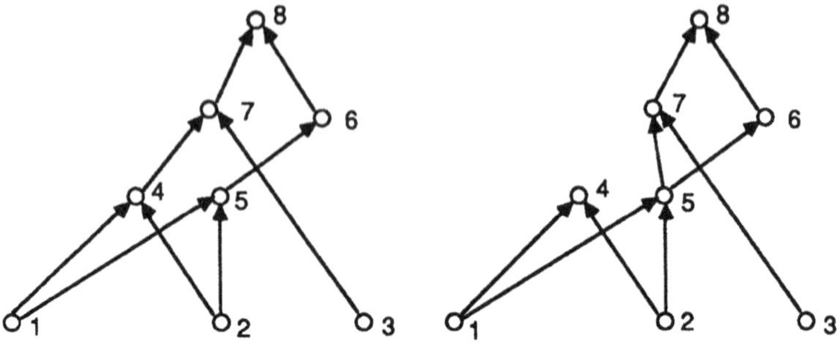

Figure 5.2: Dependency graphs on D

5.2 Computation networks

Now we will consider an alternative way to define the meaning of computational events. Instead of labelling each computational event with a complete expression, we might label it with an operator or a variable. In this case a data dependence relation cannot be derived from the precedence relation on an underlying set of expressions. It has to be given, and it must be given in such a way that it is consistent with the input demands of each computational event imposed by the labelling. The computational events now obtain the following meaning: if a computational event is labelled with a variable x, then it represents an input of x. (Of course different inputs of the same variable x must yield the same value.) If it is labelled with an operator, then it represents the application of that operator to input data. These data are produced by the immediate predecessors with respect to the given data dependence relation.

Definition 5.5: Let $\mathcal{A} = \langle \mathcal{S}; F \rangle$ be an algebra and let $(A_x \mid x \in X)$ be a family of sets such that $X \cap F = \emptyset$ and $\{ A_x \mid x \in X \} \subseteq \mathcal{S}$. *A computation network in* X *over* $\mathcal{A}$ is a triple $\langle D, R, f \rangle$, where D is a countable set, R is a binary relation on D and f is a function $D \to F \cup X$.

D is our set of computational events. R is the data dependence relation, describing the relation between inputs and outputs of computational events. f, finally, is a function that labels each computational event with an operator or a variable; it assigns a meaning to every computational event according to the above.

Definition 5.6: Let $CN = \langle D, R, f \rangle$ be a computation network in X over $\langle \mathcal{S}; F \rangle$.

1. CN is *causal* iff R is finite-downward.
2. CN is *well-formed* iff:
   wf1. If $f(\text{p}) \in X$, then $sp(\text{p}, R) = \emptyset$.
   wf2. If $f(\text{p}) \in F$ and $n(f(\text{p})) = 0$, then $sp(\text{p}, R) = \emptyset$.

wf3. If $f(\mathrm{p}) \in F$ and $n(f(\mathrm{p})) > 0$, then there is a function $s_{\mathrm{p}}: I_{f(\mathrm{p})} \to sp(\mathrm{p}, R)$ that is onto, and for all $i \in I_{f(\mathrm{p})}$ holds that $S_{f(s_{\mathrm{p}}(i))} = S_{if(\mathrm{p})}$.

See definition 2.5 of algebras for the definition of the sort functions $S_{f(s_{\mathrm{p}}(i))}$, $S_{if(\mathrm{p})}$. In the following we will consider computation networks that are causal and well-formed. The causality condition is sometimes strengthened in the literature so that $\langle D, R\rangle$ must be a tree. The well-formedness condition, on the other hand, is often completely overlooked or ignored in the literature. If it is not met, then there is a mismatch between the inputs supplied for by R and the requirements of the operator; there might be an incorrect number of inputs, for instance, or the inputs might not have the proper sort. We might not even know which input that constitutes which argument! s_{p} connects outputs of computational events to arguments of the operator exactly as the selector functions do in definition 5.4. Note the additional sort constraint. It is not necessary to have such a constraint when computational events are labelled with expressions, since the expressions themselves are already constructed according to such constraints.

Causal and well-formed computation networks can be given a particularly simple meaning, which follows from the meaning of the computational events in D and the data dependence relation R:

Definition 5.7: Let $\langle D, R, f\rangle$ be a causal and well-formed computation network in X over $\mathcal{A} = \langle \mathcal{S}; F\rangle$. For every $\mathrm{p} \in D$, the *output function* $o(\mathrm{p})$ is given by:

1. If $f(\mathrm{p}) \in X$, then $o(\mathrm{p}) = e^X_{f(\mathrm{p})}$.
2. If $f(\mathrm{p}) \in F$, then $o(\mathrm{p}) = f(\mathrm{p}) \circ \langle\, o(s_{\mathrm{p}}(i)) \mid i \in I_{f(\mathrm{p})}\,\rangle$.

It is easy to verify that o always is well-defined when CN is causal and well-formed.

Definition 5.8: Let $\langle D, R, f\rangle$ be a causal and well-formed computation network in X over $\mathcal{A} = \langle \mathcal{S}; F\rangle$. The function $\psi: D \to \mathcal{E}^{(X)}(\mathcal{A})$ is defined by:

1. If $f(\mathrm{p}) \in X$, then $\psi(\mathrm{p}) = f(\mathrm{p})$.
2. If $f(\mathrm{p}) \in F$, then $\psi(\mathrm{p}) = f(\mathrm{p}) \bullet \langle\, \psi(s_{\mathrm{p}}(i)) \mid i \in I_{f(\mathrm{p})}\,\rangle$.

Exactly as o, ψ is always well-defined when CN is causal and well-formed.

Proposition 5.2: *If $CN = \langle D, R, f \rangle$ is causal and well-formed, then $\psi(D)$ is an evaluable set of expressions.*

Proof. The only interesting case is when $\psi(\mathrm{p}) = f \bullet \langle \mathbf{p}_i \mid i \in I_f \rangle$ and $n(f) > 0$. But then, since CN is well-formed, there exists an s_{p} such that $\langle \mathbf{p}_i \mid i \in I_f \rangle = \langle \psi(s_{\mathrm{p}}(i)) \mid i \in I_{f(\mathrm{p})} \rangle$. Thus $\mathbf{p}_i = \psi(s_{\mathrm{p}}(i)) \in \psi(sp(\mathrm{p}, R)) \subseteq \psi(D)$ for all $i \in I_{f(\mathrm{p})} = I_f$. ■

We are now ready to state some facts about the correspondence between sets of computational events and computation networks.

Theorem 5.2: *Let $\mathcal{A} = \langle \mathcal{S}; F \rangle$.*

1. *For every causal and well-formed computation network $\langle D, R, f \rangle$, D is a set of computational events of $\psi(D)$ and $R \in \mathcal{R}(D, \psi, \psi(D))$.*
2. *Let $E \subseteq \mathcal{E}^{(X)}(\mathcal{A})$ be an evaluable set of expressions. Let $\psi: D \to E$ be onto. For every $\prec \in \mathcal{R}(D, \psi, E)$, $\langle D, \prec, f \rangle$ is a causal and well-formed computation network in X over $\mathcal{A}$, where f is given by:*

$$\psi(\mathrm{p}) = \begin{cases} f(\mathrm{p}), & \psi(\mathrm{p}) \in X \\ f', & \psi(\mathrm{p}) = f' \bullet \langle \mathbf{p}_i \mid i \in I_{f'} \rangle. \end{cases}$$

Proof. See appendix G. ■

Proposition 5.3: *Let $CN = \langle D, R, f \rangle$ be a causal and well-formed computation network. For every $\mathrm{p} \in D$, $o(\mathrm{p}) = \phi \circ \psi(\mathrm{p})$.*

Proof. See appendix G. ■

Let us finally make a remark on the connection between computation networks as defined here and *data flow graphs* defining data flow programs, [DK82]. Data flow graphs can to some extent be seen as computation networks with possible cycles, i.e. non-causal computation networks. *The introduction of cycles will, however, make the previous interpretation of nodes as single events impossible, and we can no more associate a simple output function $o(\mathrm{p})$ with each node* p. Such a graph can instead be seen as a recursive definition of functions between *sequences* of values, where the functions on sequences are formed by point-wise application of the operators on the elements of the input sequences. The meaning of the graph is then given by fixed-point semantics for the data type of such sequences, [Kah74,Ke77A]. Data flow graphs usually also contain nodes that perform more general functions on sequences, see for instance the "stream language" L, [PA85]. It is also possible to consider data flow graphs that define functions on other data types than sequences, e.g. trees, [Ke77B].

6. Compound events of computation

So far, we have associated exactly one expression with one event, indicating the application of the outermost operator of that expression at that event. However, it is in many cases practical to consider events that correspond to the evaluation of more than one expression. Consider for instance a so-called "inner-product step", [KLe80], an operation frequently performed by processing elements in systolic arrays for matrix operations:

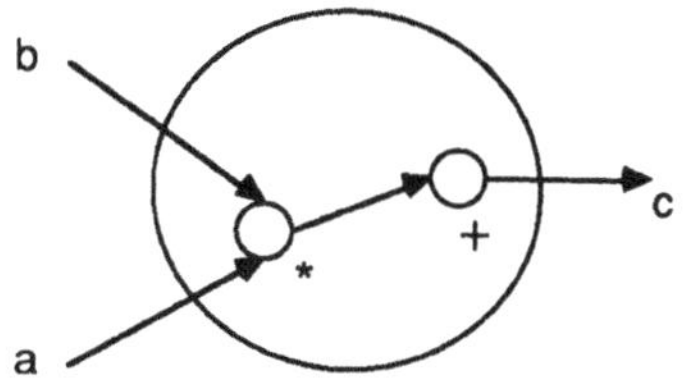

Figure 6.1: Inner-product step

This operation consists of two steps, one of multiplying a and b and the other of adding the result to c. What we want to do is to consider this operation as an *atomic* event where the inner structure is hidden. In this thesis we will consider two ways to accomplish this.

The first approach is simply to partition a set of computational events. Every block in such a partition is then an event, where all computational events in it are carried out simultaneously. A block in such a partition will be called a *cell action*. The second approach is to consider a set of equations:

$$x_1 = \mathbf{p}_1$$

$$\vdots$$

$$x_n = \mathbf{p}_n$$

$$\vdots$$

Every equation is to be interpreted as an assignment, where the variable to the left is assigned the value of the evaluated expression on the right-hand side. Such single-valued assignments might be bundled into *concurrent assignments*

$$\langle x_1, \ldots, x_k \rangle \leftarrow \langle \mathbf{p}_1, \ldots, \mathbf{p}_k \rangle.$$

Every concurrent assignment can be seen as an event of computation. We will investigate under which circumstances a set of concurrent assignments is equivalent to a set of cell actions.

6.1 Cell actions

Before giving a strict definition of cell actions, we give the following technical definition:

Definition 6.1: Let A, B be sets, let $f: A \to B$ and let ε_f be the equivalence relation on A induced by f. Let R_A be a binary relation on A and R_B a binary relation on B. $\langle B, R_B \rangle$ *is finitely inherited from* $\langle A, R_A \rangle$ *under* f, or $fi(\langle A, R_A \rangle, \langle B, R_B \rangle, f)$, iff:

fi1(f): For all $a \in A$, $[a]\varepsilon_f$ is finite.

fi2($\langle A, R_A \rangle, \langle B, R_B \rangle, f$): for all $b, b' \in B$

$$bR_Bb' \iff (b \neq b' \wedge \exists a, a' \in A[aR_Aa' \wedge f(a) = b \wedge f(a') = b']).$$

fi3(R_A, R_B): R_A^+ finite-downward $\Longrightarrow$ R_B^+ finite-downward.

The three properties fi1–fi3 have nothing to do with each other. It just turns out to be convenient to group them into one property, since they will always appear in the same context. fi1(f) is a restriction on f only. If f and R_A are given, then R_B, when it exists, is determined by fi2($\langle A, R_A \rangle, \langle B, R_B \rangle, f$). fi3($R_A, R_B$), finally, assures that we do not introduce any cycles in R_B when R_A is acyclic. The meaning of *fi* becomes more evident if the partition induced by f is considered, as shown in the following figure:

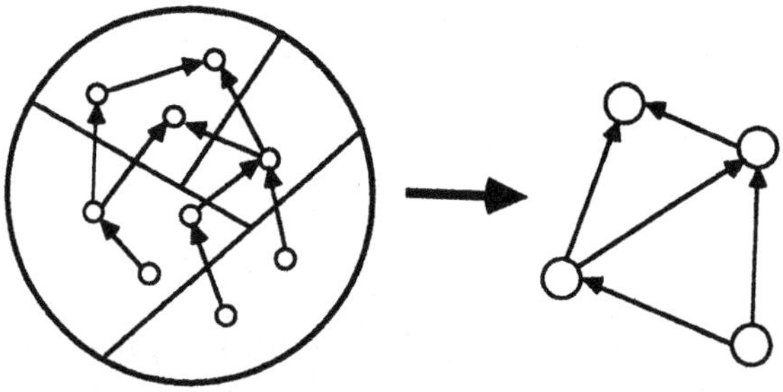

Figure 6.2: $fi(\langle A, R_A \rangle, \langle B, R_B \rangle, f)$

If we consider the structures to be directed graphs, then the effect of fi2 is to extract the "external" edges of R_A into R_B. Edges internal to a block are hidden, since $bR_Bb' \implies b \neq b'$.

Lemma 6.1: *(fi-transitivity): fi is transitive with respect to its first two arguments, that is:*

$$fi(\langle A, R_A\rangle, \langle B, R_B\rangle, f) \wedge fi(\langle B, R_B\rangle, \langle C, R_C\rangle, g) \implies fi(\langle A, R_A\rangle, \langle C, R_C\rangle, g \circ f).$$

Proof. See appendix H. ■

It is easily verified that $fi(\langle A, R_A\rangle, \langle A, R_A\rangle, i_A)$ for every structure $\langle A, R_A\rangle$, where i_A is the identity function on A. Thus, the structures $\langle A, R_A\rangle$ are objects of a category, where the morphisms are the functions $f\colon A \to B$ fulfilling $fi(\langle A, R_A\rangle, \langle B, R_B\rangle, f)$.

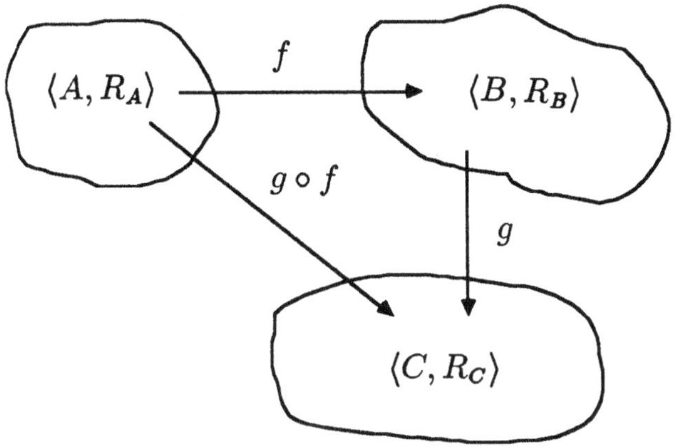

Figure 6.3: transitivity of *fi*

Lemma 6.2: *(fi-uniqueness):*
$fi(\langle A, R_A\rangle, \langle B, R_B\rangle, f) \wedge fi(\langle A, R_A\rangle, \langle B, R'_B\rangle, f) \implies R_B = R'_B$.

Proof. The lemma follows immediately from $fi2(\langle A, R_A\rangle, \langle B, R_B\rangle)$ and $fi2(\langle A, R_A\rangle, \langle B, R'_B\rangle)$ in definition 6.1, since the right-hand sides in the equivalences become equal for both R_B and R'_B. ■

So when R_A and $f\colon A \to B$ are given, then R_B, if it exists, is uniquely determined by $fi(\langle A, R_A\rangle, \langle B, R_B\rangle, f)$.

Lemma 6.3: *fi*$(\langle A, R_A\rangle, \langle B, R_B\rangle, f) \implies R_B|_{f(A)} = R_B$.

Proof. The result follows directly from fi2$(\langle A, R_A\rangle, \langle B, R_B\rangle, f)$. ∎

Lemma 6.4: *If fi*$(\langle A, R_A\rangle, \langle B, R_B\rangle, f)$, $R'_A \subseteq R_A$ *and if* R_B^+ *is finite-downward, then there exists an* R'_B *such that fi*$(\langle A, R'_A\rangle, \langle B, R'_B\rangle, f)$ *and* $R'_B \subseteq R_B$.

Proof. See appendix H. ∎

Lemma 6.5: *If* R_A *is irreflexive and if* f *is a bijection* $A \to B$, *then there exists a relation* R_B *on* B *such that fi*$(\langle A, R_A\rangle, \langle B, R_B\rangle, f)$ *and* f *is a graph isomorphism* $\langle A, R_A\rangle \to \langle B, R_B\rangle$.

Proof. See appendix H. ∎

We are now ready to define what we mean by sets of cell actions:

Definition 6.2: Let $\langle D, \prec\rangle$ be a computational event structure. Let C be a partition of D. $\langle C, \prec_c\rangle$ is a *cell action structure derived from* $\langle D, \prec\rangle$, and $\prec_c$ is the *immediate precedence relation on* C iff *fi*$(\langle D, \prec\rangle, \langle C, \prec_c\rangle, \varphi_C)$, where φ_C is the natural mapping $D \to C$. The blocks in C are called *cell actions*. C is called a *set of cell actions (SoC)*.

Cell actions model events of computation in a network of processing elements, or cells. From fi2 we can conclude, that for two cell actions c, c' holds that $c \prec_c c'$ iff $c \neq c'$ and there exist computational events $p \in c$, $p' \in c'$ such that $p \prec p'$. Since $\prec$ describes the data transmission between computational events, $\prec_c$ describes the data transmission between cell actions. Data transfers internal to a cell action are hidden. We can also note that $\prec_c^+$ is finite-downward since $\prec$ is finite-downward. Furthermore, $\langle C, \prec_c^*\rangle$ is an elementary event structure.

A cell action structure can be seen as an implementation-independent description of a restricted kind of algorithm. The cell actions are then the steps in the algorithm that are to be performed. The immediate precedence relation gives partial precedence constraints; if a cell action precedes another, then the first one must be carried out before the other, since the latter uses data computed by the former. The class of algorithms possible to describe by cell action structures is of course limited. It contains all algorithms where we always perform the same steps, with the same precedence constraints, *regardless of input data.* When such algorithms are implemented in an imperative programming language, the program typically consists of straight-line code and loops where we know the number of iterations in advance.

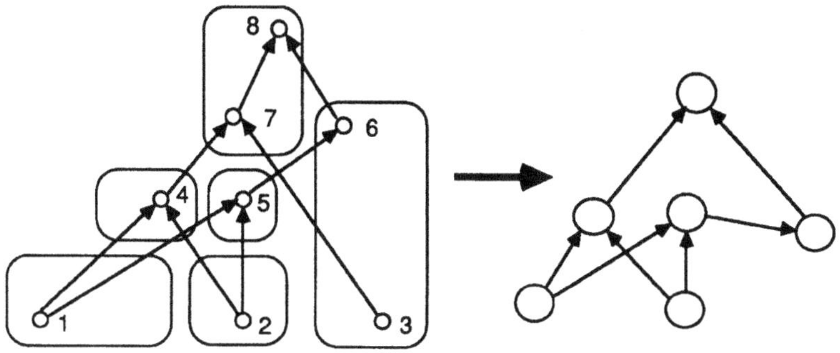

Figure 6.4: A way to partition D into $\langle C, \prec_c \rangle$.

Example 6.1: Reconsider example 5.1. A possible set of cell actions of D is the partition of D shown in figure 6.4. The resulting cell action structure $\langle C, \prec_c \rangle$ is also shown in figure 6.4. ■

Let us finally consider the semantics of a cell action; it is given by the functional semantics of the computational events it contains:

Definition 6.3: For every cell action c, its *output function* $o(c)$ is $\phi \circ \psi(c) = \{ o_\psi(\mathrm{p}) \mid \mathrm{p} \in c \}$.

Thus, every cell action defines a set of functions, namely the set of output functions of the computational events it is formed of.

6.2 Concurrent assignments

Recursion equations have by now been widely recognized as a mathematical description of static algorithms, that is suitable for extracting the possible parallelity in the given algorithm. Systems of *uniform recurrence equations* or *UREs*, [KMW69], have been used to specify algorithms suitable for implementation on regular multiprocessor systems, [Q83]. A slight extension of UREs is considered in [Raj86]. A related, more general formulation is used in [MF86]. The special case of *linear recurrences* has also been studied, [LW85]. More general formulations are based on a graph approach, [Le83,Le84,RFS82A], sets of concurrent assignments, [Li83], or recursion equations over flat data types with an undefined least element, [C85A,C85B,C86A,C86C]. All these developments have in common that only *acyclic* recursion structures are considered. Therefore, it is in fact not necessary to consider any sort of fixed-point semantics for the system of equations. Instead it is possible to interpret each step in the recurrence as an event of *assignment*. Assignments compute one or more values from previously computed values, i.e. the recursive order of computation is turned into an iterative one.

Our aim here is to relate the concept of recursion equations to the previously developed concept of cell actions. Therefore, we will give our own definition of what we mean by a system of recursion equations. We will give it in such a way that it will be easy to relate it to the previous development in this thesis.

Definition 6.4: Let $\mathcal{A}$ be an algebra and let X be a countable set of variables. A *recursion scheme* $\mathbf{F}$ *in* X *over* $\mathcal{A}$ is a partial function $X \to \mathcal{E}^{(X)}(\mathcal{A})$, for which $S(x) = S(\mathbf{p})$ for all $\langle x, \mathbf{p}\rangle \in \mathbf{F}$. $X_a(\mathbf{F})$, the *set of assigned variables of* $\mathbf{F}$, is the largest $X' \subseteq X$ such that $\mathbf{F}|_{X'}$ is total. $X_f(\mathbf{F})$, the *set of free variables of* $\mathbf{F}$, is $X \setminus X_a(\mathbf{F})$.

The recursion scheme $\mathbf{F}$ is a partial function and therefore a set of pairs. Every pair $\langle x, \mathbf{F}(x)\rangle$ in $\mathbf{F}$ is considered to be an equation $x = \mathbf{F}(x)$. (Note that $\mathbf{F}(x)$ is an arbitrary expression in $\mathcal{E}^{(X)}(\mathcal{A})$ and need not contain x itself.) $\mathbf{F}$ therefore specifies a system of equations for the variables in $X_a(\mathbf{F})$. Note that definition 6.4 recursion schemes is identical to definition 2.13 of substitutions. Thus, every recursion scheme is formally a substitution with $dom(\mathbf{F}) = X_a(\mathbf{F})$.

Definition 6.5: The relation $\prec_{\mathbf{F}}$ on the recursion scheme $\mathbf{F}$ is defined by: for all $\langle x, \mathbf{F}(x)\rangle, \langle x', \mathbf{F}(x')\rangle$ in $\mathbf{F}$,

$$\langle x, \mathbf{F}(x)\rangle \prec_{\mathbf{F}} \langle x', \mathbf{F}(x')\rangle \iff x \in varset(\mathbf{F}(x')).$$

Lemma 6.6: *$\langle x, \mathbf{F}(x)\rangle$ is minimal with respect to $\prec_{\mathbf{F}}$ iff $varset(\mathbf{F}(x)) \subseteq X_f(\mathbf{F})$.*

Proof. See appendix H. ■

Definition 6.6: The recursion scheme $\mathbf{F}$ is *causal* if and only if $\prec_{\mathbf{F}}^{+}$ is finite-downward.

If $\mathbf{F}$ is causal, then every pair $\langle x, \mathbf{F}(x)\rangle$ can be given a special meaning. It can be seen as a *single-assignment* $x \leftarrow \mathbf{F}(x)$ where the variable x is assigned the value of the expression $\mathbf{F}(x)$, which is evaluated with the values of the variables in $varset(\mathbf{F}(x))$ substituted for. Since $\prec_{\mathbf{F}}^{+}$ is finite-downward, every assigned variable is assigned a value after a finite number of steps. A free variable can be seen as an input to the system; it is not known in advance but becomes instantiated to some value when it first occurs in a right-hand side of an assignment.

The meaning of a causal recursion scheme will now be formalized. In such a scheme every single-assignment can be seen as the computation of a polynomial. This polynomial corresponds to a formal expression that is built up according to the variable dependencies expressed by the precedence relation $\prec_{\mathbf{F}}$.

Definition 6.7: Let $\mathbf{F}$ be a causal recursion scheme in X. For any $x \in X$, the *output function of x under* $\mathbf{F}$, $o_{\mathbf{F}}(x)$, is defined by:

$$o_{\mathbf{F}}(x) = \begin{cases} e_x^X, & x \in X_f(\mathbf{F}) \\ \phi(\mathbf{F}(x)) \circ \langle\, o_{\mathbf{F}}(x') \mid x' \in \mathit{varset}(\mathbf{F}(x)) \,\rangle & x \in X_a(\mathbf{F}) \end{cases}$$

and the *output expression of x under* $\mathbf{F}$, $\psi_{\mathbf{F}}(x)$, is defined by

$$\psi_{\mathbf{F}}(x) = \begin{cases} x, & x \in X_f(\mathbf{F}) \\ \psi_{\mathbf{F}}(\mathbf{F}(x)), & x \in X_a(\mathbf{F}). \end{cases}$$

Note that $\psi_{\mathbf{F}}$ is a partial function $X \to \mathcal{E}^{(X)}(\mathcal{A})$ and thus formally a substitution in $\mathcal{E}^{(X)}(\mathcal{A})$.

Theorem 6.1: *Let* $\mathbf{F}$ *be a causal recursion scheme in* X. *For any* $x \in X$, $o_{\mathbf{F}}(x) = \phi \circ \psi_{\mathbf{F}}(x)$.

Proof. See appendix H. ■

So for a causal recursion scheme, all output functions are polynomials in $\mathcal{P}^{(X)}(\mathcal{A})$ since they are images of expressions in $\mathcal{E}^{(X)}(\mathcal{A})$. But as a matter of fact they will only be dependent of the free variables in $X_f(\mathbf{F})$. The reason is that the expressions $\psi_{\mathbf{F}}(x)$ always will be in $\mathcal{E}^{(X_f(\mathbf{F}))}(\mathcal{A})$. This is stated in the following theorem:

Theorem 6.2: *Let* $\mathbf{F}$ *be a causal recursion scheme in* X. *For any* $x \in X$,

$$\psi_{\mathbf{F}}(x) \in \mathcal{E}^{(X_f(\mathbf{F}))}(\mathcal{A}).$$

Proof. See appendix H. ■

This result makes clear what the semantics of a causal recursion scheme is. Every assignment in such a scheme defines a function, that is a polynomial which is dependent of the free variables only.

Definition 6.8: Let M be a partition of the recursion scheme $\mathbf{F}$. If

$$\mathit{fi}(\langle \mathbf{F}, \prec_{\mathbf{F}} \rangle, \langle M, \prec_M \rangle, \varphi_M),$$

then M is a *set of concurrent single-assignments (cs-assignments) derived from* $\mathbf{F}$, and $\prec_M$ is *the immediate precedence relation on* M. $\langle M, \prec_M \rangle$ is a *cs-assignment structure derived from* $\langle \mathbf{F}, \prec_{\mathbf{F}} \rangle$. If $\mathbf{F}$ is causal, then M is called a causal set of concurrent single-assignments (CCSA-set) and $\langle M, \prec_M \rangle$ is called a CCSA structure.

Every cs-assignment is formally a substitution. Therefore the definition of *dom* and *range* applies to cs-assignments as well. For causal sets of concurrent assignments the cs-assignments are, as we shall se later, the counterparts to cell actions in the framework of causal recursion schemes. Each cs-assignment m, in a set of concurrent assignments derived from $\mathbf{F}$, can be seen as a single event where every variable x in $dom(m)$ is assigned the value of $\mathbf{F}(x)$. Thus, $range(m)$ can be seen as the inputs to m and $dom(m)$ as the outputs. A convenient notation for the cs-assignment $\{\langle x_1, \mathbf{F}(x_1)\rangle, \dots, \langle x_k, \mathbf{F}(x_k)\rangle\}$ is

$$\begin{gathered} x_1 \leftarrow \mathbf{F}(x_1) \\ \vdots \\ x_k \leftarrow \mathbf{F}(x_k). \end{gathered}$$

Definition 6.9: Let M be a CCSA-set derived from $\mathbf{F}$. For any $m \in M$, the *output function of* m *under* $\mathbf{F}$, $o_{\mathbf{F}}(m)$ is given by

$$o_{\mathbf{F}}(m) = \{\, o_{\mathbf{F}}(x) \mid \langle x, \mathbf{F}(x)\rangle \in m \,\}.$$

So every cs-assignment in a CCSA-set defines a multiple-valued function, which in turn is given by the recursively defined output functions for the assignments in it.

Not surprisingly, the precedence relation $\prec_M$ can be determined by matching input and output variables of cs-assignments:

Theorem 6.3: *Let* $\langle M, \prec_M\rangle$ *be a cs-assignment structure. For all* m, m' *in* M,

$$m \prec_M m' \iff m \neq m' \wedge dom(m) \cap range(m') \neq \emptyset.$$

Proof. See appendix H. ∎

A variable in $dom(m) \cap range(m')$ represents a value that is produced by m and used by m'. Thus, $\prec_M$ is really a data dependence relation. There is, however, a peculiarity that should be noted. We must require $m \neq m'$ in the right-hand side of the equivalence in theorem 6.3 above. The reason is that $dom(m) \cap range(m)$ might be nonempty even though $m \not\prec_M m$. The following example of a cs-assignment illustrates this:

$$m = \begin{matrix} x \leftarrow f(y) \\ y \leftarrow g(z) \end{matrix}$$

Here $dom(m) = \{x, y\}$ and $range(m) = \{y, z\}$, so $dom(m) \cap range(m) = \{y\}$. But still the data dependence structure is acyclic. y should be seen as a quantity internal to m that does not affect the data dependency. As a matter of fact we can convert m into a cs-assignment m', such that $o_{\mathbf{F}'}(m') = o_{\mathbf{F}}(m)$ and $dom(m) \cap range(m) = \emptyset$, by substituting $g(z)$ for y in the first assignment:

$$m = \begin{array}{l} x \leftarrow f(g(z)) \\ y \leftarrow g(z) \end{array}$$

It would be nice to get rid of the complication that *dom* and *range* sometimes have variables in common which do not indicate a data dependency. This is the motivation for the following definition.

Definition 6.10: The cs-assignment m is *pure* iff $dom(m) \cap range(m) = \emptyset$. The set M of cs-assignments is pure iff all $m \in M$ are pure. The cs-assignment structure $\langle M, \prec_M \rangle$ is pure iff M is pure.

Proposition 6.1: *For any cs-assignment structure* $\langle M, \prec_M \rangle$

$$\forall m, m' \in M[m \prec_M m' \iff dom(m) \cap range(m') \neq \emptyset] \iff M \textit{ is pure.}$$

Proof.

$\Longrightarrow$: Assume that $m \prec_M m' \iff dom(m) \cap range(m') \neq \emptyset$ for any $m, m' \in M$. For all $m \in M$ holds that $m \not\prec_M m$. It follows that $dom(m) \cap range(m) = \emptyset$ for all $m \in M$.

$\Longleftarrow$: Follows directly from theorem 6.3. ■

Corollary 6.1: *If M is pure, then $\prec_M$ is irreflexive.*

In the example above, m was converted into the pure cs-assignment m' with the same output function by substituting $g(z)$ for y. In fact it is *always* possible to transform a CCSA-set into an equivalent pure CCSA-set in this manner. Such a transformation is defined more formally in the following definition.

Definition 6.11: For any set M of cs-assignments derived from a causal recursion scheme $\mathbf{F}$ in X, the new recursion scheme $p(\mathbf{F})$ and the function $p: \mathbf{F} \to p(\mathbf{F})$ are defined by: for all $\langle x, \mathbf{p} \rangle \in \mathbf{F}$,

$$p(\langle x, \mathbf{p} \rangle) = \langle x, p(\varphi_M(\langle x, \mathbf{p} \rangle))(\mathbf{p}) \rangle$$

and

$$p(\mathbf{F}) = \{ p(y) \mid y \in \mathbf{F} \}.$$

We also define

$$P(M) = \{ p(m) \mid m \in M \}.$$

This definition deserves some discussion. φ_M is the natural mapping $\mathbf{F} \to M$. For any $\langle x, \mathbf{p}\rangle \in \mathbf{F}$, denote $\varphi_M(\langle x, \mathbf{p}\rangle)$ by m. Definition 6.11 is recursive; since $p(\langle x, \mathbf{p}\rangle) = \langle x, \mathbf{q}\rangle$ for some $\mathbf{q}$ it holds that $dom(p(m)) = dom(m)$. Thus, it follows, for those $\langle x, \mathbf{p}\rangle$ for which $varset(\mathbf{p}) \cap dom(m) = \emptyset$, that $p(\langle x, \mathbf{p}\rangle) = \langle x, \mathbf{p}\rangle$ and the images of the other assignments in $\mathbf{F}$ can be determined recursively following $\prec_{\mathbf{F}}|_m$, the restriction of $\prec_{\mathbf{F}}$ to m. This relation is, for every $\langle x, \mathbf{p}\rangle$, a subrelation to the well-founded ordering $\prec_{\mathbf{F}}$ and thus well-founded itself. Therefore, $p(\mathbf{F})$ is properly defined by the above. Since these restricted relations are not connected for different m, every $p(\langle x, \mathbf{p}\rangle)$ will be uniquely determined by the above. Thus, p is well-defined. Since $p(\langle x, \mathbf{p}\rangle) = \langle x, \mathbf{q}\rangle$ for some expression $\mathbf{q}$, $x \neq x'$ implies $p(\langle x, \mathbf{p}\rangle) \neq p(\langle x', \mathbf{p}'\rangle)$. Therefore p is 1-1 (if $\langle x, \mathbf{p}\rangle \in \mathbf{F}$ there is no $\langle x, \mathbf{p}'\rangle \in \mathbf{F}$ since $\mathbf{F}$ is a partial function). By definition, p is onto. Thus, p is a bijection $\mathbf{F} \to p(\mathbf{F})$. We can also note that $p(\mathbf{F})$ is a recursion scheme in X, as $\mathbf{F}$.

Theorem 6.4: *For any CCSA structure $\langle M, \prec_M\rangle$ derived from $\mathbf{F}$, $\langle P(M), \prec_{P(M)}\rangle$ is an isomorphic pure CCSA structure derived from $p(\mathbf{F})$. p is an isomorphism $M \to P(M)$, and $o_{p(\mathbf{F})}(p(m)) = o_{\mathbf{F}}(m)$ for all $m \in M$.*

Proof. See appendix H. ∎

In this section we have defined CCSA structures by partitioning causal recursion schemes. As a matter of fact it is sometimes possible to go the other way around; for certain sets M of substitutions it is possible to derive a causal recursion scheme such that M is a set of cs-assignments derived from it.

Theorem 6.5: *Let M be a subset of $\mathcal{S}^{(X)}(\mathcal{A})$, such that all $m \in M$ are finite and $m \neq m' \implies dom(m) \cap dom(m') = \emptyset$ for all $m, m' \in M$. Let the relation $\prec_M$ on M be given by*

$$m \prec_M m' \iff dom(m) \cap range(m') \neq \emptyset$$

for all $m, m' \in M$. If $\prec_M^+$ is finite-downward, then $\mathbf{F}_M = \bigcup(\, m \mid m \in M\,)$ is a causal recursion scheme in X over $\mathcal{A}$ and $\langle M, \prec_M\rangle$ is a pure CCSA structure derived from $\langle \mathbf{F}_M, \prec_{\mathbf{F}_M}\rangle$.

Proof. See appendix H. ∎

6.3 Equivalence of cell actions and cs-assignments

Now we will investigate under which circumstances there is a set of cell actions that is equivalent to a given CCSA-set and vice versa. First we will see how to construct a set of cell actions that is equivalent to a given CCSA-set.

Informally, a cs-assignment m will be interpreted in the following way: it will correspond to a set of events $c(m)$, where for every $\langle x, \mathbf{p}\rangle \in m$ there is an event of computation for $\mathbf{p}$ and every subexpression of $\mathbf{p}$ that is not an assigned variable. The latter represent values produced by previous computational events and should not be seen as such events themselves. Thus, $c(m)$ contains all computational events necessary to evaluate the expressions of the variables assigned by m, given the inputs from other cs-assignments. $\{\, c(m) \mid m \in M \,\}$ will be our set of cell actions corresponding to M.

There is, however, a complication that should not be overlooked. Consider an assignment $x \leftarrow x'$, where x' is an assigned variable. Apparently we cannot find any computational events corresponding to it according to above, since that construction does require that some evaluation should take place of an expression that is not an assigned variable. This leads to the following definition of *guarded* assignments, where the name is chosen in analogy with process algebra:

Definition 6.12: $\langle x, \mathbf{p}\rangle$ in the recursion scheme $\mathbf{F}$ is guarded iff $\mathbf{p} \notin X_a(\mathbf{F})$. $\mathbf{F}$ is guarded iff all assignments in $\mathbf{F}$ are guarded. The set M of cs-assignments is guarded iff it is derived from a guarded recursion scheme. The cs-assignment structure $\langle M, \prec_M\rangle$ is guarded iff M is guarded.

How, then, should an assignment $x \leftarrow x'$, where $x' \in X_a(\mathbf{F})$, be treated? The natural way to think of it, is to consider it as an event where the value of x' is propagated unaffected. Such events are modelled by applications of the identity function $i_{S(x)}$. So if we add this function as an operator to our algebra, then we can replace the assignment $x \leftarrow x'$ by the guarded assignment $x \leftarrow i_{S(x)} \bullet x'$. More formally, we define the following transformation on recursion schemes:

Definition 6.13: Let $\mathcal{A}$ be an algebra. For any recursion scheme $\mathbf{F}$ over $\mathcal{A}$, the recursion scheme $g(\mathbf{F})$ over $i(\mathcal{A})$ and the function $g\colon \mathbf{F} \to g(\mathbf{F})$ are defined by: for every $\langle x, \mathbf{p}\rangle \in \mathbf{F}$,

$$g(\langle x, \mathbf{p}\rangle) = \begin{cases} \langle x, \mathbf{p}\rangle, & \mathbf{p} \notin X_a(\mathbf{F}) \\ \langle x, i_{S(x)} \bullet \mathbf{p}\rangle, & \mathbf{p} \in X_a(\mathbf{F}) \end{cases}$$

and

$$g(\mathbf{F}) = \{\, g(y) \mid y \in \mathbf{F} \,\}.$$

It is easy to see that g is a bijection, that $X_f(g(\mathbf{F})) = X_f(\mathbf{F})$ and that $X_a(g(\mathbf{F})) = X_a(\mathbf{F})$. It should also be intuitively clear that $o_{g(\mathbf{F})}(x) = o_{\mathbf{F}}(x)$ for all $x \in X$, since the application of the identity function $i_{S(x)}$ does not change the output function. This is formally stated in the theorem below.

Theorem 6.6: *For every causal recursion scheme* $\mathbf{F}$ *over* $\mathcal{A}$, $g(\mathbf{F})$ *is a causal guarded recursion scheme over* $i(\mathcal{A})$ *such that* $o_{g(\mathbf{F})}(x) = o_{\mathbf{F}}(x)$ *for all* $x \in X_a(\mathbf{F})$. *Furthermore,* g *is a graph isomorphism* $\langle \mathbf{F}, \prec_{\mathbf{F}} \rangle \to \langle g(\mathbf{F}), \prec_{g(\mathbf{F})} \rangle$.

Proof. See appendix H. ■

The following corollary follows immediately from theorem 6.6:

Corollary 6.2: *For every set* M *of cs-assignments derived from a causal recursion scheme* $\mathbf{F}$ *over* $\mathcal{A}$, *the set of cs-assignments* $G(M) = \{ g(m) \mid m \in M \}$ *is derived from the causal guarded recursion scheme* $g(\mathbf{F})$ *over* $i(\mathcal{A})$. $G: M \to G(M)$, *defined by* $G(m) = g(m)$ *for all* $m \in M$, *is a graph isomorphism* $\langle M, \prec_M \rangle \to \langle G(M), \prec_{G(M)} \rangle$ *and for all* $m \in M$ *holds that* $o_{g(\mathbf{F})}(G(m)) = o_{\mathbf{F}}(m)$.

Having stated this we can restrict our attention to guarded sets of assignments, knowing that every set of cs-assignments can be converted into a guarded one. For a guarded CCSA-set we can construct a cell action structure as sketched in the beginning of this section, without the complication of finding computational events for assigned variables. However, since a cell action structure is derived from a computational event structure, which in turn is derived from an evaluable set of expressions, we must also construct these sets.

Definition 6.14: For any guarded CCSA structure $\langle M, \prec_M \rangle$, derived from a recursion scheme $\mathbf{F}$ in X over $\mathcal{A}$, define:

$$C(m) = \{ \langle m, \mathbf{p} \rangle \mid \exists \langle x', \mathbf{p}' \rangle \in m[\mathbf{p} \in ec(\mathbf{p}') \setminus X_a(\mathbf{F})] \} \qquad \text{(c1)}$$

for any $m \in M$, and furthermore

$$D(M) = \bigcup (C(m) \mid m \in M), \qquad \text{(c2)}$$

$$C(M) = \{ C(m) \mid m \in M \}. \qquad \text{(c3)}$$

For any $\langle m, \mathbf{p} \rangle \in D(M)$, define

$$\psi_M(\langle m, \mathbf{p} \rangle) = \psi_{\mathbf{F}}(\mathbf{p}) \qquad \text{(c4)}$$

and

$$E(M) = \psi_M(D(M)). \qquad \text{(c5)}$$

Finally, for every $\langle m, \mathbf{p} \rangle \in D(M)$ such that $\mathbf{p} = f \bullet \langle \mathbf{p}_i \mid i \in I_f \rangle$, define $s_{\langle m,\mathbf{p} \rangle}: I_f \to D(M)$ by: for all $i \in I_f$,

$$s_{\langle m,\mathbf{p} \rangle}(i) = \begin{cases} \langle m, \mathbf{p}_i \rangle, & \mathbf{p}_i \notin X_a(\mathbf{F}) \\ \langle m', \mathbf{F}(\mathbf{p}_i) \rangle, & \mathbf{p}_i \in X_a(\mathbf{F}),\ \mathbf{p}_i \in dom(m'). \end{cases} \tag{c6}$$

and

$$SF(M) = (\, s_{\langle m,\mathbf{p} \rangle} \mid \langle m, \mathbf{p} \rangle \in D(M) \text{ and } \mathbf{p} \text{ is compound}\,). \tag{c7}$$

The idea of definition 6.14 is to construct, for each cs-assignment, a cell action containing one computational event for each subexpression of the right-hand side expressions of the assignments, except for the possible assigned variables which correspond to computational events in other cell actions. The following results show that the concepts above are well-defined. We assume below that $\langle M, \prec_M \rangle$ is a guarded CCSA structure, derived from a recursion scheme $\mathbf{F}$ in X over $\mathcal{A}$, as required in definition 6.14.

Lemma 6.7: *$E(M)$ is an evaluable subset of $\mathcal{E}^{(X_f(\mathbf{F}))}(\mathcal{A})$. $D(M)$ is a set of computational events of $E(M)$. $SF(M)$ is a family of selector functions with respect to $D(M)$ and ψ_M.*

Proof. See appendix H. ■

Lemma 6.8: *There is a binary relation $\prec_{C(M)}$ on $C(M)$ such that $\langle C(M), \prec_{C(M)} \rangle$ is a cell action structure derived from $\langle D(M), \prec_{SF(M)} \rangle$. C is a graph isomorphism $\langle M, \prec_M \rangle \to \langle C(M), \prec_{C(M)} \rangle$.*

Proof. See appendix H. ■

The final result is stated in the following theorem:

Theorem 6.7: *For any guarded CCSA structure $\langle M, \prec_M \rangle$, derived from a recursion scheme $\mathbf{F}$ in X over $\mathcal{A}$, $\langle C(M), \prec_{C(M)} \rangle$ is a cell action structure. Furthermore, C is a graph isomorphism $\langle M, \prec_M \rangle \to \langle C(M), \prec_{C(M)} \rangle$ such that $o_{\mathbf{F}}(m) \subseteq o(C(m))$ for all $m \in M$.*

Proof. See appendix H. ■

Corollary 6.2 and theorem 6.7 now immediately give the following theorem as a corollary:

Theorem 6.8: *For any CCSA structure $\langle M, \prec_M \rangle$, derived from a recursion scheme $\mathbf{F}$ in X over $\mathcal{A}$, $\langle C(G(M)), \prec_{C(G(M))} \rangle$ is a cell action structure. Furthermore, $C \circ G$ is a graph isomorphism $\langle M, \prec_M \rangle \to$*

$\langle C(G(M)), \prec_{C(G(M))}\rangle$ *such that* $o_{\mathbf{F}}(m) \subseteq o(C(G(m)))$ *for all* $m \in M$. $E(G(M))$ *is an evaluable subset of* $\mathcal{E}^{(X_f(\mathbf{F}))}(i(\mathcal{A}))$.

This completes the first part of our investigation concerning equivalence between CCSA-sets and sets of cell actions. For every CCSA-set M, there is a set of cell actions $C(G(M))$ that subsumes M in the sense that $o_{\mathbf{F}}(m) \subseteq o(C(G(m)))$ for every $m \in M$. The fact that $o_{\mathbf{F}}(m)$ not necessarily is equal to $o(C(G(m)))$ is not of a deep nature. Rather, it is due to the way we have chosen to define output functions for cell actions and cs-assignments, respectively. Every computational event in a cell action c adds an output function to the total output function of c. A cs-assignment, on the other hand, can be seen as equivalent to a cell action where some computational events are designated as special "output" events, distinguished by that they define an assigned variable. An alternative definition of output function for cell actions could include a set of "output" events from which the set of functions is formed, instead of from the whole cell action. This would, however, complicate matters unnecessarily much. Therefore this approach is not taken here.

Let us now turn to the opposite problem: given a cell action structure, find an isomorphic CCSA structure such that the output function of every cs-assignment is equal to the corresponding cell action. This will turn out to be somewhat easier than the previously treated problem. The idea is to construct, for every cell action c, a cs-assignment consisting of one assignment for each $\mathrm{p} \in c$. The construction is given in the following definition:

Definition 6.15: Let $\langle C, \prec_c\rangle$ be a cell action structure, derived from a computational event structure $\langle D, \prec\rangle$ of a set of expressions $E \subseteq \mathcal{E}^{(X)}(\mathcal{A})$ under ψ. Assume that $D \cap X = \emptyset$. For all $c \in C$, *the set of input computational events to c*, *ice*(c), is defined by the following:

$$ice(c) = \{\, \mathrm{p}' \mid \mathrm{p}' \in D \wedge \mathrm{p}' \notin c \wedge \exists \mathrm{p} \in c[\mathrm{p}' \prec \mathrm{p}]\,\}. \qquad (ice)$$

Assume that $\prec$ is given by the family of selector functions $\{\, s_{\mathrm{p}} \mid \mathrm{p} \in comp(D)\,\}$. For every cell action $c \in C$ and for all computational events $\mathrm{p} \in c \cup ice(c)$, we define the *associated expression* $ae_c(\mathrm{p})$ by:

ae1. For all $\mathrm{p} \in ice(c)$, $ae_c(\mathrm{p}) = \mathrm{p}$.
ae2. For all $\mathrm{p} \in c$ such that $\psi(\mathrm{p}) \in X$, $ae_c(\mathrm{p}) = \psi(\mathrm{p})$.
ae3. For all $\mathrm{p} \in c$ such that $\psi(\mathrm{p}) = f \bullet \langle\, \mathbf{p}_i \mid i \in I_f\,\rangle$,
$ae_c(\mathrm{p}) = f \bullet \langle\, ae_c(s_{\mathrm{p}}(i)) \mid i \in I_f\,\rangle$.

Now, for every $c \in C$, we define

$$M(c) = \{\, \mathrm{p} \leftarrow ae_c(\mathrm{p}) \mid \mathrm{p} \in c\,\}, \qquad (m1)$$

and furthermore

$$M(C) = \{ M(c) \mid c \in C \}, \qquad (m2)$$
$$\mathbf{F}_C = \bigcup(M(c) \mid c \in C). \qquad (m3)$$

The associated expressions of computational events in c are built up with the input computational events of c as variables representing the values they provide for c. This is the reason why we require $D \cap X = \emptyset$, we do not want to mix up assigned and free variables. This is not a serious restriction. If $D \cap X \neq \emptyset$, then we can equally well consider a set D' that is in 1-1-correspondence with D and disjoint from X.

The results below hold for any cell action structure $\langle C, \prec_c \rangle$, that is derived from a computational event structure $\langle D, \prec \rangle$ of a set of expressions $E \subseteq \mathcal{E}^{(X)}(\mathcal{A})$.

Lemma 6.9: *$\mathbf{F}_C$ is a causal recursion scheme in $X \cup D$ over $\mathcal{A}$ and $X_f(\mathbf{F}_C) = X$. $M(C)$ is a CCSA-set derived from $\mathbf{F}_C$.*

Proof. See appendix H. ∎

Theorem 6.9: *M is a graph isomorphism $\langle C, \prec_c \rangle \to \langle M(C), \prec_{M(C)} \rangle$. For all $c \in C$ holds that $o(c) = o_{\mathbf{F}_C}(M(c))$.*

Proof. See appendix H. ∎

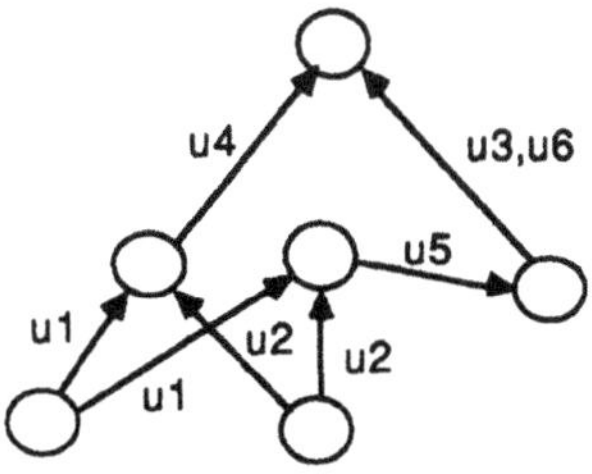

Figure 6.5: $\langle M, \prec_M \rangle$

With the result above in mind we can conclude the following: for every CCSA structure $\langle M, \prec_M \rangle$ there is an equivalent cell action structure $\langle C(G(M)), \prec_{C(G(M))} \rangle$. Furthermore, for every cell action structure $\langle C, \prec_c \rangle$, there is an equivalent CCSA structure $\langle M(C), \prec_{M(C)} \rangle$. Cell action structures and CCSA structures are therefore interchangeable concepts, and we might use any of these to represent algorithms of the restricted kind we consider here. If the algorithms are to be specified directly step-by-step rather than from a functional output specification, then CCSA structures are clearly to prefer. If, on the other hand, the algorithms are to be derived from an output specification, then cell action structures are more suited as algorithm descriptions.

Example 6.2: Reconsider once more the examples 5.1 and 6.1. The following CCSA structure $\langle M(C), \prec_{M(C)}\rangle$ is easily seen to be isomorphic with $\langle C, \prec_c\rangle$:

$$u_1 \leftarrow x \qquad u_2 \leftarrow y \qquad \begin{array}{l} u_3 \leftarrow z \\ u_6 \leftarrow \sin(u_5) \end{array}$$

$$u_4 \leftarrow u_1 + u_2 \qquad u_5 \leftarrow u_1 + u_2 \qquad \begin{array}{l} u_7 \leftarrow u_3 \cdot u_4 \\ u_8 \leftarrow u_3 \cdot u_4 + u_6 \end{array}$$

Here every assigned variable u_i corresponds to the computational event i. By recursively substituting output expressions for assigned variables, according to definition 6.7, it is easy to see that the output expression for every u_i is equal to $\psi(i)$. Thus, the output function is the same for every cell action and the corresponding cs-assignment. In figure 6.5 $\langle M, \prec_M\rangle$ is drawn as a graph. Every edge is labelled with the assigned variable(s) that causes the corresponding dependence. Cf. figure 6.4. ∎

7. A hardware model

In this chapter we consider a simple hardware model. It is quite abstract, in the sense that it does not contain any means of describing geometrical and electrical properties of circuits and layouts. Instead the emphasis is on such issues as timing and connectivity; we consider points in a *space* which are connected by communication lines in some fashion. These points are dimensionless and they are possible locations for *cells*, or processing elements. The communication lines are connections in an abstract sense and are thought of as being able of carrying any sort of information that may be sent (boolean values, floating-point numbers, ...).

The next step is to connect hardware and algorithm. By "algorithm" we here simply mean a cell action structure. Each cell action corresponds to an *event*. Therefore we need some corresponding notion of events in the hardware model. The solution is to add a global *time*, that is common to all cells. The result is a set of pairs of cells and times, where each pair $\langle t, r \rangle$ represents the unique event taking place in cell r at time t. The introduction of a global time means that we restrict our attention to *synchronous* systems with a global clock signal, where the synchronization makes it possible to speak of simultaneous events. The essential idea in this thesis is to connect events of cell actions and events in the hardware by a *schedule*, telling where and when each cell action is to be performed.

This approach leads to a kind of "high-level" model where much detail is hidden. The advantage is, besides the mathematical simplicity, that the description becomes more implementation-independent. The internal details of a cell are irrelevant. We do not care about the technology used to build a cell, or even if its actions are hardwired or programmed, as long as it performs the cell actions assigned to it at the proper times. What we obtain is a specification for each cell of what it should do at certain times and a set of input/output data relationships between events in the cells. From this the communication lines necessary to support the communication requirements of the schedule can be derived. This can be seen as a 'topological floorplanning" where the placements of the cells and the routing of wires between them is determined.

7.1 Space-time

Space-time is a model of the events in a network of cells; every space point is a possible location for a cell and at every time something (for instance a cell action) may take place at a cell.

Definition 7.1: (Space-time)

st1. Any countable nonempty set R is a *space.*

st2. For any *least time* $t_0 \in Z$, $T = \{ n + t_0 \mid n \in N \}$ is a *set of times.*

st3. If T is a set of times and R is a space, then $T \times R$ is a *space-time.*

Every set of times T starts with a least time t_0 and it is isomorphic with N. This means that the less-than relation "$<$" on T is well-founded. There is no reason but convenience to have a least time that is nonzero. In the rest of this paper we will w.l.o.g. assume that $T = N$, if nothing else is explicitly stated.

Discrete time, as introduced here, has the following physical interpretation: one unit of discrete time corresponds to a physical time that is long enough to allow any cell to finish what it is doing. The simple definition of a space-time as the cartesian product of a set of times and a space, is a model of the events in a synchronous, clocked system. In such a system one unit of discrete time naturally corresponds to one clock cycle. If the cells are built from combinatorial logic, then all levels have to settle before the next clock tick.

7.2 Communication constraints

Definition 7.2: The binary relation $\prec_s$ on the space-time S is a *communication ordering* on S if and only if:

co1. For all $\langle t, r \rangle$, $\langle t', r' \rangle$ in S, $\langle t, r \rangle \prec_s \langle t', r' \rangle \implies t \leq t'$.

co2. $\prec_s^+$ is finite-downward.

$\langle S, \prec_s \rangle$ is called a *communication structure.*

Communication orderings are used to describe the communication constraints that we want to pose on the resulting hardware. $s \prec_s s'$ means that communication may take place between the events s and s'.

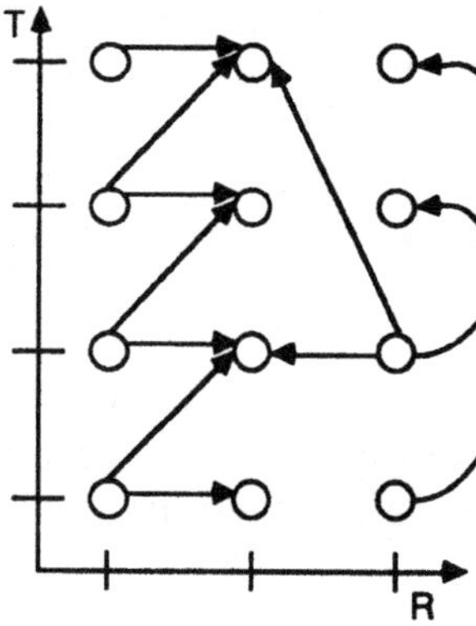

Figure 7.1: A communication ordering.

Col is a time causality criterion; communication cannot be directed backwards in time. It is, however, possible for communicating events to take place at the same time. The interpretation of this, in a clocked system, is that the signal carrying the information does not pass any latch on its way between the cells where the events take place. Since the clock cycle time is supposed to be long enough to allow everything to settle, both events will have taken place before the next discrete time. Thus, they can be considered as having taken place simultaneously, and the communication between them can be seen as instant. From co2 follows that $\prec_s^+$ is a strict partial order. This implies that there can be no cycles in the communication structure. Since col prohibits cycles others than such going through events taking place at the same time, co2 prohibits cycles with no delay in the communication structure.

Communication orderings can themselves be ordered by set inclusion; $\prec_s \subseteq \prec_s'$ means that $\prec_s'$ can support all communication that $\prec_s$ supports. We are interested in set operations on communication orderings, since they express how the communication constraints changes when the corresponding communication orderings are combined in different ways. $\prec_s \cap \prec_s'$, for instance, is the relation expressing that the constraints given by both $\prec_s$ and $\prec_s'$ are to hold. $\prec_s \cup \prec_s'$, on the other hand, gives the relaxed constraints obtained when adding the communication capabilities given by $\prec_s$ and $\prec_s'$.

Let us denote the set of all communication orderings over S by $\mathcal{C}(S)$. Then we can state the following:

Proposition 7.1: $\emptyset \in \mathcal{C}(S)$. *If* $\prec_s, \prec_s' \in \mathcal{C}(S)$, *then* $\prec_s \cap \prec_s' \in \mathcal{C}(S)$.

Proof. See appendix I. ■

Thus, $\langle \mathcal{C}(S), \cap \rangle$ is a semilattice since the intersection of two communication orderings always is a communication ordering itself and since "$\cap$" is idempotent, commutative and associative. There is also a least element $\emptyset$.

Proposition 7.2: $\prec_s \in \mathcal{C}(S) \wedge \prec \subseteq \prec_s \implies \prec \in \mathcal{C}(S)$.

Proof. See appendix I. ■

Proposition 7.3: $\prec \notin \mathcal{C}(S) \wedge \prec \subseteq \prec' \implies \prec' \notin \mathcal{C}(S)$.

Proof. We prove the equivalent reversed implication $\prec' \in \mathcal{C}(S) \implies \prec \in \mathcal{C}(S) \vee \prec \not\subseteq \prec'$ instead. Assume $\prec' \in \mathcal{C}(S)$. If $\prec \not\subseteq \prec'$ the implication is trivially true, so assume that $\prec \subseteq \prec'$. But then $\prec \in \mathcal{C}(S)$ follows from proposition 7.2. ■

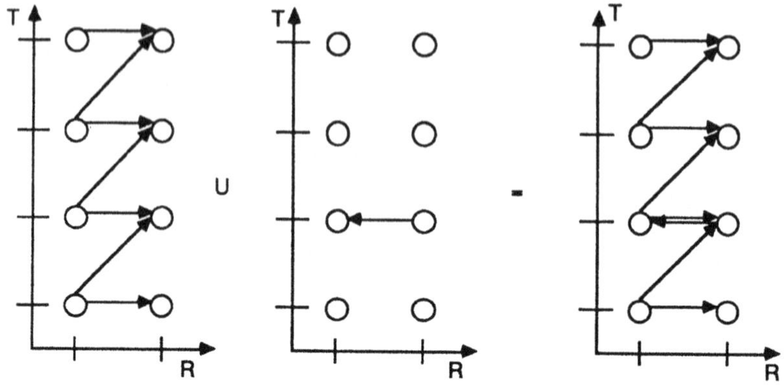

Figure 7.2: The union of two communication orderings

Note, that the union of two communication orderings is not always a communication ordering. A counterexample is shown in 7.2.

The fact that the union of two communication orderings is not always a communication ordering is due to the fact that we allow *ripple*, that is: communication between events taking place at the same discrete time (which corresponds to a signal rippling through several cells without being interrupted by latches). So if the union of two communication orderings with ripple is formed, then zero-delay cycles may be introduced. An important subclass of communication orderings is the one where ripple is not allowed.

Definition 7.3: The relation $\prec$ on the space-time S is *ripple-free* iff:

$$\forall \langle t, r \rangle, \langle t', r' \rangle \in S[\langle t, r \rangle \prec \langle t', r' \rangle \implies t < t'].$$

The set of all ripple-free relations over S is denoted by $\mathcal{R}(S)$.

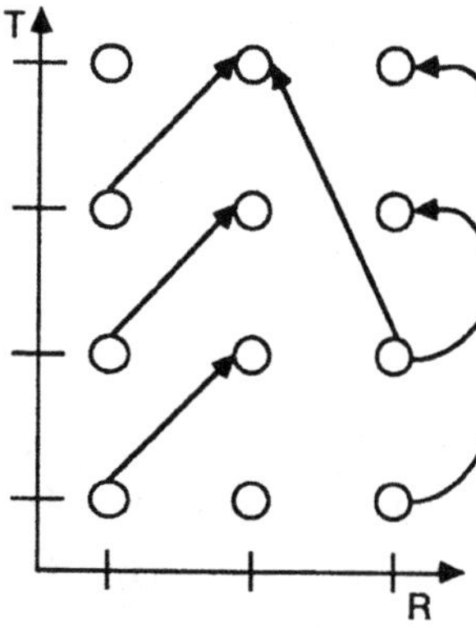

Figure 7.3: A ripple-free space-time relation.

Lemma 7.1: *For any space-time S, all $\prec \in \mathcal{R}(S)$ are well-founded.*

Proof. Define the mapping $f: S \rightarrow T$ by $f(\langle t, r\rangle) = t$ for all $\langle t, r\rangle \in S$. Since R is ripple-free, it follows that f is monotone: $sRs' \implies f(s) < f(s')$ for all $s, s' \in S$. $<$ is well-founded on T, thus R must be well-founded on S. ■

Proposition 7.4: *For any space-time S, $\langle \mathcal{R}(S); \cup, \cap \rangle$ is a lattice. $\emptyset$ is a least element. $\prec_T$, defined by $\langle t, r\rangle \prec_T \langle t', r'\rangle$ iff $t < t'$, belongs to $\mathcal{R}(S)$ and is a greatest element.*

Proof. See appendix I. ■

We are particularly interested in the ripple-free relations that also are communication orderings. The following proposition gives a condition for when this is the case.

Proposition 7.5: *Let S be a space-time. If $\prec \in \mathcal{R}(S)$, then $\prec \in \mathcal{C}(S)$ iff for all $s \in S$ $sp(s, \prec)$ is finite.*

Proof. See appendix I. ■

Proposition 7.6: *For any space-time S, $\langle \mathcal{R}(S) \cap \mathcal{C}(S); \cup, \cap \rangle$ is a lattice with $\emptyset$ as a least element.*

Proof. See appendix I. ■

It is not always the case that $\prec_T$ belongs to $\mathcal{R}(S) \cap \mathcal{C}(S)$, and there are situations when there is no greatest element in $\mathcal{R}(S) \cap \mathcal{C}(S)$. The following proposition gives the exact conditions for when there is a greatest element:

Proposition 7.7: *For any space R, $\prec_T \in \mathcal{R}(T \times R) \cap \mathcal{C}(T \times R)$ iff R is finite. If $\prec_T \notin \mathcal{R}(T \times R) \cap \mathcal{C}(T \times R)$, then there is no greatest element in $\mathcal{R}(T \times R) \cap \mathcal{C}(T \times R)$.*

Proof. See appendix I. ■

The following corollary follows immediately from propositions 7.4 and 7.7:

Corollary 7.1: *For any finite space R holds that $\mathcal{R}(T \times R) \subseteq \mathcal{C}(T \times R)$.*

Let us finally consider the degenerate case when $|R| = 1$. Note that this case is of some interest. The space-time $T \times R$, where R has a single element, describes the events in a uniprocessor.

Proposition 7.8: *For any space R, where $|R| = 1$, $\mathcal{R}(T \times R) = \mathcal{C}(T \times R)$.*

Proof. See appendix I. ∎

7.3 Schedules of sets of cell actions

Definition 7.4: Let A be a set, let $\prec$ be a relation on A and let S be a space-time. For any $F: A \rightarrow S$, we define *the mapped relation* $\prec_F$ on S by $\mathit{fi}(\langle A, \prec\rangle, \langle S, \prec_F\rangle, F)$ if it exists.

By lemma 6.2 $\prec_F$, must be unique if it exists. Therefore it is well-defined. A function $F: A \rightarrow T \times R$ will sometimes be seen as two functions, $t_F: A \rightarrow T$ and $r_F: A \rightarrow R$ given by $F(a) = \langle t_F(a), r_F(a)\rangle$ for all $a \in A$. A convenient way to denote this relationship is $F = \langle t_F, r_F\rangle$.

We now consider functions from a set of cell actions, with precedence relation $\prec_c$, to a space-time. Then the *mapped precedence relation* $\prec_{cF}$ is of interest. The interpretation of F is that the cell action c is performed at the event $F(c)$. Thus, $\prec_{cF}$ gives the communication requirements: if $F(c) \prec_{cF} F(c')$, then output data from c is used as input to c' and therefore this data must be sent between the events $F(c)$ and $F(c')$. Let us now define what kind of mappings from set of cell actions to space-time we allow:

Definition 7.5: Let $\langle C, \prec_c\rangle$ be a cell action structure and let $\langle S, \prec_s\rangle$ be a communication structure. W is a *weakly correct mapping (wcm)* $\langle C, \prec_c\rangle \rightarrow \langle S, \prec_s\rangle$ iff it is a function $C \rightarrow S$ such that:

$$\prec_{cW} \text{ exists and } \prec_{cW} \subseteq \prec_s.$$

W is a *correct mapping (cm)* $\langle C, \prec_c\rangle \rightarrow \langle S, \prec_s\rangle$ iff it is a weakly correct mapping $\langle C, \prec_c\rangle \rightarrow \langle S, \prec_s\rangle$ and 1-1.

The schedule of cell actions defined by a weakly correct mapping can be visualized in a *space-time diagram*, that shows the mapped precedence relation in space-time and possibly the communication ordering in question. A simple example is shown in figure 7.4.

The existence of $\prec_{cW}$ is a finiteness constraint. For instance it guarantees that only a finite number of cell actions are mapped to any point in space-time. $\prec_{cW} \subseteq \prec_s$ is a monotonicity constraint. It assures that the communication requirements of $\langle C, \prec_c\rangle$ can be met by $\langle S, \prec_s\rangle$ under the schedule given by W.

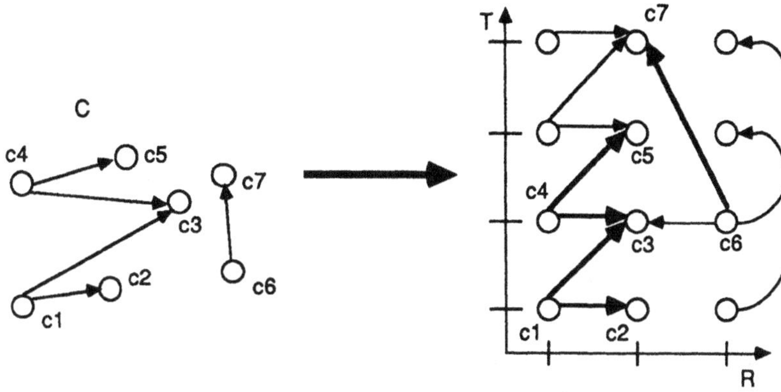

Figure 7.4: A cell action structure and a space-time diagram for a correct mapping.

Proposition 7.9: *If W is a weakly correct mapping $\langle C, \prec_c \rangle \to \langle S, \prec_s \rangle$, then, for all $\prec$ such that $\prec_{cW} \subseteq \prec \subseteq \prec_s$, holds that W is a weakly correct mapping $\langle C, \prec_c \rangle \to \langle S, \prec \rangle$.*

Proof. Assume that W is a weakly correct mapping $\langle C, \prec_c \rangle \to \langle S, \prec_s \rangle$ and that $\prec_{cW} \subseteq \prec \subseteq \prec_s$. By proposition 7.2, $\prec$ is a communication ordering. Since W is a weakly correct mapping $\langle C, \prec_c \rangle \to \langle S, \prec_s \rangle$, it follows that $\prec_{cW}$ exists. By the assumption $\prec_{cW} \subseteq \prec$. ∎

A communication ordering $\prec_s$ sometimes describes communication constraints given by an existing processor configuration. A mapped precedence relation $\prec_{cW}$ describes the communication requirements of the cell action schedule given by W. There are two cases where the model developed so far can be applied:

1. *Scheduling*: in this case the communication ordering is given in advance which corresponds to an existing multiprocessor system. The task is to find a schedule of the cell actions for which the communication requirements can be met by the given communication capabilities.
2. *Synthesis*: in the case of synthesis there is no given communication ordering. The first step in the synthesis procedure is instead to find a function W from cell actions C to space-time S, such that the mapped precedence relation $\prec_{cW}$ is a communication ordering *itself*. Since the mapped precedence relation contains itself, W is then a weakly correct mapping $\langle C, \prec_c \rangle \to \langle S, \prec_{cW} \rangle$. In this case it is really the general rules for communication orderings that give the constraints, rather than a given communication ordering. This will be referred to as "free synthesis". A sacond step in the procedure of synthesis is to find a multiprocessor configuration which supports at least the

communication capabilities given by the mapped precedence relation. This issue will be treated later.

There are also "mixed" cases that fall between the two extremes described above. A given communication ordering may for instance describe desired constraints on the structure of a system to be synthesized, rather than the communication capabilities of an existing system. This will be referred to as "restricted synthesis".

An important special case is when the set of cell actions is finite. Then the conditions for a function to be a weakly correct mapping can be relaxed:

Proposition 7.10: *Let $\langle C, \prec_c \rangle$ be a cell action structure where C is finite and let $\langle S, \prec_s \rangle$ be a communication structure. Let W be a function $C \rightarrow S$ and let $\prec_{cw}$ be given by* $\mathrm{fi2}(\langle C, \prec_c \rangle, \langle S, \prec_{cw} \rangle, W)$. Then W is a weakly correct mapping $\langle C, \prec_c \rangle \rightarrow \langle S, \prec_s \rangle$ iff $\prec_{cw} \subseteq \prec_s$.

Proof. See Appendix I. ■

For a *correct* mapping $M: \langle C, \prec_c \rangle \rightarrow \langle S, \prec_s \rangle$ there is a *unique* cell action for each space-time point in $M(C)$. This is often desirable, since then every cell action c directly describes the action performed at the event $M(c)$. If a mapping is weakly correct but not correct, then more than one cell action is mapped onto the same space point. The interpretation of this is that all these cell actions are performed at the same event, as one big cell action. There are situations when this freedom is useful, but it also leads to pathological cases. If C is finite, then a weakly correct mapping from C may for instance map all cell actions onto one single space-time point! This case is of course extremely uninteresting, since such a weakly correct mapping hides the inner structure of the set of cell actions completely and we are mainly interested in the interplay between the dependence structure of the set of cell actions and the hardware structure.

There is also another reason to avoid weakly correct mappings that map too many cell actions onto one single space-time point. An implicit assumption about weakly correct mappings is, that *everything that is mapped onto one space-time point must be possible for one processing element to perform in one time unit.* A discrete integer time step corresponds to a physical time interval. If the resulting cell actions grow too complex, then the physical time interval must be longer, or the processing element must be more powerful, or both. Especially there has to be some upper bound on the amount of work needed to carry out a resulting cell action. If not, then there will be no processor powerful enough and no physical time interval long enough to allow for all cell actions to be performed by one processor during one discrete time step. Thus, there must be a bound on the number of cell actions mapped to the same space-time point. There must also be a bound on the number of computational events contained in one cell action. Finally, there must be a bound on the amount of work needed to carry out one computational event.

This last restriction implies that there must be a bound on the work required to perform one operation in the algebra we are dealing with. This essentially means that all sorts must be finite. Permissible algebras are for instance such over booleans, floating-point numbers, finite character sets and integers of limited size. Algebras over, say, lists and character strings are not allowed, since there is no upper bound in general on the size of the representation of such objects. (This does *not* mean that we cannot describe, say, character-wise operations on strings. But then the basic operations considered are defined over characters rather than strings.)

When a function $F: C \to S$ is 1-1, then the conditions for it to be a correct mapping can be simplified:

Proposition 7.11: *Let* $\langle C, \prec_c \rangle$ *be a cell action structure and let* $\langle S, \prec_s \rangle$ *be a communication structure. Let* F *be a function* $C \to S$ *that is 1-1. Then* F *is a correct mapping* $\langle C, \prec_c \rangle \to \langle S, \prec_s \rangle$ *iff* $c \prec_c c' \implies F(c) \prec_s F(c')$ *for all* $c, c' \in C$.

Proof. See Appendix I. ■

It is not surprising, that given a weakly correct mapping that maps different cell actions onto the same space-time point, it is possible to derive a new set of cell actions where each cell action contains exactly the computational events that takes place at one event in space-time. There will also exist a *correct* mapping from this set of cell actions to the space-time, which is equivalent to the old one in the sense that it will result in the same scheduling of computational events. This is stated more formally in the following theorem:

Theorem 7.1: *Let* $\langle C, \prec_c \rangle$ *be a cell action structure derived from the computational event structure* $\langle D, \prec \rangle$. *Let* $\langle S, \prec_s \rangle$ *be a communication structure. For every weakly correct mapping* $W: \langle C, \prec_c \rangle \to \langle S, \prec_s \rangle$ *it holds that* $\langle C', \prec_c' \rangle$, *where* $C' = D/\varepsilon_{W \circ \varphi_C}$, *is a cell action structure derived from* D, *and there exists a correct mapping* $M: \langle C', \prec_c' \rangle \to \langle S, \prec_s \rangle$ *such that* $W \circ \varphi_C = M \circ \varphi_{C'}$. *The immediate precedence relation* $\prec_c'$ *on* C' *is given by: For all* $\mathrm{p}, \mathrm{p}' \in D$,

$$\varphi_{C'}(\mathrm{p}) \prec_c' \varphi_{C'}(\mathrm{p}') \iff W \circ \varphi_C(\mathrm{p}) \prec_{cW} W \circ \varphi_C(\mathrm{p}').$$

Proof. See Appendix I. ■

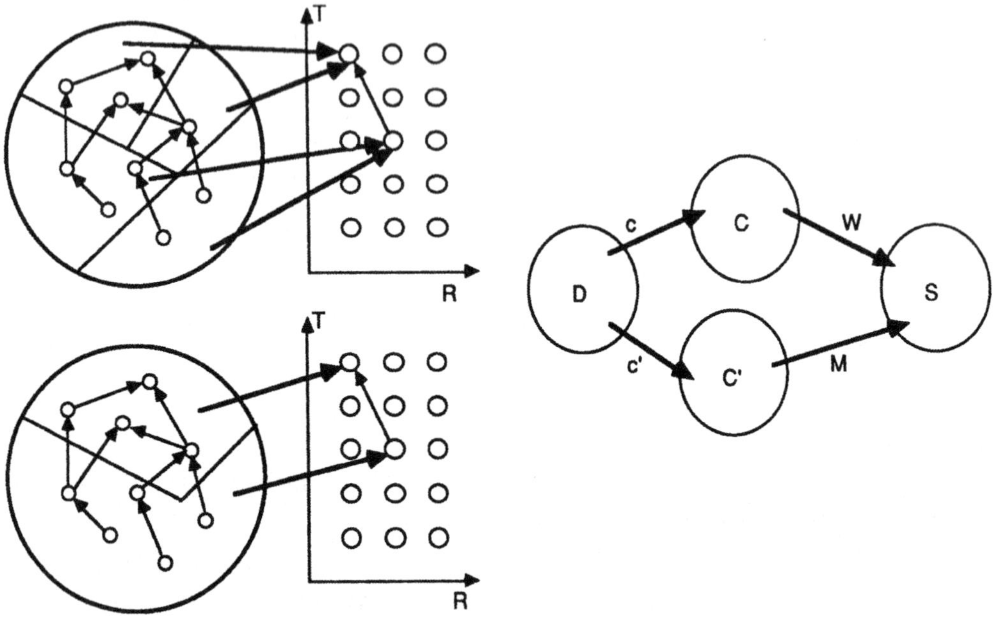

Figure 7.5: Merge of cell actions according to theorem 7.1 and a commuting diagram illustrating the theorem.

An interpretation of theorem 7.1 is that all the old cell actions mapped to the same point in space-time are merged into a new, bigger cell action. This is illustrated in figure 7.5.

7.4 Free synthesis

We will now address the issue of free synthesis: given a function from a cell action structure to a space-time, will the mapped precedence relation be a communication ordering? First there is a result about the existence of such a function:

Theorem 7.2: *For every cell action structure $\langle C, \prec_c \rangle$ and for every space-time S, there exists a correct mapping $M: \langle C, \prec_c \rangle \to \langle S, \prec_{cM} \rangle$ such that $\prec_{cM}$ is ripple-free.*

Proof. See Appendix I. ■

Next we consider the case where the function is given.

Proposition 7.12: *For every cell action structure $\langle C, \prec_c \rangle$, for every space-time S and for every function $W: C \rightarrow S$, $\prec_{cW}$ is a communication ordering iff it exists and $\langle t,r \rangle \prec_{cW} \langle t',r' \rangle \implies t \leq t'$ for all $\langle t,r \rangle, \langle t',r' \rangle \in S$.*

Proof.

$\Longrightarrow$: We prove this direction by proving that the negated consequence implies that $\prec_{cW}$ is not a communication ordering. If $\prec_{cW}$ does not exist, then it can of course not be a communication ordering. If, for some $\langle t,r \rangle, \langle t',r' \rangle \in S$, it is not the case that $\langle t,r \rangle \prec_{cW} \langle t',r' \rangle \implies t \leq t'$, then col in definition 7.2 is violated.

$\Longleftarrow$: col follows directly. Since $\prec_{cW}$ exists, it is given by $fi(\langle C, \prec_c \rangle, \langle S, \prec_{cW} \rangle, W)$. Then, since $\prec_c$ is finite-downward, it follows that $\prec_{cW}$ must be finite-downward. ■

Proposition 7.13: *For every cell action structure $\langle C, \prec_c \rangle$, for every space-time S and for every function $W: C \rightarrow S$ holds that if W is 1-1 and if $\langle t,r \rangle \prec_{cW} \langle t',r' \rangle \implies t \leq t'$ for all $\langle t,r \rangle, \langle t',r' \rangle \in S$, then $\prec_{cW}$ is a communication ordering.*

Proof. Directly from lemma 6.5, since W is a bijection $C \rightarrow W(C)$, and from proposition 7.12. ■

Corollary 7.2: *For every cell action structure $\langle C, \prec_c \rangle$, for every space-time S and for every function $W: C \rightarrow S$ holds that if W is 1-1 and if $\langle t,r \rangle \prec_{cW} \langle t',r' \rangle \implies t < t'$ for all $\langle t,r \rangle, \langle t',r' \rangle \in S$, then $\prec_{cW}$ is a ripple-free communication ordering.*

In related work, the premise in corollary 7.2 is usually chosen as the basic restriction on functions to a space-time, [Mo82,Li83,C86B,Raj86].

7.5 Schedules of sets of concurrent single-assignments

As stated in chapter 6, CCSA structures and cell action structures are equivalent concepts. Therefore, it is of some interest to define weakly correct mappings also from CCSA structures:

Definition 7.6: Let $\langle M, \prec_M \rangle$ be a CCSA structure and let $\langle S, \prec_s \rangle$ be a communication structure. $W: M \rightarrow S$ is a *weakly correct mapping (wcm)* $\langle M, \prec_M \rangle \rightarrow \langle S, \prec_s \rangle$ if and only if:

$$\prec_{MW} \text{ exists and } \prec_{MW} \subseteq \prec_s .$$

W is a *correct mapping (cm)* $\langle M, \prec_M \rangle \rightarrow \langle S, \prec_s \rangle$ iff it is a weakly correct mapping $\langle M, \prec_M \rangle \rightarrow \langle S, \prec_s \rangle$ and 1-1.

For any CCSA structure $\langle M, \prec_M \rangle$, there is by theorem 6.8 an isomorphic cell action structure $\langle C(G(M)), \prec_{C(G(M))} \rangle$, with graph isomorphism $C \circ G$ from $\langle M, \prec_M \rangle$ to $\langle C(G(M)), \prec_{C(G(M))} \rangle$. The following proposition follows immediately from the isomorphism:

Proposition 7.14: *Let $\langle M, \prec_M \rangle$ be a CCSA structure and let $\langle S, \prec_s \rangle$ be a communication structure. W is a weakly correct mapping $\langle M, \prec_M \rangle \to \langle S, \prec_s \rangle$ if and only if $W \circ (C \circ G)^{-1}$ is a weakly correct mapping $\langle C(G(M)), \prec_{C(G(M))} \rangle \to \langle S, \prec_s \rangle$.*

The isomorphism between M and $C(G(M))$ implies that all results proved here about weakly correct mappings from cell action structures carries over to weakly correct mappings from CCSA structures.

The equivalence between cell action structures and CCSA structures of course also works the other way. For any cell action structure $\langle C, \prec_c \rangle$, there is by theorem 6.9 an isomorphic CCSA structure $\langle M(C), \prec_{M(C)} \rangle$, with graph isomorphism M from $\langle C, \prec_c \rangle$ to $\langle M(C), \prec_{M(C)} \rangle$. The following proposition follows immediately from the isomorphism:

Proposition 7.15: *Let $\langle C, \prec_c \rangle$ be a cell action structure and let $\langle S, \prec_s \rangle$ be a communication structure. W is a weakly correct mapping $\langle C, \prec_c \rangle \to \langle S, \prec_s \rangle$ iff $W \circ M^{-1}$ is a weakly correct mapping $\langle M(C), \prec_{M(C)} \rangle \to \langle S, \prec_s \rangle$.*

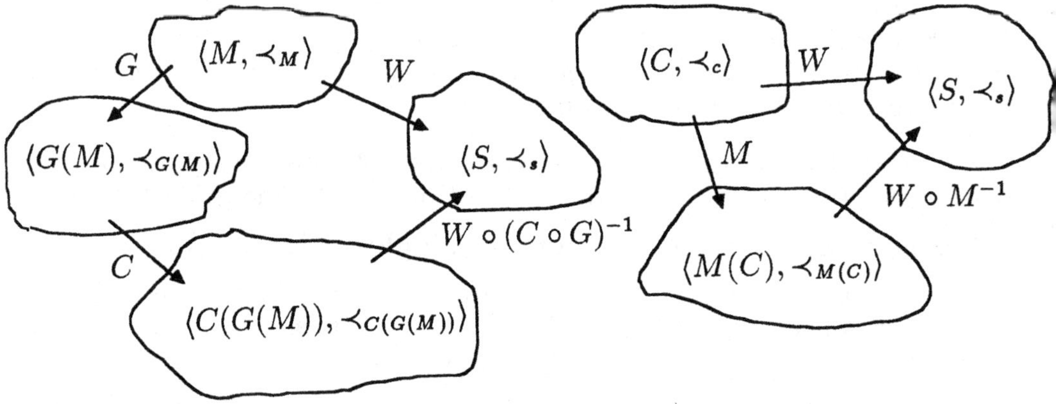

Figure 7.6: Commuting diagrams.

The commuting diagrams in figure 7.6 illustrate the contents of propositions 7.14 and 7.15.

7.6 Space-time transformations

Some proposed design methodologies work by turning existing designs into new ones, see for instance [JWCD81,WD81] or [KLi83]. In these works, a formalism for describing processor networks by expressions of delay operators applied to sequences is developed and new designs are obtained by applying correctness-preserving transformations to such expressions. In the framework developed here, the transformation of old designs to new ones is most conveniently seen as a *transformation of space-time*:

Definition 7.7: Let $\langle S, \prec_s \rangle$ and $\langle S', \prec'_s \rangle$ be communication structures. $A: S \to S'$ is a *weak space-time transformation (wstt)* $\langle S, \prec_s \rangle \to \langle S', \prec'_s \rangle$ iff:

$$\prec_{sA} \text{ exists and } \prec_{sA} \subseteq \prec'_s.$$

A weak space-time transformation $\langle S, \prec_s \rangle \to \langle S', \prec'_s \rangle$ is a *space-time transformation (stt)* $\langle S, \prec_s \rangle \to \langle S', \prec'_s \rangle$ iff it is 1-1.

The semantics of weak space-time transformations can be seen in terms of weakly correct mappings giving schedules of cell actions. A weakly correct mapping W from $\langle C, \prec_c \rangle$ to $\langle S, \prec_s \rangle$ gives a schedule of $\langle C, \prec_c \rangle$ on $\langle S, \prec_s \rangle$. If this schedule is given, then a weak space-time transformation $A: \langle S, \prec_s \rangle \to \langle S', \prec'_s \rangle$ gives a new schedule of $\langle C, \prec_c \rangle$ on $\langle S', \prec'_s \rangle$. According to the following theorem, this schedule will always fulfil our conditions regarding finiteness and causality.

Theorem 7.3: *Let $\langle S, \prec_s \rangle$ and $\langle S', \prec'_s \rangle$ be communication structures and let A be a weak space-time transformation $\langle S, \prec_s \rangle \to \langle S', \prec'_s \rangle$. Let $\langle C, \prec_c \rangle$ be a cell action structure. Then, for every weakly correct mapping $W: \langle C, \prec_c \rangle \to \langle S, \prec_s \rangle$, $A \circ W$ is a weakly correct mapping $\langle C, \prec_c \rangle \to \langle S', \prec'_s \rangle$.*

Proof. See Appendix I. ∎

Corollary 7.3: *If W is a correct mapping $W: \langle C, \prec_c \rangle \to \langle S, \prec_s \rangle$ and if A is a space-time transformation $\langle S, \prec_s \rangle \to \langle S', \prec'_s \rangle$, then $A \circ W$ is a correct mapping $\langle C, \prec_c \rangle \to \langle S', \prec'_s \rangle$.*

Thus, weak space-time transformations give correct schedules on the new communication structure for all cell action structures correctly scheduled by a weakly correct mapping on the first communication structure. Similar results appear in [CF84], where a more graph-theoretic approach is taken. Space-time transformations that are not weak are of special interest, since they alter neither the internal structure of the cell actions nor the dependence structure between them:

Definition 7.8: The binary relation $\stackrel{stt}{\rightarrow}$ between communication structures is defined by:
$\langle S, \prec_s \rangle \stackrel{stt}{\rightarrow} \langle S', \prec_s' \rangle$ iff there exists a space-time transformation $\langle S, \prec_s \rangle \rightarrow \langle S', \prec_s' \rangle$.
$\stackrel{stt}{\leftrightarrow}$ is defined by: $\langle S, \prec_s \rangle \stackrel{stt}{\leftrightarrow} \langle S', \prec_s' \rangle$ iff $\langle S, \prec_s \rangle \stackrel{stt}{\rightarrow} \langle S', \prec_s' \rangle$ and $\langle S', \prec_s' \rangle \stackrel{stt}{\rightarrow} \langle S, \prec_s \rangle$.

Note that there must be a space-time transformation between the communication structures *that is 1-1*, if the relation defined above is to hold between them. A definition based on weak space-time transformations instead of space-time transformations would yield a totally uninteresting relation, since such a transformation may alter the structure of the communication ordering. Every finite communication structure $\langle S, \prec_s \rangle$ would for instance be related to every other communication structure $\langle S', \prec_s' \rangle$, since there is a weak space-time transformation that maps all points in S to one single point in S'!

If $\langle S, \prec_s \rangle \stackrel{stt}{\rightarrow} \langle S', \prec_s' \rangle$, then all sets of cell actions executable on $\langle S, \prec_s \rangle$ can be executed on $\langle S', \prec_s' \rangle$ without altering the partial order of execution. $\langle S, \prec_s \rangle \stackrel{stt}{\leftrightarrow} \langle S', \prec_s' \rangle$ means that all sets of cell actions executable on one of the communication structures can be executed on the other, and vice versa. Cf. (m, k)-*simulation*, [CF84].

Proposition 7.16: $\stackrel{stt}{\rightarrow}$ *is transitive and reflexive.*

Proof. See appendix I. ■

Theorem 7.4: $\stackrel{stt}{\leftrightarrow}$ *is an equivalence relation.*

Proof. Transitivity, reflexivity follows from the corresponding properties for $\stackrel{stt}{\rightarrow}$. Symmetry follows directly from the definition. ■

Proposition 7.17: *If there exists a space-time transformation* A: $\langle S, \prec_s \rangle \rightarrow \langle S', \prec_s' \rangle$ *that is a bijection* $S \rightarrow S'$ *and if* $fi(\langle S, \prec_s \rangle, \langle S', \prec_s' \rangle, A)$, *then* $\langle S, \prec_s \rangle \stackrel{stt}{\leftrightarrow} \langle S', \prec_s' \rangle$.

Proof. Assume that A is a bijection $S \rightarrow S'$ and that $fi(\langle S, \prec_s \rangle, \langle S', \prec_s' \rangle, A)$ holds.
$\stackrel{stt}{\rightarrow}$: Follows directly.

$\overset{stt}{\leftarrow}$: Since $\prec_s$ is a communication ordering it is irreflexive. By lemma 6.5, it follows that A is a graph isomorpism $\langle S, \prec_s\rangle \rightarrow \langle S', \prec_s'\rangle$. This implies that A^{-1} is a graph isomorpism $\langle S', \prec_s'\rangle \rightarrow \langle S, \prec_s\rangle$. $\prec_s'$ is also a communication ordering and thus irreflexive. Lemma 6.5, applied once more, now implies $fi(\langle S', \prec_s'\rangle, \langle S, \prec_s\rangle, A^{-1})$. Since A is a bijection, A^{-1} is 1-1. $\langle S, \prec_s\rangle \overset{stt}{\leftarrow} \langle S', \prec_s'\rangle$ follows. ■

It is easy to believe that $\langle S, \prec_s\rangle \overset{stt}{\leftrightarrow} \langle S', \prec_s'\rangle$ implies that the graphs $\langle S, \prec_s\rangle$ and $\langle S', \prec_s'\rangle$ are isomorphic. This is however not the case, as can be seen by the following example:

Example 7.1: Consider the space-time $S = \langle N, \{0\}\rangle$. Define the communication orderings $\prec_0$, $\prec_1$ on S by:
$\prec_0$: For all $n \geq 0$ and $m \in Z^+$, $\langle m \cdot 2^n, 0\rangle \prec_0 \langle (m+1) \cdot 2^n, 0\rangle$.
$\prec_1$: For all $n \geq 1$ and $m \in Z^+$, $\langle m \cdot 2^n, 0\rangle \prec_1 \langle (m+1) \cdot 2^n, 0\rangle$.

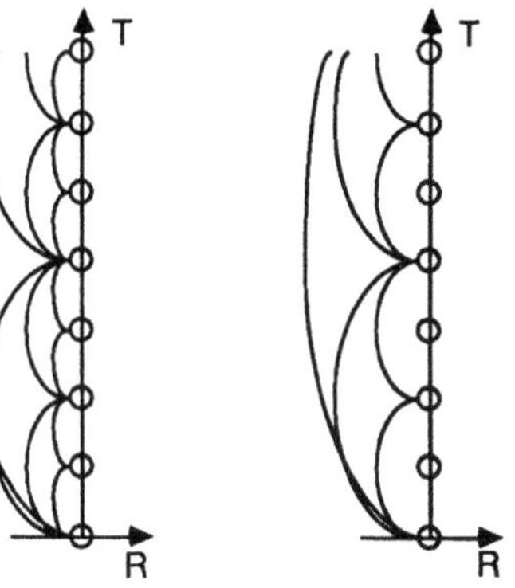

Figure 7.7: $\langle S, \prec_0\rangle$, $\langle S, \prec_1\rangle$.

Consider the function $A: S \rightarrow S$ defined by $A(\langle t, 0\rangle) = \langle 2t, 0\rangle$ for all $t \in N$. Clearly A is a space-time transformation both from $\langle S, \prec_0\rangle$ to $\langle S, \prec_1\rangle$ and from $\langle S, \prec_1\rangle$ to $\langle S, \prec_0\rangle$. Thus, $\langle S, \prec_0\rangle \overset{stt}{\leftrightarrow} \langle S, \prec_1\rangle$. But $\langle S, \prec_0\rangle$ and $\langle S, \prec_1\rangle$ are not isomorphic when considered as graphs, since $\langle S, \prec_0\rangle$ is connected while $\langle S, \prec_1\rangle$ is not. ■

7.7 Fixed hardware structures

We now consider a simple model of fixed hardware, with cells connected by wires with integer delays:

Definition 7.9: Let R be a space. Then $\Delta \subseteq R \times R \times N$ is a *fixed hardware structure on R*.

Figure 7.8: Graphical representation of a wire $\langle \delta, r, r' \rangle$ in a fixed hardware structure.

Cf. *computation networks*, [CF84]. $\langle r_1, r_2, \delta \rangle \in \Delta$ should be interpreted that there is a wire with integer delay δ between the cells located at r_1 and r_2. A way to accomplish this is to let the wire be broken by δ clocked latches. A fixed hardware structure can also be seen as a labelled directed graph where each edge is labelled with its delay. This is a direct equivalent of the delay operator formalism used in [JWCD81,KLi83], or [WD81]. The model is also very close to the one used in [LS81]. An especially interesting part is the subgraph with zero delay:

Definition 7.10: For every fixed hardware structure Δ on a space R, the binary relation Δ_0 on R is defined by: for all $r_1, r_2 \in R$, $r_1 \Delta_0 r_2 \iff \langle r_1, r_2, 0 \rangle \in \Delta$.

We want to pose certain restrictions on the subgraph with zero delay. In particular zero-delay cycles are undesirable, and it is also necessary to demand finiteness for fanin, fanout and the length of paths with zero delay. The following definitions provides this:

Definition 7.11: The fixed hardware structure Δ on R has:

1. *finite fanin*, if and only if for all $r \in R$, $\{ r' \mid \exists \delta \in N[\langle r', r, \delta \rangle \in \Delta] \}$ is finite.
2. *finite fanout*, if and only if for all $r \in R$, $\{ r' \mid \exists \delta \in N[\langle r, r', \delta \rangle \in \Delta] \}$ is finite.
3. *no infinite zero-delay paths* if and only if there are no infinite sequences $\langle r_z \mid z \in Z \rangle$ where $r_z \Delta_0 r_{z+1}$ for all $z \in Z$.

Note that 3 implies that if Δ has no infinite zero-delay paths, then Δ_0 is a well-founded relation on R (which, among other things, implies that there are no cycles in $\langle R, \Delta_0 \rangle$).

Every fixed hardware structure Δ on a space R defines an ordering on $T \times R$. This ordering gives the communication constraints between the possible events of computation taking place in Δ.

Definition 7.12: For every fixed hardware structure Δ on a space R, the *space-time ordering* $\prec(\Delta)$ *generated by* Δ is defined by:

$$\forall r_1, r_2 \in R \forall \delta \in N[\langle r_1, r_2, \delta \rangle \in \Delta \iff \forall t \in T[\langle t, r_1 \rangle \prec(\Delta) \langle t + \delta, r_2 \rangle]]$$

Cf. *unrollings*, [CF84]. It is interesting to see which fixed hardware structures that generate communication orderings. The following theorem gives an answer:

Theorem 7.5: *If* Δ *on* R *has finite fanin and no infinite zero-delay paths, then* $\prec(\Delta)$ *is a communication ordering on* $T \times R$.

Proof. See appendix I. ■

So far we have treated the case when a communication order is generated by a fixed hardware structure. This communication order gives the monotonicity constraint when scheduling sets of cell actions on the structure. It is interesting to consider the reverse problem; given a communication ordering $\prec_s$, find an in some sense "minimal" fixed hardware structure Δ that supports the communication capabilities of $\prec_s$:

Definition 7.13: For any communication ordering $\prec_s$ on $T \times R$, the *fixed hardware projection* $\Delta(\prec_s)$ *of* $\prec_s$ *on* R is given by: for all $r_1, r_2 \in R$ and $\delta \in N$,

$$\exists t_1, t_2 \in T[\langle t_1, r_1 \rangle \prec_s \langle t_2, r_2 \rangle \wedge t_2 - t_1 = \delta] \iff \langle r_1, r_2, \delta \rangle \in \Delta(\prec_s).$$

Theorem 7.6: *For any communication ordering* $\prec_s$ *holds that* $\prec_s \subseteq \prec(\Delta(\prec_s)))$. *Moreover,* $\Delta(\prec_s)$ *is the least fixed hardware structure with this property.*

Proof.

1. $\prec_s \subseteq \prec(\Delta(\prec_s)))$:

$$\begin{aligned} \langle t_1, r_1 \rangle \prec_s \langle t_2, r_2 \rangle &\implies \langle r_1, r_2, t_2 - t_1 \rangle \in \Delta(\prec_s) \\ &\implies \forall t \in T[\langle t, r_1 \rangle \prec(\Delta(\prec_s)) \langle t + t_2 - t_1, r_2 \rangle] \\ &\implies (\text{choose } t = t_1) \\ &\implies \langle t_1, r_1 \rangle \prec(\Delta(\prec_s)) \langle t_2, r_2 \rangle. \end{aligned}$$

2. minimality of $\Delta(\prec_s)$: Consider any $\Delta' \subset \Delta(\prec_s)$. Then there is a $\langle r_1, r_2, \delta \rangle$ such that (1): $\langle r_1, r_2, \delta \rangle \in \Delta(\prec_s)$ and (2): $\langle r_1, r_2, \delta \rangle \notin \Delta'$. From (1) follows that there exists a $t \in T$ such that $\langle t, r_1 \rangle \prec_s \langle t+\delta, r_2 \rangle$. From (2) follows that $\langle t, r_1 \rangle \not\prec(\Delta') \langle t + \delta, r_2 \rangle$. ■

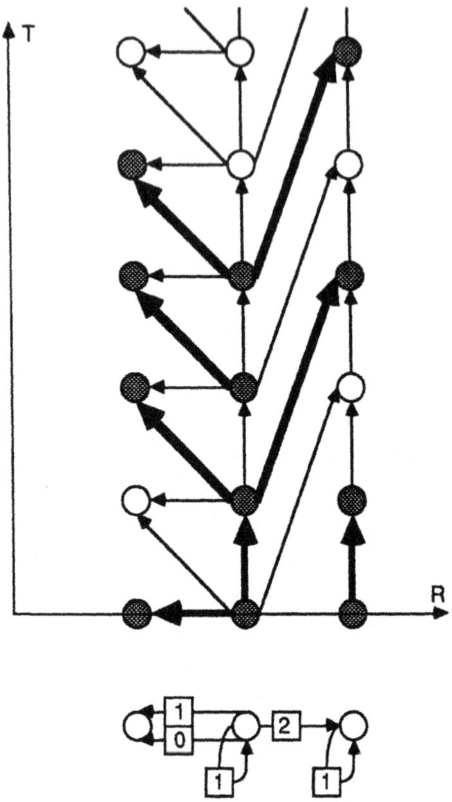

Figure 7.9: $\prec_s$, $\Delta(\prec_s)$ and $\prec(\Delta(\prec_s))$.

Figure 7.9 shows an example of a communication ordering on a space-time $T \times R$, its fixed hardware projection and the space-time ordering generated by the projection.

The two applications of the space-time mapping methodology, scheduling and synthesis, can now be extended to fixed hardware structures. The first step in scheduling on a fixed hardware structure is to generate its space-time ordering. Then, provided that the space-time ordering is a communication ordering, the scheduling proceeds as usual. In the case of free synthesis the mapped precedence relation is projected to a fixed hardware structure. By theorem 7.6 this fixed hardware structure is capable of supporting the resulting communication requirements. It is also possible to perform restricted synthesis with respect to fixed hardware structures by generating the space-time ordering, find a weakly correct

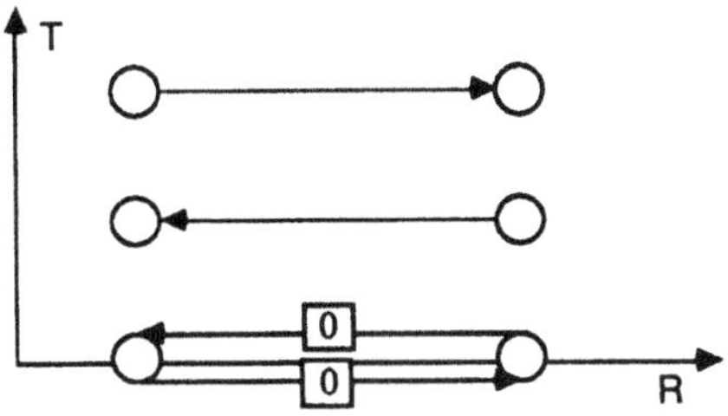

Figure 7.10: $\prec_s$ for which $\Delta(\prec_s)$ contains zero-delay cycles.

mapping with respect to that ordering and finally form the fixed hardware projection of the mapped precedence relation.

It would be nice if the fixed hardware projections of communication orderings always generated communication orderings in turn. Unfortunately this is not true. Figure 7.10 shows a simple example, where the projection of a communication ordering contains zero-delay cycles. The communication ordering in figure 7.10 may, however, very well be executed on a piece of hardware with dynamic routing.

Let us finally consider an important subclass of fixed hardware structures, namely those that do not contain any zero-delay lines at all:

Definition 7.14: The fixed hardware structure Δ is *systolic* iff for all $\langle r_1, r_2, \delta \rangle$ in Δ holds that $\delta > 0$.

The name "systolic" is chosen in accordance with the terminology in [LS81,CF84]. Systolic fixed hardware structures are closely connected with ripple-free space-time relations:

Proposition 7.18: *Any systolic fixed hardware structure Δ generates a ripple-free space-time ordering.*

Proof. See appendix I. ■

From proposition 7.18 and corollary 7.1 we obtain the following corollary:

Corollary 7.4: *If the space R is finite, then every systolic fixed hardware structure on R generates a communication ordering.*

Proposition 7.19: *For any ripple-free communication ordering $\prec_s$ on a space-time $T \times R$, $\Delta(\prec_s)$ is systolic.*

Proof. We prove the equivalent implication $\Delta(\prec_s)$ not systolic $\Longrightarrow \prec_s$ not ripple-free:

$$\begin{aligned}
\Delta(\prec_s) \text{ not systolic } &\Longrightarrow \\
\exists r_1, r_2 \in R[\langle r_1, r_2, 0\rangle \in \Delta(\prec_s)] &\Longrightarrow \\
\exists r_1, r_2 \in R[\exists t_1, t_2 \in T[\langle t_1, r_1\rangle \prec_s \langle t_2, r_2\rangle \wedge t_2 - t_1 = 0]] &\Longrightarrow \\
\exists \langle t_1, r_1\rangle, \langle t_2, r_2\rangle \in T \times R[\langle t_1, r_1\rangle \prec_s \langle t_2, r_2\rangle \wedge t_1 = t_2] &\Longrightarrow \\
\prec_s \text{ not ripple-free.}
\end{aligned}$$

■

8. Schedules and performance criteria

In this chapter we will consider *schedules*. A schedule is usually thought of as distributing the occurrence of events or actions in space-time according to some causal restrictions. Here, where the events represent evaluations of expressions, we will extend the meaning of "schedule" to include the selection of expressions for the given polynomials to compute, to choose the different events where each expression is going to be evaluated, and to choose for each event which other events that will provide it with necessary data. That is, a schedule σ will consist of a set of expressions E, a set of computational events D of E, a function $\psi: D \to E$ labelling each event in D with an expression, a data dependence relation $\prec \in \mathcal{R}(D, \psi, E)$, a set of cell actions C derived from D and finally a wcm W, distributing the cell actions of C in space-time.

In practical cases it is usually not enough to find a correct schedule of an output specification. It is also desirable to find the in some sense best, or at least a good solution to the scheduling problem. In order to determine if a schedule is better than another, some sort of *performance criterion* is needed. In this chapter we will define some. Most will be quite crude. On the other hand they will be simple to understand and evaluate. (*Optimization* with respect to these criteria will, however, be very hard or impossible in general.)

We will also define some more sophisticated concepts, which are clearly of importance for the simplicity and performance of the resulting implementation of the schedule. In order to formulate some of these criteria, we have to introduce constraints on schedules that involve the functions possible to compute by a cell at a certain time, not only the communication requirements as for the communication orderings. The formulation of these concepts will reveal problems of implementation previously not considered in the literature.

8.1 Schedules

Definition 8.1: Let O be an output specification satisfiable in $\mathcal{A}$ with domain A and let η be a representation function of O w.r.t. $\mathcal{A}$. Let $\langle S, \prec_s \rangle$ be a communication structure. $\sigma = \langle E_\sigma, \psi_\sigma, D_\sigma, \prec_\sigma, C_\sigma, W_\sigma \rangle$ is a *schedule of O on $\langle S, \prec_s \rangle$* if and only if:

sc1. E_σ is a set of expressions such that $\eta(O, A) \subseteq \phi(E_\sigma)$.

sc2. $\psi_\sigma: D_\sigma \to ec(E_\sigma)$ is onto.

sc3. $\prec_\sigma \in \mathcal{R}(D_\sigma, \psi_\sigma, E_\sigma)$.

sc4. $\langle C_\sigma, \prec_c \rangle$ is a cell action structure derived from $\langle D_\sigma, \prec_\sigma \rangle$.

sc5. W_σ is a weakly correct mapping $\langle C_\sigma, \prec_c \rangle \to \langle S, \prec_s \rangle$.

Note that $\prec_c$ is not included in the schedule. The reason is that it is uniquely determined since $\langle C_\sigma, \prec_c\rangle$ is finitely inherited from $\langle D_\sigma, \prec_\sigma\rangle$. Thus, all information is already present.

A schedule contains information about all decisions made on the way from output specification to weakly correct mapping. It tells what expressions we have chosen to represent the output specification and it tells where and when to perform each step necessary to evaluate these expressions.

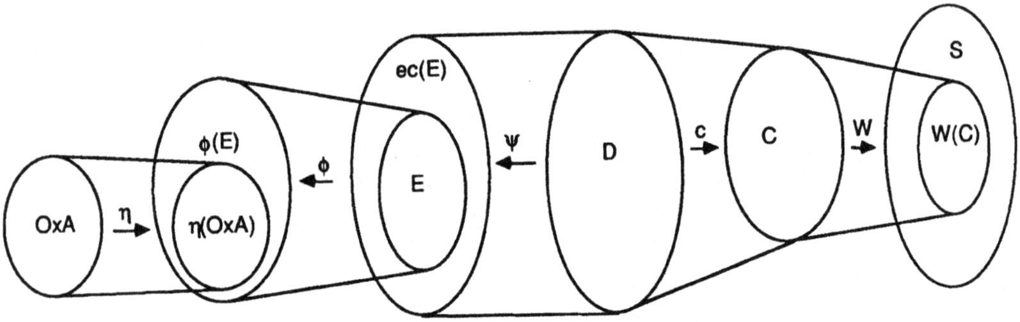

Figure 8.1: Diagram of a schedule σ.

Schedules may be schematically depicted as in figure 8.1. The synthesis problem has an intuitively appealing interpretation in this diagram; *synthesis can be seen as determining the components of a schedule from left to right.* Of course, the task of synthesis will not be that straightforward in practice. Any reasonable synthesis procedure will probably include the evaluation of some performance criterion when a schedule is found and then some feedback to earlier stages in the procedure, where components are recomputed with guidance from the earlier results.

Perhaps less evident is that the problem of *verification* also can be attacked using the concept of schedules; *verification of design correctness can be seen as determining the components of a schedule from right to left.* In this case the starting point is a system description of some kind. The task of verification is then the following: first to derive, from this description, a communication structure, a weakly correct mapping and a cell action structure, a computational event structure and a set of expressions evaluated by the system, and then to check that there is a representation function η such that the polynomials given by those expressions include $\eta(O, A)$. If this is the case, then the design is correct with respect to the output specification O. Note that this only means that our system always computes the functions in O. It tells *nothing* about other correctness criteria such as: "result x is to be available at processor r at time t", "response time of the system should be less than t" or "the system is deadlock-free". Some correctness criteria of this kind may be checked by a closer examination of the schedule. Others may be hard to check at all, simply because they do not really apply to the kind of systems we are able to describe by schedules as defined here.

As mentioned in chapter 6, CCSA structures may be chosen as a starting point for synthesis rather than output specifications. Definitions 6.14 and 8.1 provide tools to incorporate this into the formalism:

Proposition 8.1: *Let $\langle M, \prec_M \rangle$ be a CCSA structure derived from* $\mathbf{F}$. *Let $\langle S, \prec_s \rangle$ be a communication structure. If W is a weakly correct mapping $\langle M, \prec_M \rangle \to \langle S, \prec_s \rangle$, then*

$$\langle E(G(M)), \psi_{G(M)}, D(G(M)), \prec_{SF(G(M))}, C(G(M)), W \circ (C \circ G)^{-1} \rangle$$

is a schedule of $\phi(E(G(M)))$ on $\langle S, \prec_s \rangle$.

Proof. All functions in $\phi(E(G(M)))$ are, by theorem 6.6 and lemma 6.7, polynomials in $\mathcal{P}^{(X_f(\mathbf{F}))}(i(\mathcal{A}))$. Therefore, $\eta(\phi(E(G(M))), (A_x \mid x \in X_f(\mathbf{F})\,)) = \phi(E(G(M)))$. Thus, sc1 follows. Sc2, sc3 follows from lemma 6.7. Sc4 follows from lemma 6.8. Sc5 follows from proposition 7.14. ∎

8.2 Simple performance criteria

In this section we will consider some simple performance criteria for schedules σ. In the definitions below $W_\sigma = \langle t_\sigma, r_\sigma \rangle$.

Definition 8.2: (Help quantities) For any schedule σ on a space-time $T \times R$ we define:

For any $\tau \in T$

$$C_\sigma(\tau) = \{\, c \mid t_\sigma(c) \leq \tau \,\}$$

and for any $C' \subseteq C_\sigma$

$$T_\sigma(C') = \{\, t \mid \exists t', t'' \in t_\sigma(C')[t' \leq t \leq t''] \,\}.$$

$C_\sigma(\tau)$ is the set of cell actions scheduled up to the time τ. $T_\sigma(C')$ is the number of discrete times from the time of the first scheduled cell action(s) of C' to the time of the last one(s).

Definition 8.3: For any schedule σ on a space-time $T \times R$ we define:

the *execution time*

$$T_\sigma = \lim_{\tau \to \infty} |T_\sigma(C_\sigma(\tau))|$$

and the *number of cells*

$$n_\sigma^R = \lim_{\tau \to \infty} |r_\sigma(C_\sigma(\tau))|.$$

When C_σ is infinite, then T_σ and n_σ^R may be infinite. When C_σ is finite they are not, and the definitions of T_σ and n_σ^R can be simplified. The result is stated in the following proposition, that follows immediately from the definitions above:

Proposition 8.2: *If C_σ is finite, then*

$$T_\sigma = |T_\sigma(C_\sigma)|$$

and

$$n_\sigma^R = |r_\sigma(C_\sigma)|\,.$$

We can now formulate a first optimization problem:

MINIMAL TIME SCHEDULING:

Given an output specification O and a communication structure $\langle S, \prec_s \rangle$, find a schedule σ of O on $\langle S, \prec_s \rangle$ such that T_σ is minimized.

Some results regarding minimal time scheduling are known. Consider the problem to minimize T_σ on a communication structure $\langle N \times R, \prec_\mathrm{T} \rangle$ where R is finite, say $|R| = k$. Assume further that C_σ is finite, say $|C_\sigma| = n$, and fixed. Let us finally restrict ourselves to consider correct mappings W_σ only. The corresponding decision problem (given a time t, is there a correct mapping $W_\sigma = \langle t_\sigma, r_\sigma \rangle: \langle C_\sigma, \prec_c \rangle \to \langle N \times R, \prec_\mathrm{T} \rangle$ such that $t_\sigma(c) \leq t$ for all $c \in C_\sigma$?) is easily seen to be equivalent to the PRECEDENCE CONSTRAINED SCHEDULING problem, [GJ79,U75], which is known to be *NP*-complete when $k > 2$. This special case of the minimal time scheduling problem is thus *NP*-hard, so minimal time scheduling is *NP*-hard in general.

An interesting question is if there are non-trivial communication orderings other than $\prec_\mathrm{T}$, for which the restricted minimal time scheduling problem above is no more *NP*-hard. It does not seem unlikely that this can be the case for certain highly regular communication orderings. This is an interesting field for future research which seems to have quite immediate applications, for instance in compiling techniques for regularly connected multiprocessor systems.

Another special case of minimal time scheduling, where some results are known, is scheduling of arithmetic expression evaluation. This can be expressed as follows in our model: the search for an optimal schedule is restricted to schedules σ where $D_\sigma = E_\sigma$, the blocks in C_σ are singletons (so every cell action contains exactly one arithmetic operation) and W_σ is a correct mapping. On the other hand we are now allowed to change E_σ as long as the output specification is not violated. We assume that E_σ is finite and that the scheduling is done on a communication structure $\langle N \times R, \prec_\mathrm{T} \rangle$ where $|R| \geq |E_\sigma|$, so that the number of processors does not restrict the schedules. The resulting optimization methods will merely consist of algebraic rewriting techniques intended to minimize the maximal height of the expressions. In [Be72] an optimization algorithm is given for the restricted case where expressions are treelike and rewriting is done using associativity only. In the thesis of Muraoka, [Mu71], an algorithm is presented which utilizes distributivity to obtain better results. Muraoka also treats the problem of scheduling the evaluation of a polynomial $\sum_{i=0}^{n} a_i x^i$ on m processors, where m not necessarily is greater than the number of operations. Algorithms for scheduling the evaluation of expressions on a limited number of processors often consist of an interesting mix of algebraic rewriting and task scheduling methods. In the unlimited case it is possible to give upper bounds on the best solution. Such can be found for rewrite methods using associativity and commutativity, [KM74], and for methods also using distributivity, [Br74].

There are many cases where the minimal time criterion considered above is clearly not suitable. Expression schedules that are optimal with respect to execution time, for instance, often require many processors during a few steps but only a few at later stages. This leads to a poor utilization of the hardware. It is therefore interesting to define a performance criterion that measures the average hardware utilization.

Definition 8.4: For any schedule σ on a space-time $N \times R$ and any $C' \subseteq C_\sigma$ we define the *space-time hull*

$$H_\sigma(C') = T_\sigma(C') \times R_\sigma(C').$$

$R_\sigma(C')$ is the set of locations where processors must be placed to support the schedule of C' given by σ. $T_\sigma(C')$ is the set of times needed by σ to schedule C'. $H_\sigma(C')$ is thus the total set of space-time events taking place in the hardware supporting the schedule during the time of execution.

Definition 8.5: For any schedule σ on a space-time $N \times R$, the *cell utilization* U_σ is defined as

$$U_\sigma = \lim_{\tau \to \infty} |W(C_\sigma(\tau))| \,/\, |H_\sigma(C_\sigma(\tau))|$$

when this limit exists.

There are cases when this limit does not exist. If C_σ is infinite, then there might for instance be some finite τ for which $W(C_\sigma(\tau))$ becomes infinite. Even when $W(C_\sigma(\tau))$ always is finite the limit may not exist. When C_σ is *finite*, however, we immediately obtain the following result from the definition above:

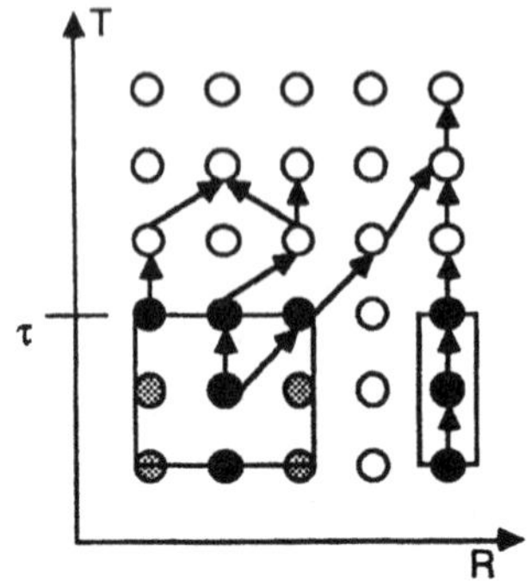

Figure 8.2: $W(C_\sigma(\tau))$, $H_\sigma(C_\sigma(\tau))$.

Proposition 8.3: *For any schedule σ where C_σ is finite* $U_\sigma = |W(C_\sigma)| / |H(C_\sigma)|$.

A concept similar to cell utilization is the "efficiency" E_p, [HJ81]. It is defined relative to a fixed expression and gives the cell utilization for a schedule of the expression on p processors.

We can now formulate an optimization problem based on cell utilization:

MAXIMAL CELL UTILIZATION SCHEDULING:

Given an output specification O and a communication structure $\langle S, \prec_s \rangle$, find a schedule σ of O on $\langle S, \prec_s \rangle$ such that U_σ is maximized.

The simple performance criteria defined in this section have one major deficiency: *they do not take into account the complexity of the resulting cell actions.* If two schedules give resulting cell actions with different complexity, then they are not comparable by these criteria. The following is an extreme example of this: when C_σ is finite there is always a weakly correct mapping that maps all cell actions to one single point in space. The resulting schedule σ will always be an optimal solution to the two problems formulated above. ($T_\sigma = 1$ and $U_\sigma = 1$. Both values are optimal.) But of course we gain nothing by such a schedule. We end up with the problem to design hardware for the resulting cell action. This design problem is equivalent to the original scheduling problem.

Thus, any reasonable optimization must be carried out over a restricted class of schedules that preserve cell action complexity. The two cases considered for the minimal time scheduling problem, namely scheduling of fixed C_σ with correct mappings and scheduling of arithmetic expression evaluation, are examples of this.

8.3 Reducing fanout

Fanout occurs when a value must be sent on more than one output line. If there are many such lines this will cause electrical problems, since a high capacitance then must be charged which may slow the circuit down. This problem is aggravated by the fact that there can be only a limited number of "nearest neighbours", so if there are many lines going out from one cell it is likely that some of them are long. This can also cause routing problems.

It is thus desirable to keep the fanout low. We will now give a definition of fanout of a *value*. Note that this is not the same as fanout of fixed hardware structures, as defined in chapter 7. That definition deals with the total number of lines going out from a cell. The definition here deals with the number of lines a particular value must be sent to.

From now on we will only consider correct mappings from pure CCSA structures to communication structures, since this will turn out to be the most convenient. By theorems 7.1, 6.4 and proposition 7.15 we can find an equivalent pure CCSA structure and correct mapping for any cell action structure and weakly correct mapping. No generality is therefore lost.

Definition 8.6: For any pure CCSA-set M derived from $\mathbf{F}$ in X and for any $x \in X_a(\mathbf{F})$, the *fanout* of x is $|\{ m \mid x \in range(m) \}|$.

The fanout of a variable x is thus the number of cs-assignments using it as input. This definition is in accordance with the intuitive meaning of fanout because of the following three facts: first, that we consider correct mappings only. Thus, two cs-assignments using the same variable can never be mapped to the same point in space-time. Second, that we assume that we need one distinct line for any data transfer with either different delay or different destination. So if a variable is to be sent to n different events in space-time there must be n lines supporting the transfers. Third, that we only consider pure CCSA-sets, that is, we do not count data transfers that are local to a cs-assignment.

We will now describe a method to transform a pure CCSA-set, with fanout greater than one for some variable, to a new, functionally equivalent CCSA-set where this variable has been replaced with a set of fresh variables, each having a fanout of one. Methods of this kind have been described before ("pipelining", "adding missing loop indices", [Mo82, MiWi84,Raj86], "broadcast avoidance", [Li83] or "reducing fan-out degrees", [C86B]). The aim of the description here is to give a deeper understanding of these methods. We will also answer some basic questions about existence and limitations of these fanout-reducing transformations.

Let us first introduce some new notation. For any set M of cs-assignments and any variable x, the relation $\prec_{Mx}$ is defined by

$$m \prec_{Mx} m' \iff x \in dom(m) \wedge x \in range(m')$$

for all $m, m' \in M$. $\prec_{Mx}$ thus describes the data dependencies caused by the transfers of the variable x. For any set of variables X' we define

$$\prec_{MX'} = \bigcup(\prec_{Mx} \mid x \in X').$$

Obviously, if M is derived from the recursion scheme $\mathbf{F}$, then $\prec_M = \prec_{MX_a(\mathbf{F})}$. That is, the total data dependence relation on M is the union of all data dependencies given by the assigned variables of $\mathbf{F}$. In the following we will consider bijections between sets of cs-assignments. For any bijection $f: M \to M'$ and relation R on M, we will violate the notational rules slightly and write R for the relation R' on M', defined by $f(m)R'f(m') \iff mRm'$ for all $m, m' \in M$.

For notational convenience we also introduce the operation "$*$" on substitutions by

$$\sigma_1 * \sigma_2 = \{\, x \leftarrow \sigma_1\sigma_2(x) \mid x \in dom(\sigma_2)\,\}$$

for all $\sigma_1, \sigma_2 \in \mathcal{S}^{(X)}(\mathcal{A})$. Note that $\sigma_1 * \sigma_2 \neq \sigma_1\sigma_2$ in general.

Definition 8.7: Let M be a pure set of cs-assignments and let x be a variable. If there exists an $m \in M$ such that $x \in dom(m)$, then I is a *causal x-enumeration in M* iff:

ce1: I is a bijection $dom(I) \to \{m\} \cup \{\, m' \mid x \in range(m')\,\}$, where $dom(I) = \{0, \ldots, n(I)\}$ and $n(I) = \big|\{\, m' \mid x \in range(m')\,\}\big|$.

ce2: $m_0 = m$.

ce3: For all $i, j \in dom(I)$ holds that $i < j \implies m_j \not\prec_M m_i$.

Note that possibly $n(I) = \infty$ in the definition above.

Proposition 8.4: *For any pure CCSA-set M derived from $\mathbf{F}$ and for any variable $x \in X_a(\mathbf{F})$ there always exists a causal x-enumeration in M.*

Proof. Since $x \in X_a(\mathbf{F})$ there is a $m \in M$ such that $x \in dom(m)$. For any causal x-enumeration it must hold that $m_0 = m$. For no m' such that $x \in range(m')$ it can be the case that $m' \prec_M m$, since $m \prec_M m'$ for all such m'. Thus, ce2 and ce3 are never contradictory. From proposition 7.11 and theorem 7.2 follows that there exists a correct mapping $W: M \to S$ (for some space-time S) such that

$$m \prec_M m' \implies t_W(m) < t_W(m'). \tag{1}$$

This total ordering with respect to time induces a bijective enumeration I of $\{m\} \cup \{ m' \mid x \in range(m') \}$, such that for all $j, i \in \{0, \ldots, n(I)\}$

$$j < i \iff t_W(m_j) < t_W(m_i). \tag{2}$$

(1) and (2) now yields

$$i \leq j \implies m_i \not\prec_M m_j$$

and since always $m_i \not\prec_M m_i$ we obtain

$$i < j \implies m_i \not\prec_M m_j.$$

Finally we can note that $m_0 = m$, since for all m' such that $x \in range(m')$ holds that $m \prec_M m'$. Thus, I is a causal x-enumeration in M. ■

Definition 8.8: Let M be a pure set of cs-assignments, derived from the recursion scheme $\mathbf{F}$ in X over $\mathcal{A}$, and let x be a variable in $X_a(\mathbf{F})$. Let I be a causal x-enumeration in M. Let X_I be a set of fresh variables ($X_I \cap X = \emptyset$), bijectively enumerated from $dom(I)$ ($i \neq j \implies x_i \neq x_j$), such that $S(x_i) = S(x)$ for all $i \in dom(I)$. We now define the function $f_{Ix}: M \to \mathcal{S}^{((X \setminus \{x\}) \cup X_I)}(\mathcal{A})$ by:

$$\begin{aligned}
f_{Ix}(m_0) &= (m_0 \setminus \{x \leftarrow m_0(x)\}) \cup \{x_0 \leftarrow m_0(x)\},\\
f_{Ix}(m_i) &= (\{x \leftarrow x_{i-1}\} * m_i) \cup \{x_i \leftarrow x_{i-1}\}, \text{ for } 0 < i < n(I),\\
f_{Ix}(m_{n(I)}) &= \{x \leftarrow x_{i-1}\} * m_i, \text{ if } n(I) < \infty,\\
f_{Ix}(m) &= m \text{ for all other } m \in M.
\end{aligned}$$

Furthermore we define $MIx = f_{Ix}(M)$.

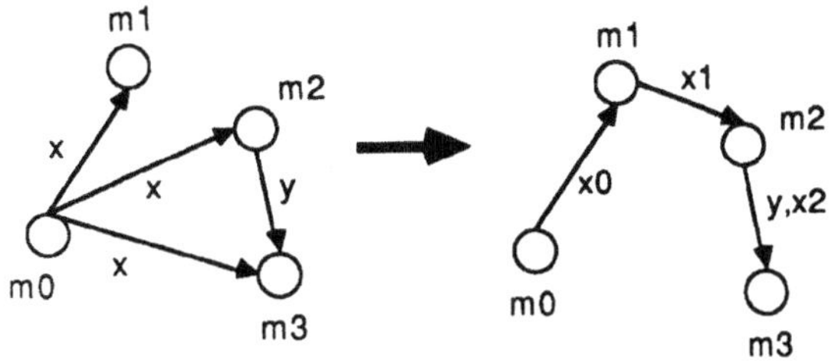

Figure 8.3: Causal x-enumeration. Transformation that reduces the fanout of x.

It is easily seen that f_{Ix} is a bijection $M \to MIx$. What are the properties of MIx? In MIx the variable x is replaced with a set of variables X_I, in a way that eliminates the possibly large fanout of x. Instead of m_0 sending x to all other m_i, $f_{Ix}(m_0)$ sends x_0, with the same value as x, to $f_{Ix}(m_1)$. Then, for all $i > 0$, $f_{Ix}(m_i)$ uses x_{i-1} instead of x and passes the value on to $f_{Ix}(m_{i+1})$, through the variable x_i. Thus, there will be $n(I)$ variables x_i, each with fanout one, instead of one variable x with fanout $n(I)$. This reduction in fanout is bought at the expense of the introduction of new data dependencies. In MIx it holds that $f_{Ix}(m_i) \prec_{MIx} f_{Ix}(m_{i+1})$ for $0 \leq i < n(I)$. In M it is not necessarily the case that $m_i \prec_M m_{i+1}$. A simple example is shown in figure 8.3.

MIx will always be a proper CCSA-set. The functions computed by MIx will be the same as the functions computed by M. This is stated more formally in the following theorem.

Theorem 8.1: *For any pure CCSA-set M derived from a recursion scheme $\mathbf{F}$ in X, for any variable x in $X_a(\mathbf{F})$ and for any causal x-enumeration I in M, MIx is a pure CCSA-set derived from a recursion scheme $\mathbf{F}'$ in $(X \setminus \{x\}) \cup X_I$. For all $x' \in X \setminus \{x\}$ it holds that $o_{\mathbf{F}'}(x') = o_{\mathbf{F}}(x')$, and $o_{\mathbf{F}'}(x_i) = o_{\mathbf{F}}(x)$ for all $i \in dom(I)$. Furthermore, $\prec_{MIx} = (\prec_M \setminus \prec_{Mx}) \cup \prec_{MIxX_I}$.*

Proof. See appendix J. ■

Next question is how to find a causal x-enumeration. An interesting situation is when a correct mapping of the original CCSA structure into a communication structure is already found. The next theorem tells that if the cs-assignments receiving x are totally ordered by the communication ordering, then the enumeration induced by this ordering is a causal x-enumeration.

Theorem 8.2: *Let M be a pure CCSA-set and let W be a correct mapping $\langle M, \prec_M \rangle \to \langle S, \prec_s \rangle$. For any $m \in M$ and for any $x \in dom(m)$, if there is an enumeration $I' : \{1, \ldots, n(I')\} \to \{ W(m') \mid x \in range(m') \}$, where $n(I') = \left| \{ W(m') \mid x \in range(m') \} \right|$, such that $W(m'_i) \prec_s W(m'_{i+1})$ for all $i \in \{1, \ldots, n(I') - 1\}$, then I, which is I' extended with $m_0 = m$, is a causal x-enumeration in M, and $W' : MIx \to S$, defined by $W'(f_{Ix}(m)) = W(m)$ for all $m \in M$, is a correct mapping $\langle MIx, \prec_{MIx} \rangle \to \langle S, \prec_s \rangle$.*

Proof. See appendix J. ■

The definition of fanout in this section only applies to assigned variables. There may, however, be reason to also consider free variables. A free variable may very well occur in the *range* of several cs-assignments. This means that we must provide the same value as input to all these cs-assignments. This value must come from somewhere outside the system. It may have been computed at several events, or it may originate from a single source. In the latter case such a variable should really be considered to have a fanout, just as an assigned variable. It is clear that fanout of free variables can be reduced using exactly the same methods as for assigned variables. This can be understood as follows: For any free variable x we may add the cs-assignment $x' \leftarrow x$, where x' is a fresh variable. Then all old occurrences of x are replaced with x'. The resulting fanout of x' can be treated using the method described here.

Let us finally make a remark on the correspondence between CCSA-sets and sets of computational events. It is possible to define the fanout of a computational event p as the number of cell actions, where the value computed by p is used as input. In the context of cs-assignments fanout is reduced by introducing new variables x_i and assignments of the form $x_i \leftarrow x_{i-1}$, to avoid usage of any variable at more than one event. In the context of cell actions the corresponding operation is to add computational events, labelled with expressions with *identity operators.* Consider the following simple example.

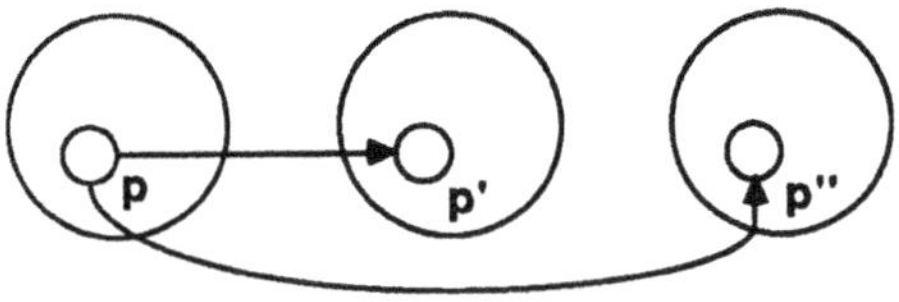

Figure 8.4: Simple cell action structure.

We have three cell actions, each containing one computational event. For simplicity we will now identify computational events and expressions. The leftmost cell action contains $\mathbf{p}$, the one in the middle $\mathbf{p}'$ and the rightmost $\mathbf{p}''$. $\mathbf{p}'$ and $\mathbf{p}''$ use the values from $\mathbf{p}$ as input, so this expression has fanout two. Furthermore, it must be the case that $\mathbf{p} \prec_e \mathbf{p}'$ and $\mathbf{p} \prec_e \mathbf{p}''$. Let us now reduce the fanout by introducing a new expression $id_{S(\mathbf{p})} \bullet \mathbf{p}$. In $\mathbf{p}''$ we substitute this expression for $\mathbf{p}$ and obtain a new expression $\mathbf{p}'''$. It is immediately clear that $\mathbf{p}'''$ computes the same function as $\mathbf{p}''$, since $\phi(id_{S(\mathbf{p})} \bullet \mathbf{p}) = \phi(\mathbf{p})$. Now the situation is that $\mathbf{p} \prec_e \mathbf{p}'$, $\mathbf{p} \prec_e id_{S(\mathbf{p})} \bullet \mathbf{p}$ and $id_{S(\mathbf{p})} \bullet \mathbf{p} \prec_e \mathbf{p}'''$.

Instead of $\mathbf{p}''$ receiving its input from $\mathbf{p}$, we now have the value-equivalent expression $\mathbf{p}'''$ receiving the input from $id_{S(\mathbf{p})} \bullet \mathbf{p}$. Since this expression is placed in the same cell action as $\mathbf{p}''$, the value of $\mathbf{p}$ need only be sent to this cell action. Thus, the fanout of $\mathbf{p}$ is reduced to

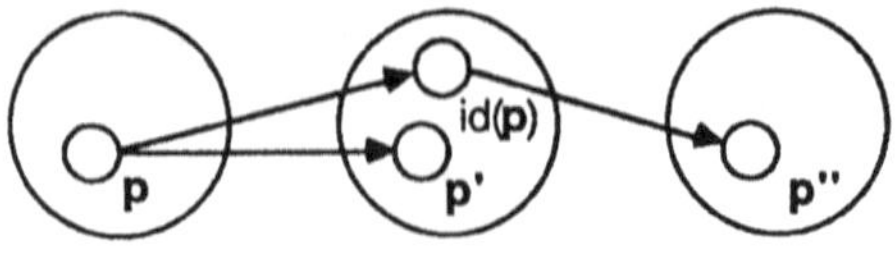

Figure 8.5: Modified cell action structure.

one. Instead we obtain a new transfer of data from $id_{S(\mathbf{p})} \bullet \mathbf{p}$ to $\mathbf{p}'''$. The situation is exactly analogous to the one for cs-assignments, where additional assignments are introduced, fanout is reduced and additional data dependencies appear.

8.4 Covering cycles of substitutions

In this section we will define a property of single points in space under a given schedule. When true, it expresses that a cell can be built using only a finite amount of hardware, in order to support a given schedule.

The property concerns what functions from inputs to outputs a certain cell must be able to compute under a given schedule. This will, of course, depend on the cs-assignments mapped to the cell at different times. Every cs-assignment $m = \langle x_1, \ldots, x_k \rangle \leftarrow \langle \mathbf{p}_1, \ldots, \mathbf{p}_k \rangle$ defines a multiple-valued function $\langle \phi(\mathbf{p}_1), \ldots, \phi(\mathbf{p}_k) \rangle$ from inputs to outputs. So if the cs-assignments $m_1, m_2, \ldots$ are mapped to the cell r at times $t_1, t_2, \ldots$, then r must be able to compute the input/output function of m_k at time t_k. This can essentially be achieved in two different ways. The first is to design r so that it computes the function of *any* m_k at *any* time. The second is to make r *programmable*, and have a program running in r that computes the function of m_k at t_k. It is also possible to think of combined cases. In any case it is clear that the more similar the functions of the cs-assignments are, the simpler the supporting cell can be designed, either with respect to hardware complexity, program complexity, or both. If the cell is made programmable, then a piece of straight-line code will always do; this is because the mapping of m_k to t_k at r determines exactly what function r must compute at time t_k. If there is some periodicity in the sequence of functions to compute, then this can be utilized to let the program loop. This will save code.

One central question must be addressed now: *what do we mean by a functional unit being able to compute a function*? Apparently the ordinary equality between functions as sets of pairs will not do. For instance we would like to consider the function $g(x, y) = x^y$ as being computable by a functional unit defined by the equation $x_{out} = z^u$, since we can match the definitions by simply renaming variables. They are, however, formally not functions in the same variables, thus they are not equal as functions! What we really would like to say is that *a function g can be* *computed by the functional unit f, iff we can obtain g by substituting the inputs of g for the inputs of f in the definition of f.* We will now develop this view further.

Consider as a simple example the following functional unit f. It has one input line x, another input line y and puts the value from line x raised to the value from line y on the line x_f.

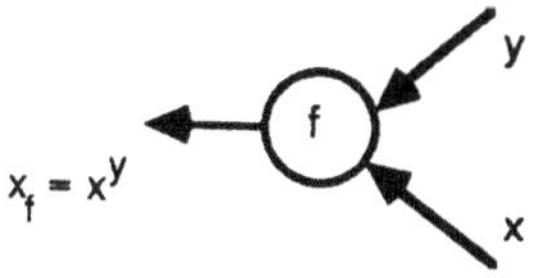

Figure 8.6: Functional unit f for x^y.

f can be defined by the assignment

$$x_f \leftarrow x^y.$$

Let us now consider the function g, defined by the assignment

$$x_g \leftarrow y^x.$$

Apparently g can be evaluated on f; it is just to switch inputs. Formally this is done by applying the substitution

$$\{x \leftarrow y, y \leftarrow x\}$$

to the expression defining f. The result is

$$\{x \leftarrow y, y \leftarrow x\}(x^y) = y^x.$$

Also the function h, defined by

$$x_h \leftarrow u^v$$

can be evaluated by f; it is just to apply the substitution

$$\{x \leftarrow u, y \leftarrow v\}$$

to the expression x^y defining f. It is also possible to compute some functions in less than two variables on f. Consider for instance the function i given by

$$x_i \leftarrow z^2.$$

This function can also be evaluated on f, since

$$\{x \leftarrow z, y \leftarrow 2\}(x^y) = z^2.$$

Thus, if we put the functional unit f at the cell r, then we can schedule the evaluation of the functions g, h and i on r by providing the proper inputs at the proper input lines of f at the proper times.

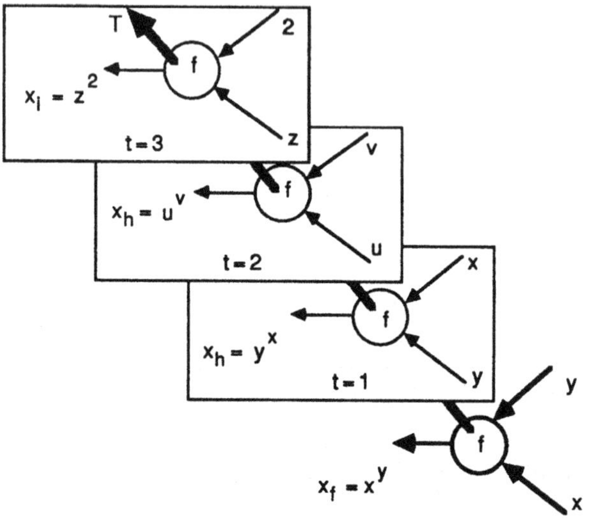

Figure 8.7: Evaluating g, h, i on f at r at times $t = 1, 2, 3$.

It is now time to introduce some formal definitions. In the light of the discussion above it seems to be convenient to use formal substitutions. Therefore we will use formal expressions to represent functions. Let us first define what we mean by substituting an input at an input line:

Definition 8.9: A *restricted substitution* in $\mathcal{E}^{(X)}(\mathcal{A})$ is a substitution $\theta \in \mathcal{S}^{(X)}(\mathcal{A})$, such that for all $x \in dom(\theta)$ holds that $\theta(x)$ is a variable or a constant. The set of all restricted substitutions in $\mathcal{E}^{(X)}(\mathcal{A})$ is denoted $s^{(X)}(\mathcal{A})$.

Second, we define what we mean by a functional unit being able to compute a function:

Definition 8.10: The relation $\geq_\phi$ on $\mathcal{E}^{(X)}(\mathcal{A})$ is for all $\mathbf{p}, \mathbf{p}' \in \mathcal{E}^{(X)}(\mathcal{A})$ defined by:

$$\mathbf{p} \geq_\phi \mathbf{p}' \iff \exists \theta \in s^{(X)}(\mathcal{A})[\mathbf{p}' =_\phi \theta(\mathbf{p})].$$

The identity $=_\phi$ is given by definition 2.16. Cf. the subsumption preorder $\leq_T$ over expressions with equational theory $\mathbf{T}$, [HO80]. The main difference is that our definition only allows *restricted* substitutions, whereas the usual definition allows substitutions in general. A less important distinction is that in the definition here the polynomials $\phi(\theta(\mathbf{p}'))$, $\phi(\mathbf{p})$ are compared, while usually the test is done on the expressions using $=_T$. By theorem 2.3 $=_\phi$ fulfils the rules for equational reasoning, so which equality to use is merely a matter of taste. Comparison of functions is chosen here since this is in closer agreement with the semantics we have adopted.

$\mathbf{p} \geq_\phi \mathbf{p}'$ means that $\mathbf{p}$ is more "general" than $\mathbf{p}$. This is the reason why we use "$\geq_\phi$" instead of "$\leq_\phi$" to denote this relation; it is intuitively appealing to consider the more general expression as "greater". We will now extend this notion of generality to cs-assignments. Observing that cs-assignments are substitutions, we will extend $\geq_\phi$ to a relation on such:

Definition 8.11: For any $\sigma_1, \sigma_2 \in \mathcal{S}^{(X)}(\mathcal{A})$, $\sigma_1 \geq_\phi \sigma_2$ if and only if

$$\exists \theta \in s^{(X)}(\mathcal{A}) \forall x \in dom(\sigma_2) \exists x' \in dom(\sigma_1)[\sigma_2(x) = \theta\sigma_1(x')].$$

Note that this extension of $\geq_\phi$ to substitutions is not similar to the corresponding extension of $\leq_T$; $\sigma_1 \leq_T \sigma_2$ iff there exists a θ such that $\theta\sigma_1 =_T \sigma_2$ (where T-equality is extended to substitutions). $\sigma_1 \geq_\phi \sigma_2$ essentially means that there is a substitution of inputs of σ_1, seen as a cs-assignment, such that for every output of σ_2 there is an assignment in σ_1 computing it.

The reason to extend $\geq_\phi$ to substitutions is the following: *a substitution σ can be seen as a specification of the relation between input and output lines of a cell.* Every variable in $range(\sigma)$ represents an input line and every variable in $dom(\sigma)$ an output line. The output line x receives the value $\phi(\sigma(x))$ computed from the values on the input lines in $varset(\sigma(x))$. Note that there is no need to restrict the set of cell-specifying substitutions to such where the domain and range are disjoint. The reason is that the only connection between inputs and outputs a substitution needs to specify is the one mentioned above. An output line is not necessarily identified with an input line by the name; output line x need not be the same as input line x. This is in contrast to cs-assignments where variables stand for *values*, not lines.

Example 8.1: Consider the two cs-assignments

$$m_1 = \begin{array}{l} x_1 \leftarrow f(y_1) \\ x_2 \leftarrow g(y_2) \end{array}$$

and

$$m_2 = \begin{array}{l} x_3 \leftarrow f(y_3) \\ x_4 \leftarrow g(y_4) \\ x_5 \leftarrow g(y_4). \end{array}$$

By applying $\theta = \{y_1 \leftarrow y_3, y_2 \leftarrow y_4\}$ to m_1 we can see that $m_1 \geq_\phi m_2$, since for every $x \in dom(m_2)$ there is an $x' \in dom(m_1)$ such that $m_2(x) = \theta m_1(x')$ (select x_1 for x_3 and x_2 for x_4, x_5.) The figure below shows a cell described by m_1 and how this cell can be used to perform m_2.

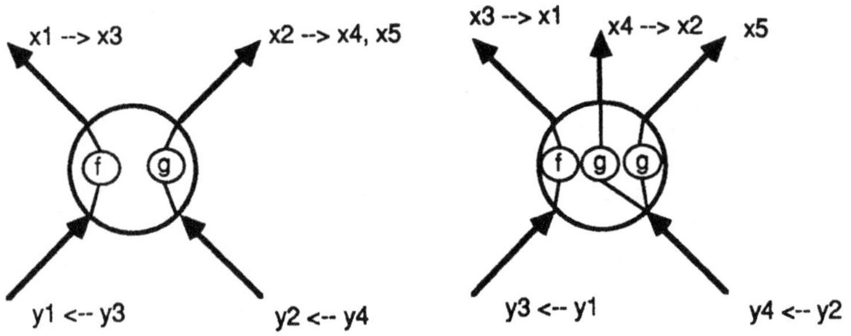

Figure 8.8: Cells given by m_1 and m_2.

On he other hand, $\theta' = \{y_3 \leftarrow y_1, y_4 \leftarrow y_2\}$ applied to m_2 shows that also $m_2 \geq_\phi m_1$. Thus, a cell specified by m_2 can perform m_1 as well.

Let us finally illustrate the fact that a cell very well can be specified by a general substitution. The substitution

$$\sigma = \{x \leftarrow f(x), y \leftarrow g(y)\}$$

specifies the same cell as m_1. It is indeed easy to verify that $\sigma \geq_\phi m_1$ and $\sigma \geq_\phi m_2$. ■

The following properties of $\geq_\phi$ follow immediately from the definition:

Proposition 8.5: *For all $\sigma \in \mathcal{S}^{(X)}(\mathcal{A})$ the following three properties hold: $\sigma \geq_\phi \sigma$, $\sigma \geq_\phi \emptyset$ and if $\emptyset \geq_\phi \sigma$, then $\sigma = \emptyset$.*

$\geq_\phi$ is thus reflexive and $\emptyset$ is a "least element".

Proposition 8.6: *$\geq_\phi$ is transitive.*

Proof. See appendix J. ■

Definition 8.12: For any $\Sigma \subseteq \mathcal{S}^{(X)}(\mathcal{A})$ and any $\sigma \in \mathcal{S}^{(X)}(\mathcal{A})$, σ is an *upper ϕ-bound of* Σ, or $\sigma \geq_\phi \Sigma$, iff for all $\sigma' \in \Sigma$ holds that $\sigma \geq_\phi \sigma'$.

This definition gives rise to some interesting questions which we will not try to answer fully here. Given an algebra $\mathcal{A}$ and a subset Σ of $\mathcal{S}^{(X)}(\mathcal{A})$, does there always exist an upper ϕ-bound σ of Σ? Does there exist a *least* upper ϕ-bound (where "least" is relative to some cost function). What about uniqueness of these bounds? What about finiteness? (The question of finiteness is especially interesting in this context; if a finite upper ϕ-bound of a set of cs-assignments is found, then this bound specifies a finite cell capable to perform all the cs-assignments.) Are there any connections with the corresponding theory of unification?

If Σ is finite, if all $\sigma \in \Sigma$ are finite and if $dom(\sigma) \cap dom(\sigma') = \emptyset$ for all $\sigma, \sigma' \in \Sigma$ (if Σ for instance a finite subset of a set of cs-assignments), then it is clear that there exists a finite upper ϕ-bound. It is simply the union of all σ in Σ. If Σ is infinite or if it contains infinite substitutions, then the existence of finite bounds is less evident. (We can, however, in the same manner as above find an *infinite* upper bound as long as the domains of the substitutions in Σ are disjoint.) The question of how to find least upper ϕ-bounds seems to be somewhat similar to the problem in unification theory of how to find a minimal complete set of unifiers for two terms in a given theory. This is known to be undecidable for many theories. It is therefore reason to suspect, that also the question about least upper ϕ-bounds is hard to answer in general.

Definition 8.13: Let $\langle M, \prec_M \rangle$ be a cs-assignment structure, let S be a space-time and let $W: M \rightarrow S$ be 1-1. The function $W^{-}: S \rightarrow M \cup \{\emptyset\}$ is defined by:

$$W^{-}(s) = \begin{cases} W^{-1}(s), & s \in W(M) \\ \emptyset, & s \notin W(M). \end{cases}$$

$W^{-}(\langle t, r \rangle)$ tells what has to be done at cell r at time t. If no cs-assignment is mapped to $\langle t, r \rangle$, then nothing has to be done there, i.e. $\emptyset$.

Definition 8.14: Let $\langle M, \prec_M \rangle$ be a CCSA structure derived from a recursion scheme in X' over $\mathcal{A}$. Let $\langle T \times R, \prec_s \rangle$ be a communication structure and let W be a correct mapping $\langle M, \prec_M \rangle \rightarrow \langle T \times R, \prec_s \rangle$. Let $n \in Z^+$, let $X \supseteq X'$ and let $\sigma_0, \ldots, \sigma_{n-1} \in \mathcal{S}^{(X)}(\mathcal{A})$.

$\langle \sigma_0, \ldots, \sigma_{n-1} \rangle$ is a *covering n-cycle of substitutions of r under W* iff σ_i is a finite upper ϕ-bound of $\{ W^{-}(\langle t, r \rangle) \mid t \in T \land t \equiv i \pmod n \}$ for every $i \in \{0, \ldots, n-1\}$.

σ is a *covering substitution of r under W* iff σ is an upper ϕ-bound of $W^{-}(T \times \{r\})$.

σ is a covering substitution iff $\langle \sigma \rangle$ is a covering 1-cycle of substitutions. For a covering n-cycle of substitutions $\langle \sigma_0, \ldots, \sigma_{n-1} \rangle$ of r under

W, σ_i defines an action of r that computes $W^-(\langle t,r\rangle)$ when $t = mn+i$, $m \in N$. For every such time there is a restricted substitution that gives information about where to provide the inputs of $W^-(\langle t,r\rangle)$ to the action defined by σ_i. $\langle\sigma_0,\ldots,\sigma_{n-1}\rangle$ specifies a *program* of length n that is powerful enough to support all cs-assignments mapped to r; the intuitive interpretation is that the actions corresponding to the substitutions are cyclically repeated so that the action corresponding to σ_i is performed at r for every time $mn+i$, $m \in N$. If σ is a covering substitution, then r only needs to perform the action of σ. In this case this action can be hardwired into r.

Note how the value of $W^-(s)$, when there is no cs-assignment mapped to s, is treated by the definition above. Since $\sigma \geq_\phi \emptyset$ for all $\sigma \in \mathcal{S}^{(X)}(\mathcal{A})$, any σ supports this "empty cs-assignment". This reflects the fact that the cell can do whatever it wants when it is idle, without affecting the outputs of the system.

8.5 Substitution fields in space-time

So far we have used communication orderings to specify constraints on weakly correct mappings. These orderings concern communication capabilities and causality; they say nothing about the capability of the cells to compute output values from inputs. We will now introduce a new concept that makes it possible to formulate what functional capability a cell has at a given time. This can be used to specify constraints on weakly correct mappings, in order to ensure that no cs-assignments are mapped to cells that cannot support them at the time given by the mapping.

Definition 8.15: For any algebra $\mathcal{A}$ and any set of variables X, a *substitution field in the space-time* $T \times R$ is a family of triples $\Phi = (\langle\sigma_s^\Phi, it_s^\Phi, ot_s^\Phi\rangle \mid s \in T \times R)$, where, for all $s \in T \times R$, σ_s^Φ is a substitution in $\mathcal{S}^{(X)}(\mathcal{A})$, $it_s^\Phi\colon range(\sigma_s^\Phi) \to Z \times R \times X$ is an *input transfer function* and $ot_s^\Phi\colon dom(\sigma_s^\Phi) \to P(Z \times R \times X)$ is an *output transfer function*. For all $\langle t,r\rangle \in T \times R$ and for all $\langle\delta,r',x'\rangle \in Z \times R \times X$ it must hold, for all $x \in dom(\sigma_{\langle t,r\rangle}^\Phi)$, that

$$\langle\delta,r',x'\rangle \in ot_{\langle t,r\rangle}^\Phi(x) \wedge t+\delta \in T \wedge x' \in range(\sigma_{\langle t+\delta,r'\rangle}^\Phi)$$
$$\Longrightarrow$$
$$it_{\langle t+\delta,r'\rangle}^\Phi(x') = \langle\delta,r,x\rangle$$

and for all $x \in range(\sigma_{\langle t,r\rangle}^\Phi)$ that

$$it^{\Phi}_{\langle t,r\rangle}(x) = \langle \delta, r', x' \rangle \wedge t - \delta \in T \wedge x' \in dom(\sigma^{\Phi}_{\langle t-\delta,r'\rangle})$$
$$\Longrightarrow$$
$$\langle \delta, r, x \rangle \in ot^{\Phi}_{\langle t-\delta,r'\rangle}(x').$$

Furthermore, for any $s \in T \times R$ and $x \in dom(\sigma^{\Phi}_s)$, it must hold that

$$\forall \langle \delta, r, x' \rangle \in ot^{\Phi}_s(x)[S(x') = S(x)]$$

and for all $x \in range(\sigma^{\Phi}_s)$

$$\langle \delta, r, x' \rangle = it^{\Phi}_s(x) \Longrightarrow S(x') = S(x).$$

A substitution field Φ in $T \times R$ is *finite* iff σ^{Φ}_s is finite for all $s \in T \times R$.

Here we use the word "field" in the same sense as in mathematical physics: a field is something that is defined for every point in space and varies with time. The interpretation is that $\sigma^{\Phi}_{\langle t,r\rangle}$ is the "most general" action the cell at r may perform at time t.

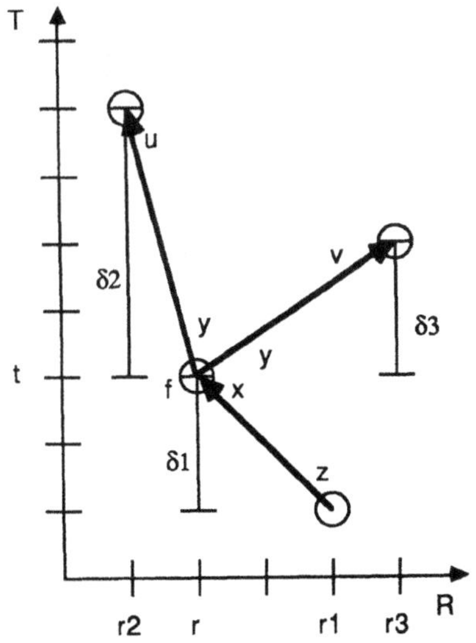

Figure 8.9: $\sigma^{\Phi}_{\langle t,r\rangle} = \{y \leftarrow f(x)\}$, $it^{\Phi}_{\langle t,r\rangle}(x) = \langle \delta_1, r_1, z \rangle$, $ot^{\Phi}_{\langle t,r\rangle}(y) = \{\langle \delta_2, r_2, u \rangle, \langle \delta_3, r_3, v \rangle\}$.

The interpretation of the input and output transfer functions of Φ is the following: every $\langle \delta, r', x' \rangle$ in $ot^{\Phi}_{\langle t,r \rangle}(x)$ represents a transfer of the value on the output port x of r at time t to the input port x' of r' at time $t + \delta$. $it^{\Phi}_{\langle t,r \rangle}(x)$ gives the source, in the form of time, place and output port, of the value received by the input port x at cell r at time t. This is why the equalities between values of certain input and output transfer functions must hold in the definition above: If a value is sent from $\langle t, r, x \rangle$ to $\langle t + \delta, r', x' \rangle$ according to $ot^{\Phi}_{\langle t,r \rangle}(x)$, then it must be received at $\langle t + \delta, r', x' \rangle$ from $\langle t, r, x \rangle$ according to $it^{\Phi}_{\langle t+\delta,r' \rangle}(x')$. Note, though, that there might be input ports for certain cells at certain times for which there is no sending port, and vice versa. The interpretation of these "loose ends" is that they can be used for communication that is not internal to the system, like receiving values of free variables and sending output values to something "outside". Finally there is a sort constraint on the input and output transfer functions; input and output ports that are connected must have the same sort.

The following exclusion principle for output transfer functions follows from the connection between input and output transfer functions, that is postulated in the definition of substitution fields:

Proposition 8.7: *(Exclusion principle for output transfer functions): If $t' + \delta' = t'' + \delta''$ and $\langle \delta', r', x' \rangle \neq \langle \delta'', r'', x'' \rangle$, then it cannot be the case that both $\langle \delta', r, x \rangle \in ot^{\Phi}_{\langle t',r' \rangle}(x')$ and $\langle \delta'', r, x \rangle \in ot^{\Phi}_{\langle t'',r'' \rangle}(x'')$.*

Proof. Let us show that the premise cannot be true if both $\langle \delta', r, x \rangle \in ot^{\Phi}_{\langle t',r' \rangle}(x')$ and $\langle \delta'', r, x \rangle \in ot^{\Phi}_{\langle t'',r'' \rangle}(x'')$. $\langle \delta', r, x \rangle \in ot^{\Phi}_{\langle t',r' \rangle}(x')$ implies that

$$it^{\Phi}_{\langle t'+\delta',r \rangle}(x) = \langle \delta', r', x' \rangle. \tag{1}$$

In the same way $\langle \delta'', r, x \rangle \in ot^{\Phi}_{\langle t'',r'' \rangle}(x'')$ implies that

$$it^{\Phi}_{\langle t''+\delta'',r \rangle}(x) = \langle \delta'', r'', x'' \rangle. \tag{2}$$

Assume that $t' + \delta' = t'' + \delta''$. (1), (2) then gives $\langle \delta', r', x' \rangle = \langle \delta'', r'', x'' \rangle$ which contradicts the premise. ■

The intuitive meaning of the exclusion principle is that the output transfer functions cannot cause two different values to be sent to the same port at the same time. As can be seen from the proof, this is an immediate consequence of that it^{Φ}_{s} is a function giving one single value in $Z \times R \times X$.

Definition 8.16: For any substitution field Φ on S, the binary relation $\preceq_\Phi$ on S is defined by:

$$\langle t,r\rangle \preceq_\Phi \langle t',r'\rangle$$
$$\Longleftrightarrow$$
$$\exists x \in dom(\sigma^\Phi_{\langle t,r\rangle})\exists x' \in X[\langle t'-t,r',x'\rangle \in ot^\Phi_{\langle t,r\rangle}(x)]\vee$$
$$\exists x' \in range(\sigma^\Phi_{\langle t',r'\rangle})\exists x \in X[\langle t'-t,r,x\rangle \in it^\Phi_{\langle t',r'\rangle}(x')]$$

for all $\langle t,r\rangle, \langle r',t'\rangle \in S$. Furthermore, we define $\prec_\Phi = \preceq_\Phi \setminus \omega_S$. Φ is *consistent* with the communication ordering $\prec_s$ iff $\prec_\Phi \subseteq \prec_s$. Φ is *causal* if it is consistent with some communication ordering.

This definition is quite natural. It says that the input and output transfers of a substitution field Φ are supported by the communication ordering $\prec_s$ when Φ is consistent with $\prec_s$. Note that we require $\prec_s$ to support not only the transfers of assigned variables with well-defined "sources" and "sinks", but also the "loose ends" mentioned above.

Definition 8.17: Let $\mathcal{A}$ be an algebra and let X be a set of variables. Let $\langle M, \prec_M\rangle$ be a cs-assignment structure and let S be a space-time. Let Φ be a substitution field on S. The function $W: M \rightarrow S$ is *a correct mapping* with respect to Φ iff it is 1-1 and there is a family of restricted substitutions $(\,\theta_m \mid m \in M\,)$, such that for all $m \in M$

$$\begin{aligned}
&\forall x \in dom(m)\exists x' \in dom(\sigma^\Phi_{W(m)})\\
&\quad [m(x) = \theta_m\sigma^\Phi_{W(m)}(x')\wedge\\
&\quad \forall m' \in M[x \in range(m') \Longrightarrow\\
&\qquad \exists x'' \in range(\sigma^\Phi_{W(m')})[\langle t_W(m') - t_W(m), r_W(m'), x''\rangle \in ot^\Phi_{W(m)}(x')\wedge\\
&\qquad\qquad \theta_{m'}(x'') = x]]].
\end{aligned}$$

This definition needs some informal explanation. Assume that W is correct with respect to Φ. Then there must be a family $(\,\theta_m \mid m \in M\,)$ of restricted substitutions, where θ_m substitutes the inputs of m for those of $\sigma^\Phi_{W(m)}$ in such a way that all outputs of m are computed by $\sigma^\Phi_{W(m)}$. Thus, m can be performed by the cell $r_W(m)$ at time $t_W(m)$ if every input of m is supplied on the proper input line at the right moment. The second conjunct concerns the transfer of variables assigned by m. It says that every receiver m' of an output variable x of m must be matched on $\sigma^\Phi_{W(m')}$, by $\theta_{m'}$, in such a way that m' receives x at the input port(s) of

$\sigma^{\Phi}_{W(m')}$, where it is sent to according to the output transfer function of $W(m)$.

The following corollary follows immediately from the definition above:

Corollary 8.1: *If $W: M \to S$ is a correct mapping with respect to Φ, then $\sigma^{\Phi}_{W(m)} \geq_{\phi} m$ for all $m \in M$.*

Theorem 8.3: *For any pure cs-assignment structure $\langle M, \prec_M \rangle$ and for any substitution field Φ on S holds that if W is a correct mapping with respect to Φ, then $\prec_{MW}$ exists and $\prec_{MW} \subseteq \prec_{\Phi}$.*

Proof. See appendix J. ∎

Assume that W is a correct mapping with respect to Φ and that Φ is consistent with $\prec_s$. By proposition 7.11 (which carries over from cell action structures to CCSA structures because of the equivalence of these concepts), W will be a correct mapping $\langle M, \prec_M \rangle \to \langle S, \prec_s \rangle$ if it is 1-1 and if $\prec_{MW} \subseteq \prec_s$. We therefore immediately obtain the following corollary:

Corollary 8.2: *For any pure CCSA structure $\langle M, \prec_M \rangle$, for any communication structure $\langle S, \prec_s \rangle$ and for any substitution field Φ on S holds that if W is a correct mapping with respect to Φ and if Φ is consistent with $\prec_s$, then W is a correct mapping $\langle M, \prec_M \rangle \to \langle S, \prec_s \rangle$.*

We can conclude that being a correct mapping with respect to a causal substitution field Φ is stronger than being a correct mapping to the corresponding communication structure $\langle S, \prec_{\Phi} \rangle$. This is not surprising, since correctness with respect to a substitution field involves the inner structure of the cs-assignments. Correctness with respect to communication structures, on the other hand, only considers the possibility to transfer data between events.

In the case of free synthesis with respect to communication orderings, the mapped precedence relation becomes the communication ordering. In the same way, a mapped cs-assignment structure defines its own substitution field:

Definition 8.18: Let M be a pure set of cs-assignments derived from $\mathbf{F}$ and let $S = T \times R$ be space-time. For any function $W: M \to S$ that is 1-1, we define the substitution field (MW) in S by: for all $\langle t, r \rangle \in S$,

$$\sigma^{(MW)}_{\langle t,r \rangle} = W^{-}(\langle t, r \rangle),$$

for all $x \in range(\sigma^{(MW)}_{\langle t,r\rangle})$

$$it^{(MW)}_{\langle t,r\rangle}(x) = \begin{cases} \langle 0, r, x\rangle & x \notin X_a(\mathbf{F}) \\ \langle t - t', r', x\rangle & \exists \langle t', r'\rangle [x \in dom(W^-(\langle t', r'\rangle))] \end{cases}$$

and for all $x \in dom(\sigma^{(MW)}_{\langle t,r\rangle})$

$$ot^{(MW)}_{\langle t,r\rangle}(x) = \{\, \langle t' - t, r', x\rangle \mid x \in range(W^-(\langle t, r\rangle))\,\}.$$

It is easily seen that W always will be correct with respect to (MW). (Set $\theta_m = \emptyset$ for all $m \in M$ in definition 8.17.) The question of causality is more interesting. Because of proposition 7.3, (MW) will be causal iff $\prec_{(MW)}$ is a communication ordering. The following theorem shows that this concept of causality is consistent with the previous concept of causality used in free synthesis with respect to communication orderings.

Theorem 8.4: *For all pure sets of cs-assignments M, for all space-times S and for all functions $W: M \to S$ that are 1-1, $\prec_{MW}$ exists and $\prec_{(MW)} = \prec_{MW}$.*

Proof. See appendix J. ■

8.6 The fixed i/o-property

We now have the means to define an important property of cells under a given schedule. It says that a cell having the property can be built in such a way that it supports the schedule with a cyclic program, where every step in the program always performs its input and output through the same lines, connected to the same ports. First we define the property relative to a substitution field:

Definition 8.19: Let Φ be a substitution field on the space-time $T \times R$. For any $n \in Z^+$, the cell $r \in R$ has *the fixed n-i/o-property in* Φ iff for all $t \in T$ holds that:

f1. $\sigma^{\Phi}_{\langle t,r\rangle}$ is finite.

f2. $\sigma^{\Phi}_{\langle t,r\rangle} = \sigma^{\Phi}_{\langle t+n,r\rangle}$.

f3. $ot^{\Phi}_{\langle t,r\rangle}(x) = ot^{\Phi}_{\langle t+n,r\rangle}(x)$ for all $x \in dom(\sigma^{\Phi}_{\langle t,r\rangle})$.

f4. $it^{\Phi}_{\langle t,r\rangle}(x) = it^{\Phi}_{\langle t+n,r\rangle}(x)$ for all $x \in range(\sigma^{\Phi}_{\langle t,r\rangle})$.

Definition 8.20: Let $\langle M, \prec_M\rangle$ be a pure cs-assignment structure, let $T \times R$ be a space-time and let W be a function $M \to T \times R$. For any $n \in Z^+$, $r \in R$ has *the fixed n-i/o property with respect to W* iff there exists a causal substitution field Φ such that W is a correct mapping with respect to Φ and r has the fixed n-i/o property in Φ.

The following corollary follows immediately from definitions 8.14, 8.19, 8.20 and corollary 8.1.

Corollary 8.3: *If r has the fixed n-i/o property with respect to W, then there exists a covering n-cycle of substitutions of r under W.*

Thus, the fixed n-i/o-property is stronger than the existence of a covering n-cycle. This is because the existence of covering n-cycles only concerns the function of each cell, but the fixed n-i/o property also involves the connections between cells. If a cell has the n-i/o property, then we know two things about it. First, the functions the cell must perform can be realized by a cyclic program of length n, where each step in the program is a substitution. Second, every specific input and output of each single step can be received from and sent to the the same lines, regardless of time. Of course, the most advantageous situation is when the cell has the fixed 1-i/o property. Then we obtain a maximum of possible hardwiring, since both the function and the connections of input and output ports of the cell are fixed in time and can be built directly into the hardware.

The kind of property described in this section has so far, at least to the knowledge of the author, been neglected in the literature. It is, however, clear that the fixed n-i/o property is of importance, if the method of synthesis based on explicit mappings to space-time is to be applied to hardware synthesis. A cell may for instance be covered by a substitution without having the fixed n-i/o property for any n at all. In such a case the design of the cell is more complex, since there must be some parts of it redirecting input and output. If the pattern of i/o redirection is complex, then these parts will probably be complex too.

8.7 Substitution fields in space

In this section we will consider time-invariant substitution fields, or substitution fields in space. That is, the substitutions and input and output transfer functions are dependent on the point in space only. This is the most detailed hardware model we will consider in this thesis. Substitution fields in space correspond to fixed hardware structures in the same way as substitution fields in space-time correspond to communication orderings.

Definition 8.21: For any algebra $\mathcal{A}$ and any set of variables X, a *substitution field in the space R* is a family of triples

$$\Phi = (\langle \sigma_r^\Phi, it_r^\Phi, ot_r^\Phi \rangle \mid r \in R),$$

where σ_r^Φ, for all $r \in R$, is a substitution in $\mathcal{S}^{(X)}(\mathcal{A})$, it_r^Φ: $range(\sigma_r^\Phi) \rightarrow N \times R \times X$ is an *input transfer function* and ot_r^Φ: $dom(\sigma_r^\Phi) \rightarrow P(N \times R \times X)$

is an *output transfer function.* For all $r \in R$, $x \in X$ and $\langle \delta, r', x' \rangle \in N \times R \times X$ it holds that

$$\langle \delta, r', x' \rangle \in ot_r^\Phi(x) \wedge x' \in range(\sigma_{r'}^\Phi) \implies it_{r'}^\Phi(x') = \langle \delta, r, x \rangle$$

and

$$it_r^\Phi(x) = \langle \delta, r', x' \rangle \wedge x' \in dom(\sigma_{r'}^\Phi) \implies \langle \delta, r, x \rangle \in ot_{r'}^\Phi(x').$$

A substitution field Φ in R is *finite* iff σ_r^Φ is finite for all $r \in R$.

The interpretation of a substitution field Φ in space is similar to the interpretation of a substitution field in space-time. Every σ_r^Φ defines the function from input to output that the fixed cell at r computes. If $\langle \delta, r', x' \rangle \in ot_r^\Phi(x)$, then there is a fixed line with delay δ from the output port x of the cell at r to r'. If $x' \in range(\sigma_{r'}^\Phi)$, then the line is connected to the input port x' of the cell at r'. If $it_r^\Phi(x) = \langle \delta, r', x' \rangle$, then there is a fixed line with delay δ to the input port x of the cell at r from r'. If $x' \in dom(\sigma_{r'}^\Phi)$, then the line is connected to the output port x' of the cell at r'.

In the same manner as a fixed hardware structure generates a relation in space-time, a substitution field in space generates a substitution field in space-time:

Definition 8.22: For any substitution field Φ in a space R, the *substitution field in $T \times R$ generated by Φ, $\hat{\Phi}$*, is the family of triples

$$(\langle \sigma_s^{\hat{\Phi}}, it_s^{\hat{\Phi}}, ot_s^{\hat{\Phi}} \rangle \mid s \in T \times R),$$

where $\sigma_{\langle t,r \rangle}^{\hat{\Phi}} = \sigma_r^\Phi$, $it_{\langle t,r \rangle}^{\hat{\Phi}} = it_r^\Phi$ and $ot_{\langle t,r \rangle}^{\hat{\Phi}} = ot_r^\Phi$ for all $\langle t, r \rangle \in T \times R$.

Proposition 8.8: *For any substitution field Φ in R, $\hat{\Phi}$ is a substitution field in $T \times R$.*

Proof. See appendix J. ∎

Proposition 8.9: *For any finite substitution field Φ in R, every $r \in R$ has the fixed 1-i/o-property in $\hat{\Phi}$.*

Proof. Trivially from definitions 8.19, 8.22. ∎

There is an exclusion principle for output transfer functions of substitution fields in space, that is similar to the one for substitution fields in space-time:

Proposition 8.10: *(Exclusion principle for output transfer functions of substitution fields in space): If $\langle \delta', r', x' \rangle \neq \langle \delta'', r'', x'' \rangle$, then it cannot be the case that both $\langle \delta', r, x \rangle \in ot^{\Phi}_{r'}(x')$ and $\langle \delta'', r, x \rangle \in ot^{\Phi}_{r''}(x'')$.*

Proof. Select t', t'' such that $t' + \delta' = t'' + \delta''$. Then, according to the exclusion principle for substitution fields in space-time,

$$\begin{aligned} \langle \delta', r', x' \rangle \neq \langle \delta'', r'', x'' \rangle &\Longrightarrow \neg(\langle \delta', r', x' \rangle \in ot^{\hat{\Phi}}_{\langle t', r' \rangle}(x') \wedge \\ &\qquad \langle \delta'', r, x \rangle \in ot^{\hat{\Phi}}_{\langle t'', r'' \rangle}(x'')) \\ &\Longrightarrow \neg(\langle \delta', r', x' \rangle \in ot^{\Phi}_{r'}(x') \wedge \\ &\qquad \langle \delta'', r, x \rangle \in ot^{\Phi}_{r''}(x'')). \end{aligned}$$

■

Any substitution field Φ in space defines a fixed hardware structure $\underline{\underline{\Delta}}(\Phi)$, in the same manner as any substitution field Φ' in space-time defines a relation $\preceq_{\Phi'}$ on the space-time.

Definition 8.23: For any substitution field Φ in space R the fixed hardware structure $\underline{\underline{\Delta}}_{\Phi}$ is defined by: For all $r, r' \in R$ and for all $\delta \in N$

$$\begin{aligned} \langle r, r', \delta \rangle \in \underline{\underline{\Delta}}_{\Phi} \iff &\exists x \in dom(\sigma^{\Phi}_{r}) \exists x' \in X[\langle \delta, r', x' \rangle \in ot^{\Phi}_{r}(x)] \vee \\ &\exists x' \in range(\sigma^{\Phi}_{r'}) \exists x \in X[\langle \delta, r, x \rangle \in it^{\Phi}_{r'}(x')], \end{aligned}$$

and Δ_{Φ} is defined by

$$\Delta_{\Phi} = \underline{\underline{\Delta}}_{\Phi} \setminus \{ \langle r, r, 0 \rangle \mid r \in R \}.$$

The definitions provided for gives two ways to extract space-time relations, that describe the communication supported between the events generated of a substitution field Φ in space. We may form either $\prec_{\hat{\Phi}}$ or $\prec(\Delta(\Phi))$. The following theorem states that these relations in fact are equal. Thus, the two ways of forming a space-time relation from a substitution field in space are consistent.

Theorem 8.5: *For any substitution field Φ in the space R, $\prec_{\hat{\Phi}} = \prec(\Delta_{\Phi})$.*

Proof. See appendix J. ■

9. Index vectors and complexity

In chapter 8 we found that even simple optimization problems for schedules can be combinatorially hard. If we want to attack realistic instances of these problems with any success, there are two basic alternatives. The first is to find an approximation algorithm that works on all instances of the problem. The second, which we will address here, is to find classes of special instances where the complexity can be reduced.

The complexity is often reduced if some form of regularity is exploited. Algorithms specified by systems of *recurrence equations* are particularly interesting from this point of view. Such systems consist of a small number of equations parameterized by indices in the variable names. Thus, a system of recurrence equations specifies a set of cs-assignments, where one cs-assignment is obtained for every combination of index values. Every cs-assignment corresponds to a single step in the recurrence. A set of cs-assignments specified by a recurrence thus contains a high degree of regularity, since many cs-assignments will differ only in the variable names.

Example 9.1: Consider the recurrence relation

$$\begin{aligned} x_i &\leftarrow a x_{i-1} \\ y_i &\leftarrow b x_{i-1} \end{aligned} \qquad i = 1, \ldots, n.$$

For every i this recurrence defines a cs-assignment m_i with input variable x_{i-1} and output variables x_i, y_i. The recurrence relation thus specifies a CCSA-set $\{ m_i \mid i = 1, \ldots, n \}$. This CCSA-set has one single free variable x_0. All other variables x_i, y_i therefore correspond to polynomials in x_0. (In this simple example it is easily seen that x_i corresponds to $a^i x_0$ and y_i to $b^i x_0$.) The regularity of the CCSA-set is reflected by the fact that for all i, j between 1 and n holds that $m_i \geq_\phi m_j$. ■

In the definitions of this chapter we will consider CCSA structures, because of their close relationship with recurrence equations sketched above. There is, however, nothing that prevents the use of cell action structures instead of CCSA-sets. Because of the equivalence between cell action structures and CCSA structures demonstrated in chapter 6, cell action structures can be considered equally well, and we will freely use the definitions and results as if they had been made for such structures whenever convenient.

9.1 Index space and index functions

Algorithms specified by recurrence equations are likely to be suitable for hardware implementation, since the repetitivity of a recurrence equation often can be translated directly into a regular hardware structure. The parametrization of the equations gives an *enumeration* of the corresponding cs-assignments. If there is more than one index in the parametrization, then the enumeration is multi-dimensional. If the regularity indicated by the enumeration is to be exploited, then it is convenient to consider the enumeration itself rather than the enumerated cs-assignments. This leads to the following definition:

Definition 9.1: (index space, index vectors, sets of index vectors): For every positive integer n, Z^n is an *index space*. Every element of an index space is an *index vector*. Every subset of an index space is a *set of index vectors* (SoIV).

Similar definitions can be found in for instance [KMC72,Le84,Li85, Mo82]. For any index space Z^n we will denote the null vector $(0,\ldots,0)$ by $\overline{0}$.

Definition 9.2: For every function G from a CCSA-set M to an index space, the relation $\prec_{MG}$ on $G(M)$ is defined by $fi(\langle M, \prec_M\rangle, \langle M(G), \prec_{MG}\rangle, G)$ if it exists. G is a *weakly correct indexing function* $\langle M, \prec_M\rangle \rightarrow \langle G(M), \prec_{MG}\rangle$ iff $\prec_{MG}$ exists.

$G(M)$ is the *set of index vectors associated with* M *by* G.

G is a *correct indexing function* $\langle M, \prec_M\rangle \rightarrow \langle G(M), \prec_{MG}\rangle$ iff it is a weakly correct indexing function $\langle M, \prec_M\rangle \rightarrow \langle G(M), \prec_{MG}\rangle$ that is 1-1.

According to lemma 6.2, $\prec_{MG}$ will be uniquely determined by G and $\prec_M$ when existent. Note that we define an indexing function to be a function *from* the "indexed" set M and *to* the "indexing" set Z^n rather than the other way around. This is not in accordance with the ordinary notion of indexing. Every cs-assignment is actually labelled with an index vector in very much the same manner as cs-assignments or cell actions are labelled with events in space-time by a weakly correct mapping. The reason for this is that we will consider functions from index space to space-time; functions from CCSA-sets to space-time can thus be formed by simply composing such functions with indexing functions. Other authors have chosen set of index vectors as a starting point; they define recurrence equations as fixed equations where the variables are functions of the index vectors, [KMW69,Q83,Raj86], or where the equation itself is a function of the index vector, [C85B,C86A,C86C]. This is actually an implicit way to define sets of cs-assignments.

Definition 9.3: Let I be a set of index vectors and let $\prec$ be a relation on I. Let $\langle S, \prec_s \rangle$ be a communication structure. F is a *weakly correct index mapping* $\langle I, \prec \rangle \to \langle S, \prec_s \rangle$ iff it is a function $I \to S$ such that:

$$\prec_F \text{ exists and } \prec_F \subseteq \prec_s.$$

W is a *correct index mapping* $\langle I, \prec \rangle \to \langle S, \prec_s \rangle$ iff it is a weakly correct index mapping $\langle I, \prec \rangle \to \langle S, \prec_s \rangle$ that is 1-1.

Theorem 9.1: *Let $\langle M, \prec_M \rangle$ be a CCSA structure and let $\langle S, \prec_s \rangle$ be a communication structure. If G is a weakly correct indexing function $\langle M, \prec_M \rangle \to \langle G(M), \prec_{MG} \rangle$ and if F is a weakly correct index mapping $\langle G(M), \prec_{MG} \rangle \to \langle S, \prec_s \rangle$, then $F \circ G$ is a weakly correct mapping $\langle M, \prec_M \rangle \to \langle S, \prec_s \rangle$.*

Proof. Let G be a weakly correct indexing function $\langle M, \prec_M \rangle \to \langle G(M), \prec_{MG} \rangle$. Then $\prec_{MG}$ exists and

$$fi(\langle M, \prec_M \rangle, \langle G(M), \prec_{MG} \rangle, G). \tag{1}$$

If F is a weakly correct index mapping $\langle G(M), \prec_{MG} \rangle \to \langle S, \prec_s \rangle$, then $\prec_{MGF}$ exists. By definition 7.4 it holds that

$$fi(\langle G(M), \prec_{MG} \rangle, \langle S, \prec_{MGF} \rangle, F). \tag{2}$$

Furthermore, $\prec_{MGF} \subseteq \prec_s$. (1), (2) and *fi*-transitivity (lemma 6.1) yields

$$fi(\langle M, \prec_M \rangle, \langle S, \prec_{MGF} \rangle, F \circ G). \tag{3}$$

By *fi*-uniqueness (lemma 6.2) we can conclude that $\prec_{MF\circ G}$ exists and that $\prec_{MF\circ G} = \prec_{MGF}$. It follows that $\prec_{MF\circ G} \subseteq \prec_s$. Thus, $F \circ G$ is a weakly correct mapping from $\langle M, \prec_M \rangle$ to $\langle S, \prec_s \rangle$. ∎

Corollary 9.1: *If G is a correct indexing function $\langle M, \prec_M \rangle \to \langle G(M), \prec_{MG} \rangle$ and if F is a correct index mapping $\langle G(M), \prec_{MG} \rangle \to \langle S, \prec_s \rangle$, then $F \circ G$ is a correct mapping $\langle M, \prec_M \rangle \to \langle S, \prec_s \rangle$.*

Given a weakly correct indexing function $G(M)$, theorem 9.1 implies that weakly correct index mappings from $G(M)$ to a space-time can be considered as well as weakly correct mappings directly from M. The composition of the weakly correct indexing function and a weakly correct index mapping is always a weakly correct mapping. As usual, a commuting diagram that illustrates this can be drawn. See figure 9.1.

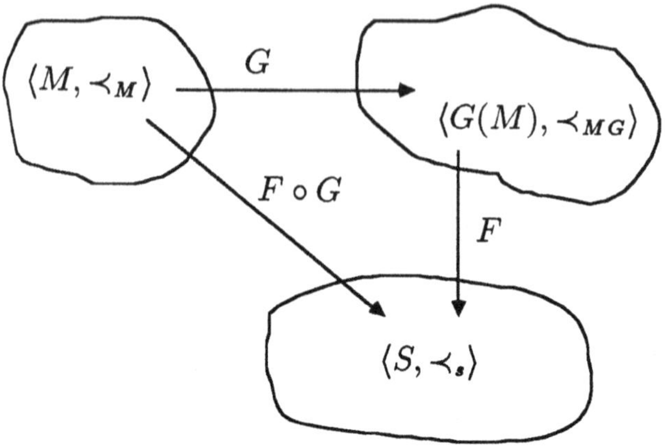

Figure 9.1: Commuting diagram.

9.2 Discrete space-time, data dependence vectors, design vectors

One reason to introduce index spaces is that they are vector spaces. For instance we can add and subtract index vectors. An important class of space-times consists of subsets of vector spaces:

Definition 9.4: A *discrete space-time* is a space-time $S = T \times R$, where T is a set of times and $R = Z^k$ for some $k \in N$. Any element of a discrete space-time is called a *space-time vector.*

Thus, a mapping from an index space to a discrete space-time is a mapping between vector spaces.

Definition 9.5: For any $k \in Z^+$, for any subset A of Z^k and for any relation R over A the *set of direction vectors of* R, $\mathcal{D}(R)$, is given by: for all $d \in Z^k$

$$d \in \mathcal{D}(R) \iff \exists a, a' \in A[d = a' - a \wedge aRa'].$$

We can speak about a direction from one point x to another point x' in a vector space. This direction is given by the difference vector $x' - x$. Thus, $\mathcal{D}(R)$ contains all directions between any two points a, a' such that aRa'. Note that direction vectors do not always belong to the set from which they are defined. But they will always belong to the embedding vector space.

Definition 9.6: Let $\langle M, \prec_M \rangle$ be a CCSA structure and let G be a weakly correct index function. Then $\mathcal{D}(\prec_{MG})$ is called *the set of data dependence vectors associated with M by G.*

A data dependence vector gives an "abstract" direction in index space, from one cs-assignment sending data to another one receiving it.

Definition 9.7: If R is a binary relation on a discrete space-time S, then $\mathcal{D}(R)$ is called *the set of design vectors of R.*

If $\prec_s$ is a communication ordering, then the design vectors of $\prec_s$ give all possible directions of communication in space-time. If $\langle \Delta t, \Delta r \rangle$ is a design vector of $\prec_s$, then there are two events $\langle t, r \rangle$, $\langle t', r' \rangle$ in S such that $t' - t = \Delta t$, $r' - r = \Delta r$ and $\prec_s$ can support a data transfer from $\langle t, r \rangle$ to $\langle t', r' \rangle$.

Proposition 9.1: *Let F be a function from a CCSA-set M to a discrete space-time S, such that $\prec_{MF}$ exists. Then, for all $m, m' \in M$,*

$$F(m) \neq F(m') \land m \prec_M m' \implies F(m) - F(m') \in \mathcal{D}(\prec_{MF}).$$

Proof. From definition 7.4 of $\prec_{MF}$ follows that if $F(m) \neq F(m')$, then $m \prec_M m' \implies F(m) \prec_{MF} F(m') \implies F(m') - F(m) \in \mathcal{D}(\prec_{MF})$. ■

Thus, $\mathcal{D}(\prec_{MF})$ contains the differences in space-time between communicating events in the schedule of M given by F. If $\langle \Delta t, \Delta r \rangle \in \mathcal{D}(\prec_{MF})$ and if we want a piece of fixed hardware to support the communication required by F, then there must somewhere exist a communication line with delay Δt and spatial direction and length Δr. A small number of design vectors implies a high regularity in the communication pattern for the scheduling, since then many data transfers will have the same length, direction and delay. This regularity is likely to be reflected in the corresponding fixed hardware structure.

9.3 Linear mappings from index space

As stated above, we can consider mappings from index sets to space-times instead of mappings directly to the space-time. The repetitivity in an algorithm specified by recurrence relations is often reflected in the low number of data dependence vectors, if a suitable index space and indexing function is selected for the recurrence equations. This property of the algorithm is preserved into the resulting space-time structure if the index mapping to the space-time is *linear*.

Since the vector spaces considered here are of the form Z^n, linear mappings can be represented by matrices. In the following we will use the same symbol for a linear mapping and the matrix representing it, when there is no risk of confusion. It should also be noted that a linear mapping is defined on the whole index space. The index mapping is, however, the restriction to the index set. When there is no risk of confusion we will use the same notation for the mapping from the whole index space and the restriction. We will also refer to the restricted function as linear, even though it really is the mapping between the vector spaces that is linear.

Theorem 9.2: *Let $\langle M, \prec_M \rangle$ be a CCSA structure, let S be a discrete space-time and let G be a weakly correct indexing function from M. If $L: G(M) \to S$ is linear and if $\prec_{L \circ G}$ exists, then*

$$\mathcal{D}(\prec_{L \circ G}) = L(\mathcal{D}(\prec_{MG})) \setminus \{\overline{0}\} = \{\, L(d) \mid d \in \mathcal{D}(\prec_{MG}) \,\} \setminus \{\overline{0}\}.$$

Proof. See appendix K. ■

Proposition 9.2: *Let $\langle M, \prec_M \rangle$ be a CCSA structure, let S be a discrete space-time and let G be a weakly correct indexing function from M. If $L: G(M) \to S$ is linear and 1-1 and if $\prec_{L \circ G}$ exists, then*

$$\mathcal{D}(\prec_{L \circ G}) = L(\mathcal{D}(\prec_{MG})) = \{\, L(d) \mid d \in \mathcal{D}(\prec_{MG}) \,\}.$$

Proof. Since $\prec_M$ is finite-downward, $\prec_{MG}$ is finite-downward and thus irreflexive. Therefore, $\overline{0} \notin \mathcal{D}(\prec_{MG})$. For a linear mapping L that is 1-1 this implies that $L(\mathcal{D}(\prec_{MG})) \setminus \{\overline{0}\} = L(\mathcal{D}(\prec_{MG}))$. The result now follows from theorem 9.2. ■

Thus, the design vectors for a communication structure obtained from a linear index mapping are exactly the mapped data dependence vectors (except for $\overline{0}$, which may occur in $\{\, L(d) \mid d \in \mathcal{D}(\prec_{MG}) \}$ when L is not 1-1). So if we consider linear mappings only, then the structure of the resulting communication ordering is to a certain extent inherent in the algorithm. The freedom that is left can still be of considerable use; examples where the method is successfully applied can be found in [CS83,Le84,Li83,MiWi84].

Another aspect is complexity. If a suitable indexing of dimensionality k' is found, then there are kk' independent parameters in a matrix representing a linear mapping to a space-time of dimension k. For a mapping in general from a CCSA-set with n elements to such a space-time nk coordinates must be found, k coordinates for each cs-assignment. Typically, for algorithms suitable for systolic implementations, $k = k' = 2$ or 3 is an appropriate choice. When n grows big, the difference in complexity between the linear and general approach increases drastically; the extreme case is infinite CCSA-sets (typically found in signal processing applications) which sometimes still can be dealt with by the linear method.

An objection against the reasoning above is that each of the matrix elements could still be any value in an infinite set; thus, the linear approach would give no real decrease in complexity. However, it is easily seen that large coefficients in the matrix will give an undesired increase of delays and spatial distances of the data transfers. Therefore, it is often possible to restrict the search to coefficients of a small magnitude. In the systolic case, with 2×2 or 3×3 matrices, it may then be feasible to do an exhaustive search in order to find the best mapping with respect to some optimality criterion.

What about existence of a set of index vectors, indexing function and linear index mapping for a given CCSA structure and a given discrete space-time? The following proposition gives the answer.

Proposition 9.3: *For every CCSA structure* $\langle M, \prec_M \rangle$ *and every discrete space-time* S*, there exists a correct indexing function* G *and a correct linear index mapping* $L\colon \langle G(M), \prec_{MG} \rangle \rightarrow \langle S, \prec_{L \circ G} \rangle$ *such that* $\prec_{L \circ G}$ *is a ripple-free communication ordering and* $L \circ G$ *is a correct mapping* $\langle M, \prec_M \rangle \rightarrow \langle S, \prec_{L \circ G} \rangle$.

Proof. Since S is a discrete space-time, it is a subset of Z^n for some $n > 1$. Thus, any nonempty subset of S is a valid set of index vectors. By theorem 7.2 (which carries over to CCSA structures), there exists a correct mapping $W\colon \langle M, \prec_M \rangle \rightarrow \langle S, \prec_{MW} \rangle$ where $\prec_{MW}$ is ripple-free. Now we can consider $W(M)$ as an index set and W as a correct indexing function. Finally we can note that the identity function I_n on the vector space that is embedding S is linear and that the restriction of it to $W(M)$ trivially is a correct index mapping $\langle W(M), \prec_{MW} \rangle \rightarrow \langle S, \prec_{MW} \rangle$. Therefore, we can select $G = W$ and $L = I_n$ which proves the result. ■

Thus, the trivial indexing given by the space-time coordinate is always a valid indexing, provided that the mapping from cs-assignments to space-time is correct. Theorem 9.2 is only a statement about existence and gives no clue how to find a good indexing. The reason is that the indexing is not given until a correct mapping is already found.

9.4 Translation-invariant communication orderings

Data dependence and design vectors are particularly relevant concepts when the communication ordering is *invariant under translation*, according to the following definition:

Definition 9.8: Let $S = T \times Z^k$ be a discrete space-time. The relation $\prec$ on S is *invariant under translation* if and only if:

$$\forall s, s' \in S \, \forall \delta \in Z^{k+1} [s + \delta, s' + \delta \in S \implies (s \prec s' \implies s + \delta \prec s' + \delta)].$$

Translation-invariant communication orderings are suitable for describing communication constraints on highly regular architectures, such as systolic array architectures. For instance, the translation-invariant communication ordering $\prec_s$ defined below describes the constraint on a two-dimensional grid, where data is sent locally and with unit delay only:

$$\forall \langle t,r\rangle, \langle t',r'\rangle \in T \times Z^2[\langle t,r\rangle \prec_s \langle t',r'\rangle \iff t' = t+1 \wedge \|r' - r\|_\infty \leq 1].$$

Here $\|r' - r\|_\infty = \max_{i=1,2} |r'_i - r_i|$. If we want to sharpen the constraint to include that no cell has any internal memory, then the condition $\|r' - r\|_\infty \leq 1$ can be changed to $\|r' - r\|_\infty = 1$. This condition implies that $\langle t,r\rangle \not\prec_s \langle t+\delta, r\rangle$ always.

Theorem 9.3: *Let $\langle M, \prec_M\rangle$ be a CCSA structure, let G be a weakly correct indexing function from M to an index space Z^n, let L be a linear mapping from $G(M)$ to a discrete space-time S and let finally $\prec_s$ be a communication ordering on S that is invariant under translation. If for all $d \in \mathcal{D}(\prec_{MG})$ holds that $Ld \in \mathcal{D}(\prec_s) \cup \{\overline{0}\}$ and if $\prec_{L\circ G}$ exists, then $L \circ G$ is a weakly correct mapping $\langle M, \prec_M\rangle \to \langle S, \prec_s\rangle$.*

Proof. By assumption $\prec_{L\circ G}$ exists. It remains to show that $\prec_{L\circ G} \subseteq \prec_s$. From the definition of $\prec_{L\circ G}$ follows, that for all $s, s' \in S$

$$\begin{gathered} s \prec_{L\circ G} s' \implies \\ s \neq s' \wedge \exists m, m' \in M[L\circ G(m) = s \wedge L \circ G(m') = s' \wedge m \prec_M m']. \end{gathered}$$

For all such m, m' it holds that $G(m') - G(m) \in \mathcal{D}(\prec_{MG})$. Since L is linear and $L(G(m') - G(m)) \in \mathcal{D}(\prec_s) \cup \{\overline{0}\}$ (theorem 9.2 and $L\circ G(m) \neq L \circ G(m')$) it follows that

$$\begin{aligned} L(G(m') - G(m)) = L\circ G(m') - L \circ G(m) \in \mathcal{D}(\prec_s) &\implies \\ \exists s'' \in S[s'' \prec_s L\circ G(m') - L\circ G(m) + s''] &\implies \\ (\text{translation invariance of } \prec_s, \text{ choose } \delta = L\circ G(m) - s'') &\implies \\ L\circ G(m) \prec_s L \circ G(m') &\implies \\ s \prec_s s'. & \end{aligned}$$

■

Thus, when L is linear and $\prec_s$ is invariant under translation, we can see if $L\circ G$ is weakly correct by checking that all mapped data dependence vectors are design vectors. When there are only a few such vectors, compared with the number of cell actions and space-time points, it is much simpler to check the correctness in this way. Theorem 9.3 is a variant of theorem 1 in [Mo82]. Note, though, that in [Mo82] it is postulated that L must have integer-valued coefficients. This does not necessarily have to be true, see [Li83,Li85] or chapter 11 for a counterexample.

9.5 Free synthesis for linear mappings

In this section we will give some results about free synthesis for linear mappings. These are merely applications of the more general results in subsection 7.4. The results are some easy-to-check conditions on the matrix elements in the matrix of the linear mapping.

Let us first recall some basic facts about linear mappings. For a vector space V, $dim(V)$ is the number of linearly independent vectors spanning V. For a matrix M, $rank(M)$ is the maximal number of linearly independent column vectors. Also, a linear mapping $M: V \to V'$ is 1-1 if and only if $dim(V) \leq dim(V')$ and $rank(M) = dim(V')$. If $dim(V) = dim(V')$, then M is 1-1 if and only if M is nonsingular, or equivalently $det(M) \neq 0$.

In the following, we will denote the time component of a space-time vector s by s_t.

Proposition 9.4: *Let $\langle M, \prec_M \rangle$ be a CCSA structure, let G be a correct indexing function to an index space Z^n and let L be a linear function from $G(M)$ to the discrete space-time $T \times Z^k$. If $n \leq k+1$, if $rank(L) = k+1$ and if for all $d \in \mathcal{D}(\prec_{MG})$ holds that $(Ld)_t \geq 0$, then $\prec_{M\,L\circ G}$ is a communication ordering on $T \times Z^k$. If $(Ld)_t > 0$ for all $d \in \mathcal{D}(\prec_{MG})$, then $\prec_{M\,L\circ G}$ is a ripple-free communication ordering on $T \times Z^k$.*

Proof. n is the dimensionality of Z^n and $k+1$ is the dimensionality of the vector space enclosing $T \times Z^k$. Thus, the condition $n \leq k+1$ and $rank(L) = k+1$ implies that L is 1-1. Then $L\circ G$ is 1-1. By proposition 9.2 follows that $\mathcal{D}(\prec_{M\,L\circ G}) = L(\mathcal{D}(\prec_{MG}))$. If for all $\langle \Delta t, \Delta r \rangle \in \mathcal{D}(\prec_{M\,L\circ G})$ holds that $\Delta t \geq 0$, then $\langle t, r \rangle \prec_{M\,L\circ G} \langle t', r' \rangle \implies t \leq t'$ for all $\langle t, r \rangle, \langle t', r' \rangle \in S$. Proposition 7.13 then implies that $\prec_{M\,L\circ G}$ is a communication ordering. If for all $\langle \Delta t, \Delta r \rangle \in \mathcal{D}(\prec_{M\,L\circ G})$ holds that $\Delta t > 0$, then for all $\langle t, r \rangle, \langle t', r' \rangle \in S$ holds that $\langle t, r \rangle \prec_{M\,L\circ G} \langle t', r' \rangle \implies t < t'$. Corollary 7.13 then gives that $\prec_{M\,L\circ G}$ is a ripple-free communication ordering. ∎

The causality condition on the mapped data dependence vectors can be easily checked by multiplying all data dependence vectors with the "time row" in the matrix L. This leads to a nice geometrical interpretation. Denote the row in L corresponding to time by l_t. This vector defines a hyperplane in the index space Z^n by the equation $l_t x = 0$ for all

x in the hyperplane. The causality condition can now be seen as follows: all data dependence vectors must lie in or above the hyperplane.

Note that $L \circ G$ may be a correct mapping even if L is not 1-1. The reason is that only the restriction of L to $G(M)$ needs to be 1-1. Dependent on the shape of $G(M)$, this restricted function may be 1-1 even if the total function $Z^n \to Z^{k+1}$ is not.

The following corollary follows immediately from proposition 9.4.

Corollary 9.2: *Let $\langle M, \prec_M \rangle$ be a CCSA structure, let G be a correct index function to an index space Z^{k+1} and let L be a linear function from $G(M)$ to the discrete space-time $T \times Z^k$. If L is nonsingular and if $(Ld)_t \geq 0$ for all $d \in \mathcal{D}(\prec_{MG})$, then $\prec_{M\,L\circ G}$ is a communication ordering on $T \times Z^k$. If for all $d \in \mathcal{D}(\prec_{MG})$ holds that $(Ld)_t > 0$, then $\prec_{M\,L\circ G}$ is a ripple-free communication ordering on $T \times Z^k$.*

This result covers most cases that are of interest. (See, however, [CS83] for an example where the dimensionalities of the vector spaces are different.)

Let us now state a result about the existence of a correct linear mapping from a given index set to a space-time. For any finite set of direction vectors $\mathcal{D}(R) = \{d_1, \dots, d_n\}$, $D(R)$ denotes a matrix $[d_1 \cdots d_n]$ with the direction vectors as columns. (Of course there will be one such matrix for every permutation of direction vectors, but in the following it will not matter which one we consider.)

Theorem 9.4: *Let $\langle M, \prec_M \rangle$ be a CCSA structure and let G be a correct index function to an index space Z^n, such that $\mathcal{D}(\prec_{MG})$ is finite. Let $S = T \times Z^k$ be a discrete space-time, where $k+1 \geq n$. If the equation*

$$D(\prec_{MG})^T x \geq \overline{0}$$

has a solution $a \in Z^n$ such that $a \neq \overline{0}$, then there exists a linear mapping $L\colon Z^n \to Z^{k+1}$ that is a correct linear mapping $\langle G(M), \prec_{G(M)} \rangle \to \langle S, \prec_{M\,L\circ G} \rangle$ whenever $L(G(M)) \subseteq S$. If $D(\prec_{MG})^T x > \overline{0}$, then $\prec_{M\,L\circ G}$ is a ripple-free communication ordering.

Proof. Consider first the case $k+1 = n$. Assume that there is an $a \in Z^n$ such that $D(\prec_{MG})^T a \geq \overline{0}$ and $a \neq \overline{0}$. Then, for every $d \in \mathcal{D}(\prec_{MG})$, holds that $d^T a \geq 0$. Furthermore, since a is a non-zero integer vector, there must exist an orthogonal $n \times n$-matrix Q with rational elements such that $Qa = (\alpha \;\; 0 \;\; \dots \;\; 0)^T$, where $\alpha > 0$. Thus, for any $d \in \mathcal{D}(\prec_{MG})$, holds that $d^T a = d^T I a = d^T (Q^T Q) a = (Qd)^T Qa = (Qd)_0 \alpha \geq 0$, where $(Qd)_0$ is the first component of Qd. Since $\alpha > 0$ it follows that $(Qd)_0 \geq 0$, or $\beta (Qd)_0 = (\beta Q d)_0 \geq 0$ for any nonnegative β. Since the elements of Q are rational, a positive β can be chosen so that

$L = \beta Q$ has integer elements only. Then L is a linear mapping $Z^n \rightarrow Z^n$ and furthermore $(Ld)_0 \geq 0$. This component of Ld can be interpreted as the time component, and when $L(G(M)) \subseteq S$ proposition 9.4 implies that $\prec_{M\,L\circ G}$ is a communication ordering on S. It follows that L is a correct index mapping $\langle G(M), \prec_{G(M)}\rangle \rightarrow \langle S, \prec_{M\,L\circ G}\rangle$.
When $D(\prec_{MG})^T a > \overline{0}$, a similar reasoning gives that $Ld > 0$ for all $d \in \mathcal{D}(\prec_{MG})$. Corollary 9.2 then implies that $\prec_{MG}$ is ripple-free.

When $k + 1 > n$ and the conditions of the theorem are fulfilled, we can use the proof above to find a correct linear index mapping $Z^n \rightarrow Z^n$ such that $G(M) \subseteq T \times Z^{n-1}$. This mapping can be trivially extended to a linear mapping $Z^n \rightarrow Z^{k+1}$, where $G(M) \subseteq T \times Z^k$. ■

Theorem 9.4 has an intuitively appealing geometrical interpretation: if we can find a hyperplane through the origin in index space such that all that dependence vectors are on one side of it, then it is possible to find a linear index mapping that preserves causality into time. Note that the condition $L(G(M)) \subseteq T \times Z^k$ is important. A linear mapping may very well fulfil the hyperplane condition but map index vectors of cs-assignments outside the selected space-time. If the index set is infinite, then it may, for a given linear index mapping, very well be the case that there is *no* space-time such that all index vectors are mapped to it.

9.6 Geometrical interpretation of linear mappings

Let us finally mention something about the geometrical interpretation of linear mappings. A linear mapping from a coordinate space to itself can be seen as a movement of the points in the space. Geometrical objects (sets of points in the space) are transformed by a 1-1 linear mapping in the following way: they are *rotated* around the origin and *stretched* or *compressed* in different directions.

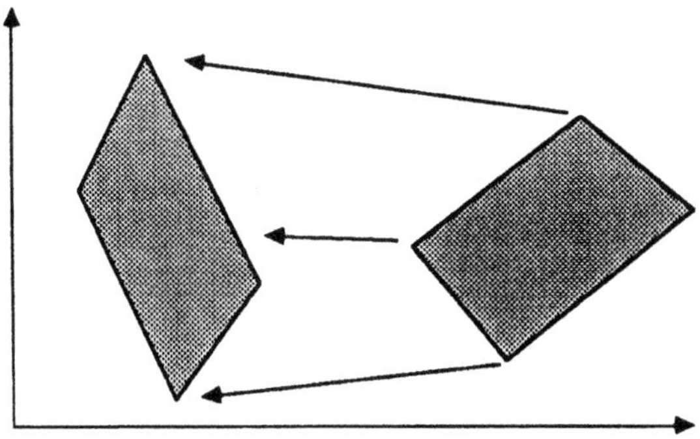

Figure 9.2: Linear transformation of geometrical object.

This interpretation can be used as a support for the intuition, when looking for linear mappings from a given index set. The index set is a set of points and can thus be seen as a geometrical object, that will be transformed according to the above. If the index space and space-time are two- or three-dimensional, then the transformed object can easily be evaluated with respect to simple performance criteria such as execution time or cell utilization.

Even in higher dimensions the execution time is simple to evaluate, when the index set is a convex polyhedron. A property of linear mappings that are 1-1 is that hyperplanes are mapped to hyperplanes. Thus, corners are mapped to corners. Therefore, we can find the execution time by checking the time coordinates of all mapped corners of the index set.

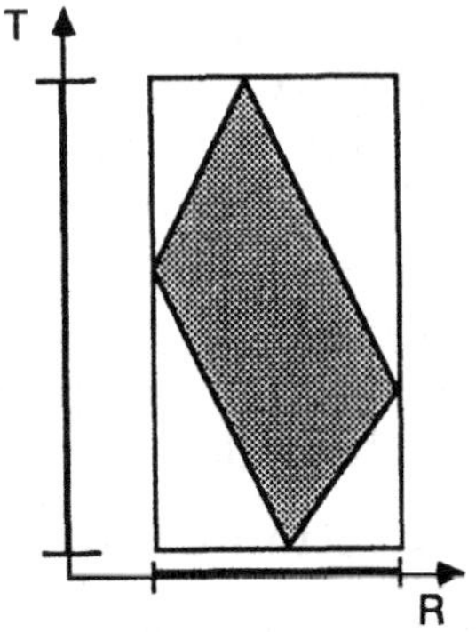

Figure 9.3: Evaluating performance criteria.

Clearly the rotation of the transformed object is of importance. Dependent on the shape of the object, different orientations will yield different execution time and cell utilization. Also of importance is the scaling effectuated by the stretchings and compressions. Note that the mapped index set may contain "holes" where no index vector is mapped. A linear mapping may very well introduce such holes even if they are not present in the index set. The reason is the scaling of the mapped object mentioned above; since the number of points where index vectors are mapped cannot be greater than the number of mapped index vectors, an expansion will introduce holes. These holes represent events where a cell is idle. Thus, it is desirable to avoid mappings that are expansions.

There is an easy way to determine whether a linear mapping is an expansion or not. Simply evaluate the *determinant* of the matrix representing it! The determinant is a scale factor (with a possible minus sign) of the linear mapping; if it is greater than one, then the mapping is an expansion, if it is less than one, then it is a contraction. The extreme case is of course when the determinant is zero: then the matrix is singular and the mapped object is collapsed into a lower dimensionality. It is thus desirable to find a suitable transformation matrix with a determinant as low (but non-zero) as possible, without violating the condition that every point in the index set is mapped to an integer lattice point in space-time.

10. Data dependent scheduling

10.1 Introduction

So far, this thesis has been concerned with the case where the schedule is determined in advance. This chapter deals with the more general case, where the schedule is dependent of the value of the inputs. Thus, the approach developed here is related to other methods for parallel scheduling on more general multiprocessor architectures. This chapter is more suggestive to its nature, as it does not lead to any concrete methods for synthesis. The aim is rather to put the contents of the rest of the thesis in a more general context, and to enhance the understanding of parallel systems in general.

10.2 The model of information

As before, we consider *output specifications* O over a domain $A = \prod(A_x \mid x \in X)$, where O is a set of functions in the variables $x \in X$ and each x is ranging over A_x. When the values a_x of all variables x are known, they form a tuple $a = \langle a_x \mid x \in X \rangle$ that is an element of A. During the evaluation of the functions in O the values of the variables will successively become known. As the value of a variable becomes known, the subset of A of possible values of a is further restricted. Therefore we adopt the following concept of information (for instance, cf. *information systems*, [Sco82]):

Definition 10.1: $\mathcal{I}(A)$, The *subsets of information* of a domain A, is $\mathcal{P}(A) \setminus \{\emptyset\}$. The *information structure* of A is the directed set $\langle \mathcal{I}(A), \supseteq \rangle$.

Every subset A' of A represents the case that a must be in A'. A itself is the least element and represents that we know nothing about a at all; it can be any element in A. The maximal elements are the singletons $\{a'\}$, $a' \in A$. $\{a'\}$ represents the case that $a = a'$, that is, we have total information about all variables. The empty set has no particular meaning and is therefore excluded.

Note that there are other ways of obtaining subsets of A than making the values of variables known. The system specified by O may for instance be used in a special environment, where there are certain restrictions on the input variables. This environment can then be specified by a subset of A.

The by far most important case is still when reading the values of variables is the only way for the system to obtain information about its environment. In this case only one type of subsets of A is interesting, namely those for which some components of the tuples are fixed and the others range over all possible values as we go through the elements of the subset:

Definition 10.2: For all $X' \subseteq X$ and for all $\langle b_x \mid x \in X' \rangle$,

$$A/\langle b_x \mid x \in X' \rangle = \{ \langle a_x \mid x \in X \rangle \mid \forall x \in X' [a_x = b_x] \}.$$

$\mathcal{H}(A)$, the *hyperplanes of* A, is the set of all such subsets of A. The operation "/", *projection*, is also defined on hyperplanes of A by

$$(A/\langle b_x \mid x \in X' \rangle)/\langle b_x \mid x \in X'' \rangle = A/\langle b_x \mid x \in X' \cup X'' \rangle.$$

(Here it is assumed that the tuples $\langle b_x \mid x \in X' \rangle$ and $\langle b_x \mid x \in X'' \rangle$ agree on b_x for $x \in X' \cap X''$).

There is a close correspondence between the relation $\supseteq$ over $\mathcal{H}(A)$ and the "less-defined" relation $\sqsubseteq$ on tuples with possibly undefined components, [Ma74]. A component of a tuple being undefined is the same as the component not being fixed in the corresponding hyperplane. If a hyperplane is a superset of another hyperplane, then the tuple corresponding to the first hyperplane is less defined than the one corresponding to the second hyperplane.

Definition 10.3: The *strict information structure* of a domain A is the directed set $\langle \mathcal{H}(A), \supseteq \rangle$.

Sometimes it is more convenient to refer to a predicate on A instead of the corresponding subset. When there is no risk of confusion, we will use the most convenient representation.

10.3 Advantages of dynamic scheduling

The aim of dynamic scheduling is to use the successively increasing information about the input a to eliminate unnecessary work, that is, to avoid evaluating expressions whose values are not needed. There are essentially two ways that expressions may become superfluous:

If there are functions in O which are data-dependent in our algebra $\mathcal{A}$, then there are subsets A', A'' of A for which $A' \supset A''$ and $\eta(O, A') \supset \eta(O, A'')$. Thus, the increased information achieved when going from A' to A'' can be used to avoid computing the polynomials in $\eta(O, A') \setminus \eta(O, A'')$.

Operators may be *non-strict* (definition 2.4), which means that for particular values of some arguments the operator becomes independent of other arguments. If a non-strict operator is involved in an expression, then some of its subexpressions may become unnecessary to evaluate, depending on the outcome of the evaluation of other subexpressions.

Example 10.1: Consider the expression $if \bullet \langle \mathbf{p}_1, \mathbf{p}_2, \mathbf{p}_3 \rangle$. If $\mathbf{p}_1$ evaluates to *false*, then $\mathbf{p}_2$ becomes unnecessary to evaluate. If it evaluates to *true*, then we don't need to evaluate $\mathbf{p}_3$. As an alternative we may consider the two expressions $\mathbf{p}_2$, $\mathbf{p}_3$ and the representation function $\eta: A \to \mathcal{P}^{(X)}(\mathcal{A})$ defined by:

$$\eta(a) = \begin{cases} \phi(\mathbf{p}_2), & \phi(\mathbf{p}_1)(a) = true \\ \phi(\mathbf{p}_3), & \phi(\mathbf{p}_1)(a) = false. \end{cases}$$

This defines a function with a data-dependent structure in an algebra without the *if*-operator, that is equal to $\phi(if \bullet \langle \mathbf{p}_1, \mathbf{p}_2, \mathbf{p}_3 \rangle)$. Which of the expressions $\mathbf{p}_2$ and $\mathbf{p}_3$ we will have to evaluate is dependent of the input a; when we know enough about a to determine the value of $\phi(\mathbf{p}_1)(a)$, then we know which one to choose. ■

10.4 Restricted polynomials

When we know that the input is in A' instead of A, it is sufficient to consider the restrictions of the output functions to A' instead of the original functions on A. The same applies to the corresponding polynomials:

Definition 10.4: Let $A' \in \mathcal{I}(A)$. For all output specifications O over A, $O|_{A'} = \{ f|_{A'} \mid f \in O \}$. $\mathcal{P}^{(X)}(\mathcal{A})|_{A'}$, *the polynomials in* X *over* $\mathcal{A}$ *restricted to* A', is the set $\{ p|_{A'} \mid p \in \mathcal{P}^{(X)}(\mathcal{A}) \}$.

It is clear that equalities that hold for polynomials in $\mathcal{P}^{(X)}(\mathcal{A})$ also hold in $\mathcal{P}^{(X)}(\mathcal{A})|_{A'}$, since functions f_1, f_2 for which $f_1(x) = f_2(x)$ for all elements x in a domain also are equal on a subset of that domain. On the other hand new equalities can be introduced, since perhaps no points in A where the values differ are in A'. These equalities can be used to find simpler expressions corresponding to a given polynomial, when the input is restricted. In order to express this formally, we make the following definition:

Definition 10.5: For all $A' \in \mathcal{I}(A)$, the *restricted natural homomorphism* $\mathcal{E}^{(X)}(\mathcal{A}) \to \mathcal{P}^{(X)}(\mathcal{A})|_{A'}$, $\phi_{A'}$, is defined by:

1. $\phi_{A'}(\mathbf{x}) = e_x^X|_{A'}$, all $x \in X$.
2. For all compound expressions $f \bullet \langle \mathbf{p}_i \mid i \in I_f \rangle$ in $\mathcal{E}^{(X)}(\mathcal{A})$, $\phi_{A'}(f \bullet \langle \mathbf{p}_i \mid i \in I_f \rangle) = f \circ \langle \phi_{A'}(\mathbf{p}_i) \mid i \in I_f \rangle$.

Consider again the expression $if \bullet \langle \mathbf{p}_1, \mathbf{p}_2, \mathbf{p}_3 \rangle$. If $\phi_{A'}(\mathbf{p}_1) = true$, then

$$\phi_{A'}(if \bullet \langle \mathbf{p}_1, \mathbf{p}_2, \mathbf{p}_3 \rangle) = \phi_{A'}(\mathbf{p}_2).$$

This can be used to avoid the evaluation of $\mathbf{p}_3$ when the input is in A'.

Definition 10.6: Let O be an output specification with domain A that is satisfiable in $\mathcal{A}$, and let η be a representation function of O with respect to $\mathcal{A}$. For all $A' \in \mathcal{I}(A)$, $\eta_{A'}: O|_{A'} \times A' \rightarrow \mathcal{P}^{(X)}(\mathcal{A})|_{A'}$ is defined by: $\eta|_{A'}(f|_{A'}, a) = \eta(f, a)|_{A'}$ for all $f \in O$, $a \in A'$.

$\eta|_{A'}$ gives the connection between the restricted functions in $O|_{A'}$ and the restricted polynomials in $\mathcal{P}^{(X)}(\mathcal{A})$. The following proposition shows that the property defining a representation function η carries over to $\eta_{A'}$:

Proposition 10.1: *Let $A' \in \mathcal{I}(A)$. for all $a \in A'$, $\eta_{A'}(f|_{A'}, a)(a) = f|_{A'}(a)$.*

Proof. For any $f \in O$ follows, from the definitions of η and $\eta_{A'}$, that $\eta_{A'}(f|_{A'}, a)(a) = \eta(f, a)|_{A'}(a) = \eta(f, a)(a) = f(a) = f|_{A'}(a)$ for all $a \in A'$. ■

Let us now formally define a function that connects restricted polynomials in $\mathcal{P}^{(X)}(\mathcal{A})|_{A'}$ with the further restricted ones in $\mathcal{P}^{(X)}(\mathcal{A})|_{A''}$:

Definition 10.7: Let $A', A'' \in \mathcal{I}(A)$ and assume that $A'' \subseteq A'$. The function $r^{A'}_{A''}: \mathcal{P}^{(X)}(\mathcal{A})|_{A'} \rightarrow \mathcal{P}^{(X)}(\mathcal{A})|_{A''}$, the *restriction function* $A' \rightarrow A''$, is defined by: $r^{A'}_{A''}(p) = p|_{A''}$ for all $p \in \mathcal{P}^{(X)}(\mathcal{A})|_{A'}$.

Note that $p|_{A''} = p|_{A'}|_{A''}$ for all $p \in \mathcal{P}^{(X)}(\mathcal{A})$. It is also clear that there is exactly one $p|_{A''}$ for every $p|_{A'}$. Thus, $r^{A'}_{A''}$ is well-defined.

Proposition 10.2: $r^{A'}_{A''} \circ \phi_{A'} = \phi_{A''}$.

Proof. See Appendix L. ■

Proposition 10.3: *For all $f \in O$ and for all $a \in A''$, $r^{A'}_{A''} \circ \eta_{A'}(f|_{A'}, a) = \eta_{A''}(f|_{A''}, a)$.*

Proof. Consider an arbitrary $f \in O$. For all $a \in A''$,

$$r^{A'}_{A''} \circ \eta_{A'}(f|_{A'}, a) = r^{A'}_{A''}(\eta_{A'}(f|_{A'}, a)) = r^{A'}_{A''}(\eta(f, a)|_{A'}) =$$
$$\eta(f, a)|_{A'}|_{A''} = \eta(f, a)|_{A''} = \text{(definition 10.6)} = \eta_{A''}(f|_{A''}, a).$$

■

The following corollary for sets follows immediately.

Corollary 10.1: *For all* $f \in O$, $r^{A'}_{A''} \circ \eta_{A'}(f|_{A'}, A'') = \eta_{A''}(f|_{A''}, A'')$.

Next corollary follows immediately from the two preceding propositions.

Corollary 10.2: *For all* $\mathbf{p} \in \mathcal{E}^{(X)}(\mathcal{A})$, $f \in O$ *and* $a \in A''$ *holds that*

$$\phi_{A'}(\mathbf{p}) = \eta_{A'}(f|_{A'}, a) \implies \phi_{A''}(\mathbf{p}) = \eta_{A''}(f|_{A''}, a).$$

Thus, the connection between an output function and a formal expression is preserved when the information increases.

If we analogously to $r^{A'}_{A''}$ define the partial function $\rho: O|_{A'} \times A' \to O|_{A''} \times A''$ by $\rho(\langle f|_{A'}, a\rangle) = \langle f|_{A''}, a\rangle$ for all $f \in O$ and all $a \in A''$, then proposition 10.3 can be expressed as $(r^{A'}_{A''} \circ \eta_{A'})|_{O|_{A'} \times A''} = \eta_{A''} \circ \rho$.

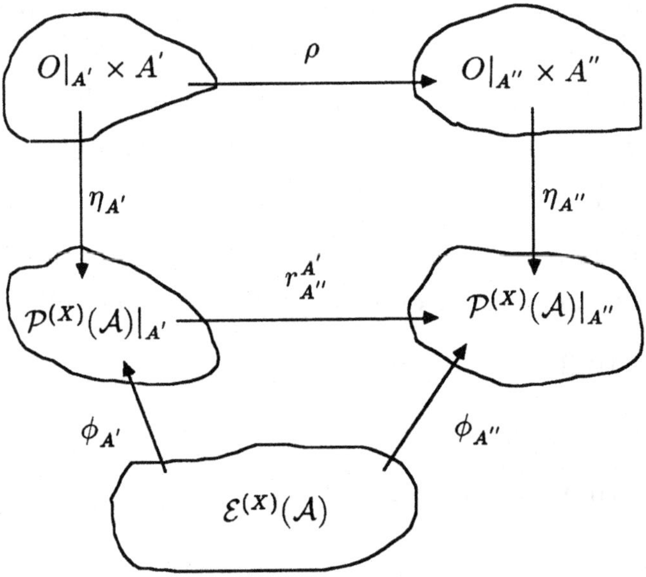

Figure 10.1: Commuting diagram.

The results in this chapter can now be summarized in a commuting diagram, as shown in figure 10.1.

10.5 Schedules

In this subchapter we will extend the definition of schedules from chapter 8 to include also *partial* schedules, where all steps necessary to perform perhaps are not scheduled. Note that a weakly correct mapping here is considered to be a function from *computational events* to space-time. That is, we disregard cell actions for the moment, since they are of no special significance here. The resulting description will still be equivalent.

Definition 10.8: Let O be an output specification satisfiable in $\mathcal{A}$ with domain A and let η be a representation function of O with respect to $\mathcal{A}$. Let $\langle S, \prec_s \rangle$ be a communication structure. For any $A' \in \mathcal{I}(A)$, $\sigma = \langle E_\sigma, \psi_\sigma, D_\sigma, \prec_\sigma, W_\sigma \rangle$ is a *partial* A'*-schedule of* O *on* $\langle S, \prec_s \rangle$, if and only if:

1. E_σ is a set of expressions such that $\phi_{A'}(E_\sigma) \cap \eta_{A'}(O|_{A'}, A') \neq \emptyset$.
2. $\psi_\sigma: D_\sigma \to ec(E_\sigma)$ is onto.
3. $\prec_\sigma \in \mathcal{R}(D_\sigma, \psi_\sigma, E_\sigma)$.
4. $W_\sigma: D_\sigma \to S$ is a weakly correct mapping $\langle D_\sigma, \prec_\sigma \rangle \to \langle S, \prec_s \rangle$.

If $\eta_{A'}(O|_{A'}, A') \subseteq \phi_{A'}(E_\sigma)$, then σ is a *total* A'*-schedule of* O *on* $\langle S, \prec_s \rangle$.

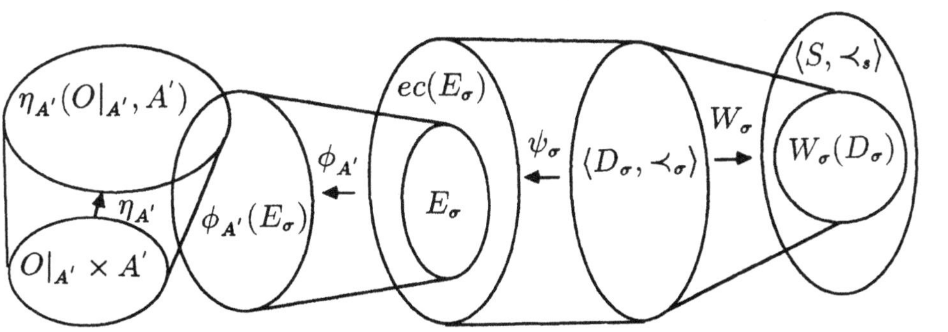

Figure 10.2: A partial A'-schedule.

Since $O|_{A'}$ and A' are non-empty, $\eta(O|_{A'}, A')$ is non-empty. Therefore, a total schedule is always a partial schedule as well. Schedules according to definition 8.1 are essentially total A-schedules. The requirement for total schedules that $\eta_{A'}(O|_{A'}, A') \subseteq \phi_{A'}(E_\sigma)$ is dropped for partial A'-schedules, which is a considerable weakening. Nevertheless, the concept of partial schedules is necessary to give an appropriate description of systems with dynamic scheduling.

Proposition 10.4: *If σ is a total A'-schedule of O on S, then it is also a total A''-schedule of O on S, for all $A'' \subseteq A'$.*

Proof.

$$\begin{aligned}\eta_{A'}(O|_{A'}, A') \subseteq \phi_{A'}(E_\sigma) &\Longrightarrow \eta_{A'}(O|_{A'}, A'') \subseteq \phi_{A'}(E_\sigma) \\ &\Longrightarrow r^{A'}_{A''} \circ \eta_{A'}(O|_{A'}, A'') \subseteq r^{A'}_{A''} \circ \phi_{A'}(E_\sigma) \\ &\Longrightarrow \eta_{A''}(O|_{A''}, A'') \subseteq \phi_{A''}(E_\sigma),\end{aligned}$$

where the last implication follows from corollary 10.1. ∎

Let us now define what we mean by dynamic scheduling. It is clear that it should be a collection of different partial schedules, each one applicable at a certain situation. What do we mean by "a certain situation"? Of course, the different subsets of the domain A represent different situations with varying knowledge of the inputs, which should affect the schedule . But this is not enough. It is reasonable to assume that also the way we reached the subset in question should play a role. Thus, there is a need for a notion of *history*:

Definition 10.9: A *history* in a domain A is a string in $\mathcal{I}(A)^+$, constructed according to the following:

1. All $A' \in \mathcal{I}(A)$ are histories in A.
2. If $A' \in \mathcal{I}(A)$ and if h is a history in A such that $A' \subset head(h)$, then $A'.h$ is a history in A.

The set of all histories in A is denoted $H(A)$. A *strict history* is a history in $\mathcal{H}(A)^+$, constructed according to the following:

1. A is a strict history.
2. If $A' \in \mathcal{H}(A)$ and if h is a strict history such that $A' \subset head(h)$, then $A'.h$ is a strict history.

The set of all strict histories in A is denoted $SH(A)$.

"." denotes string concatenation. *head*(s) gives the first character in the string s. When $h \notin \mathcal{I}(A)$ we will also use *tail*, which gives the rest of s when *head*(s) is removed. For a history h, *head*(h) will be referred to as the *present situation* and *tail*(h) as the *preceding history*. This agrees with the intuitive interpretation of histories as finite sequences of situations, where situations with increasing information are appended to the left.

It should now be quite clear what we are heading for. We want to introduce scheduling strategies where there is a schedule for each possible history. We will define them so that the schedule for a given history encompasses the schedules for all its preceding histories; a schedule will never alter the decisions made earlier. Furthermore, there must be a causality constraint; a schedule for the history h cannot take place before its present situation occurs, and the computational events not scheduled by the preceding history cannot be scheduled at times preceding the present situation.

Thus, for every present situation there is a time when exactly the information it represents is avaliable. What is this time dependent of? Obviously it must be a function of the preceding history, since its schedule determines the times of variables to be input which increases the information available to the system. If we are considering *strict* histories, then this is the only way the system obtains information. Non-strict histories, on the other hand, represent developments where information becomes available in other ways. Information about the input might for instance be known à priori, which gives rise to a history not starting in A. Information might also arrive in a fashion not determined by the schedule during the execution. These cases can be summarized in the concept of an *environment* which we do not specify further. We will assume that the time of a present situation is a function of its preceding history and the environment.

Definition 10.10: Let A be a domain, let EN be a set of environments and let T be a set of times. A *scheduling time function* is a function $t_s\colon H(A) \times EN \to T$ that is monotonic with respect to prefixing, that is: for all $h \in H(A)$, $A' \in \mathcal{I}(A)$ such that $A'.h \in H(A)$ and for all $e \in EN$, $t_s(h,e) < t_s(A'.h,e)$.

Now, finally, we are ready to define what we mean by a dynamic scheduling strategy:

Definition 10.11: Let O be an output specification with domain A. Let $\langle T \times R, \prec_s \rangle$ be a communication structure and let EN be a set of environments. Let t_s be a scheduling time function $H(A) \times EN \to T$. A *dynamic scheduling strategy* for O with respect to t_s is a family $\Sigma = (\sigma(h,e) \mid \langle h,e \rangle \in H(A) \times EN)$ of partial *head*(h)-schedules of O on $\langle T \times R, \prec_s \rangle$, fulfilling:

1. For all $e \in EN$ and all $A' \in \mathcal{I}(A)$, $\sigma(A',e)$ is a partial A'-schedule of O on $\langle T \times R, \prec_s \rangle$.
2. (inclusion) For all $e \in EN$, if $h \notin \mathcal{I}(A)$ then $E_{\sigma(h,e)} \supseteq E_{\sigma(tail(h),e)}$, $\psi_{\sigma(h,e)} \supseteq \psi_{\sigma(tail(h),e)}$, $D_{\sigma(h,e)} \supseteq D_{\sigma(tail(h),e)}$, $\prec_{\sigma(h,e)} \supseteq \prec_{\sigma(tail(h),e)}$ and $W_{\sigma(h,e)} \supseteq W_{\sigma(tail(h),e)}$.
3. (causality) Let $W_{\sigma(h,e)} = \langle t_{\sigma(h,e)}, r_{\sigma(h,e)} \rangle$. For all $e \in EN$ and for all $\mathrm{p} \in D_{\sigma(h,e)} \setminus D_{\sigma(tail(h),e)}$, $t_{\sigma(h,e)}(\mathrm{p}) > t_s(h,e)$.

Here we have allowed the schedules to be functions also of the environment. Note that we consider the functions $\psi_{\sigma(h,e)}$, $W_{\sigma(h,e)}$ to be sets of pairs, which enables us to use set inclusion in the definition. The inclusion property guarantees that decisions made by earlier schedules will not be altered by later schedules. This will be further developed below. The causality property expresses the natural fact that we cannot schedule according to a given situation before it occurs. This is a sort of "weakest" causality constraint, it does not take into account that it might take time to determine the new schedule when a new situation occurs.

As a matter of fact, the definition allows for scheduling strategies which are hardly realizable at all. The reason is the view of new information as instantly available everywhere, and that we know all consequences of the increased information as soon as it exists. Consider for instance a polynomial $p(x_1, \ldots, x_n)$. If the knowledge represented by the present situation A' is such that the values of $x_1, \ldots, x_n$ are known, then there is a constant polynomial c such that $p(x_1, \ldots, x_n)|_{A'} = c|_{A'} = c$. Thus, we instantly know the value of p when A' occurs, even without calculating it according to some expression for $p(x_1, \ldots, x_n)$! The new schedule can use this equality to schedule the constant c instead of some more complicated expression for p. This is admittedly not exactly in accordance with the spirit in the rest of this thesis, where information is considered to be transmitted through input and computed values only. On the other hand, the definition enables us to describe systems of a more general nature and it contains the total schedules, as defined in chapter , as a special case. Thus, we can use it to compare the properties of such systems with the properties of more general systems.

A dynamic scheduling scheme does not guarantee termination at all. Since the schedules are partial, it might just happen that certain output functions never become scheduled at all. Of course termination is a very interesting property. Therefore we now give definitions that formalizes the idea of termination in the context developed here.

Definition 10.12: f is *completely scheduled* by $\sigma(h, e)$ if and only if

$$\phi_{head(h)}(E_{\sigma(h,e)}) \supseteq \eta_{head(h)}(f|_{head(h)}, head(h)).$$

If f is completely scheduled by $\sigma(h, e)$, then there are expressions in $E_{\sigma(h,e)}$ for all restricted polynomials that may give a value for f. There is at least one computational event for each expression and subexpression in $ec(E_{\sigma(h,e)})$, and each computational event has a point in space-time given by $W_{\sigma(h,e)}$. Thus, everything needed to calculate f is scheduled. An interesting question is now if this property always is preserved by a dynamic scheduling strategy, when new situations are appended to the history. The following theorem states that this is the case.

Theorem 10.1: *If f is completely scheduled by $\sigma(h,e)$, then, for all h' such that $h'.h \in H(A)$, f is completely scheduled by $\sigma(h'.h,e)$.*

Proof. We first prove that f is completely scheduled by $\sigma(A'.h,e)$, for all $A' \in \mathcal{I}(A)$ such that $A'.h \in H(A)$. Since f is completely scheduled by $\sigma(h,e)$ it holds that

$$\phi_{head(h)}(E_{\sigma(h,e)}) \supseteq \eta_{head(h)}(f|_{head(h)}, head(h)).$$

Now, $A' \subseteq head(h)$ and $E_{\sigma(A'.h,e)} \supseteq E_{\sigma(h,e)}$. Thus,

$$\phi_{head(h)}(E_{\sigma(A'.h,e)}) \supseteq \eta_{head(h)}(f|_{head(h)}, A')$$

which implies

$$r_{A'}^{head(h)}(\phi_{head(h)}(E_{\sigma(A'.h,e)})) \supseteq r_{A'}^{head(h)}(\eta_{head(h)}(f|_{head(h)}, A')).$$

According to proposition 10.2 and corollary 10.1 this is the same as

$$\phi_{A'}(E_{\sigma(A'.h,e)}) \supseteq \eta_{A'}(f|_{A'}, A')$$

which is the desired result for $\sigma(A'.h,e)$. The same result for $\sigma(h'.h,e)$ now follows from a simple inductive proof over strings. ■

Thus, complete scheduling is preserved under concatenation of situations to histories. Also, by corollary 10.2, when the input is further restricted, the same expressions will always be tied to the new restriction of a certain output function in the same way as to the old one. Since the sets of a new schedule always are supersets of the sets of the old one, the evaluation of these expressions will be distributed in space-time in exactly the same way as before. Thus, the computation of the output function will proceed unaffected of new schedules. Every polynomial corresponding to it will be computed at some time which will not be altered in the future. Therefore, complete scheduling is really a termination property.

Let us now compare the schedules considered in chapter 8 with the more general scheduling strategies considered here.

Proposition 10.5: *Let σ be a total A-schedule. A family $(\sigma(h,e) \mid \langle h,e\rangle \in H(A) \times EN)$ of partial $head(h)$-schedules of O on $\langle S, \prec_s\rangle$, for which $\sigma(h,e) = \sigma$ for all $\langle h,e\rangle \in H(A) \times EN$, is always a dynamic scheduling strategy for O with respect to any scheduling time function.*

Proof. We must show that the three conditions in definition 10.11 are met by σ.

1. We must prove that for all $A' \in \mathcal{I}(A)$, σ is a partial A'-schedule. The only condition in definition 10.8 affected by A' is 1. Since all A' in $\mathcal{I}(A)$ are subsets of A and since σ is a total A-schedule, proposition 10.4 gives that σ is a total A'-schedule and thus also a partial A'-schedule for all $A' \in \mathcal{I}(A)$.
2. The set inclusions all hold with equality.
3. Since $\sigma(h,e) = \sigma(tail(h),e)$ for all $h \notin \mathcal{I}(A)$, $D_{\sigma(h,e)} \setminus D_{\sigma(tail(h),e)} = \emptyset$. Thus, there are no computational events for which the causality property must hold and it holds vacuously. ■

A dynamic scheduling strategy as in the proposition above will be referred to as a *static scheduling strategy.* The schedules considered in chapter 8 can be seen as static schedule strategies.

Theorem 10.2: *All $f \in O$ are scheduled for all histories by any static scheduling strategy σ for O.*

Proof. Since σ is a total A-schedule, it is total for all $A' \in \mathcal{I}(A)$ according to proposition 10.4. Thus, for all $A' \in \mathcal{I}(A)$, $\eta_{A'}(O|_{A'}, A') \subseteq \phi_{A'}(E_\sigma)$ which implies that for all $f \in O$ holds that $\eta_{A'}(f|_{A'}, A') \subseteq \phi_{A'}(E_\sigma)$, that is: All $f \in O$ are totally scheduled, for all histories $A' \in \mathcal{I}(A)$. The theorem now follows immediately from theorem 10.1. ■

Theorem 10.2 shows one important aspect of static scheduling strategies: *the evaluation of all functions in the output specification is guaranteed to terminate.* On the other hand this safety is bought at the cost of a possible overhead, if output functions have a data-dependent structure or if there are non-strict operators. This overhead is caused by the fact that a static strategy must schedule every possible case in advance, and it cannot use any increased information to prune away evaluations that become superfluous. The main use of static scheduling strategies must therefore be in the case when this overhead is low or none, say, for scheduling the evaluation of polynomials over an algebra with strict operators only.

10.6 Strict schedules

In the preceding subchapter we considered systems, which could possibly obtain information from the environment by other means than through the value of input variables. Now we restrict our attention to systems which do not have this property. This means that we consider strict histories only. Also, we will consider schedules and scheduling time functions which are dependent of the history only. Under these circumstances the scheduling time function for a history will be totally determined by the schedule of the previous history, since that schedule determines when the next unknown variables will be input. When these variables are input, more information about the total input is available and a new present situation occurs.

Definition 10.13: Let O be an output specification in X with domain A. Let $\langle T \times R, \prec_s \rangle$ be a communication structure. A *strict dynamic scheduling strategy* for O is a family $\Sigma = (\sigma(h) \mid h \in SH_\Sigma(A))$ of partial $head(h)$-schedules of O on $\langle T \times R, \prec_s \rangle$. *The strict Σ-histories*, $SH_\Sigma(A)$, *the Σ-scheduling time function* t_Σ: $SH_\Sigma(A) \to T$ and help quantities D_Σ, $X(D)$, D_{ne}, X_{ni}, D_{ni}, D_{next} are defined recursively below (where $W_{\sigma(h)} = \langle t_{\sigma(h)}, r_{\sigma(h)} \rangle$ for all $t \in T$):
Help quantities: for all $h \in SH_\Sigma(A)$,

$$D_\Sigma(t,h) = \{\mathrm{p} \mid \mathrm{p} \in D_{\sigma(h)} \wedge t_{\sigma(h)}(p) \leq t\}.$$

$$X(D) = \psi_{\sigma(h)}(D) \cap X \text{ for all } D \subseteq D_{\sigma(h)}.$$

$$D_{ne}(h) = D_{\sigma(h)} \setminus D_\Sigma(t_\Sigma(h), h).$$

$$X_{ni}(h) = X \setminus X(D_\Sigma(t_\Sigma(h), h)).$$

$$D_{ni}(h) = \{\mathrm{p} \mid \mathrm{p} \in D_{ne}(h) \wedge \exists x \in X_{ni}(h)[\psi_{\sigma(h)}(\mathrm{p}) = x]\}.$$

$$D_{next}(h) = \{\mathrm{p} \mid \mathrm{p} \in D_{ni}(h) \wedge \neg \exists \mathrm{p}' \in D_{ni}(h)[t_{\sigma(h)}(\mathrm{p}') < t_{\sigma(h)}(\mathrm{p})]\}.$$

Recursive definition of $SH_\Sigma(A)$, t_Σ:
sh1. $A \in SH_\Sigma(A)$. $t_\Sigma(A) < \min(t_{\sigma(A)}(\mathrm{p}) \mid p \in D_{\sigma(A)})$.
sh2. If $h \in SH_\Sigma(A)$, then for all $a \in \prod(A_x \mid x \in X(D_{next}(h)))$

$$h' = (head(h)/a).h \in SH_\Sigma(A)$$

and $t_\Sigma(h') = t_{\sigma(h)}(\mathrm{p})$ for some $\mathrm{p} \in D_{next}(h)$.
Now $\Sigma = (\sigma(h) \mid h \in SH_\Sigma(A))$ must fulfil the following:

1. $\sigma(A)$ is a partial A-schedule of O on $T \times R$.
2. (inclusion) If $h \neq A$, then $E_{\sigma(h)} \supseteq E_{\sigma(tail(h))}$, $\psi_{\sigma(h)} \supseteq \psi_{\sigma(tail(h))}$, $D_{\sigma(h)} \supseteq D_{\sigma(tail(h))}$, $\prec_{\sigma(h)} \supseteq \prec_{\sigma(tail(h))}$ and $W_{\sigma(h)} \supseteq W_{\sigma(tail(h))}$.
3. (causality) For all $\mathrm{p} \in D_{\sigma(h)} \setminus D_{\sigma(tail(h))}$, $t_{\sigma(h)}(\mathrm{p}) > t_\Sigma(h)$.

There is an obvious need for an informal explanation of the concepts introduced in the definition above:

- $D_\Sigma(t,h)$ is the set of computational events that occur before or at the time t, under the schedule given by the history h.
- $X(D)$ are the variables x in X for which there are computational events in D labelled with them. Thus, these variables will be input under the partial schedule that D is part of.
- $D_{ne}(h)$ is the set of computational events in $D_{\sigma(h)}$ that will occur after $t_\Sigma(h)$, the scheduling time of h.

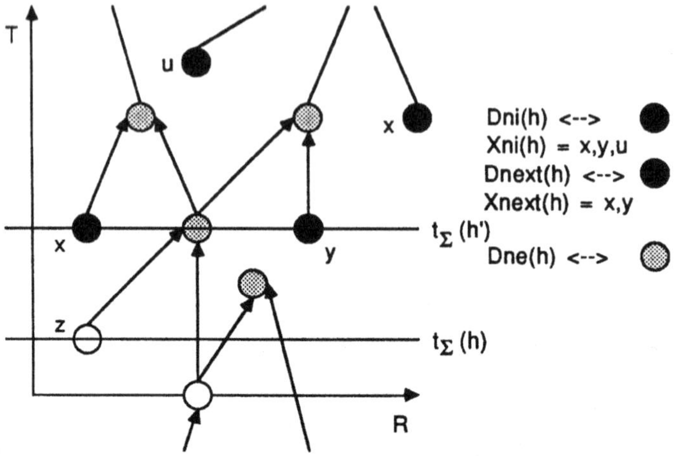

Figure 10.3: Illustration of some of the concepts in the definition of strict dynamic scheduling strategies.

- $X_{ni}(h)$ is the set of variables, scheduled by $\sigma(h)$, which have not yet been input at $t_\Sigma(h)$.
- $D_{ni}(h)$ is the set of computational events which have not yet occurred and that corresponds to variables which have not yet been input at $t_\Sigma(h)$.
- $D_{next}(h)$ are the computational events corresponding to the first new variables that are input after $t_\Sigma(h)$.

The histories that can develop under the dynamic scheduling strategy Σ are the ones in $SH_\Sigma(A)$. Every strict Σ-history starts off with A, which means that there is no à priori information about the input. $\sigma(A)$ determines $X(D_{next}(A))$, the first variables to be input, and the possible values of them in turn give the possible hyperplanes $A' = A/\langle\, x = a_x \mid x \in X(D_{next}(A))\,\rangle$ that represent the increased information about the total input. $\sigma(A)$ also gives the time $t_\Sigma(A'.A)$ for the possible histories $A'.A$. Now, when A' is known, $\sigma(A'.A)$ can be applied. All new computational events not scheduled by $\sigma(A)$ must be scheduled at times greater than $t_\Sigma(A'.A)$. $\sigma(A'.A)$ determines which new variables that will be input next and when they are to be input. This gives a new set of possible histories, from the new possible hyperplanes that represents the further increased information about the total input. This process of history development goes on infinitely or until D_{next} is empty, in which case no new variables are input by the latest schedule and no more information can be obtained.

It is easily shown, by induction on strings, that $SH_\Sigma(A) \subseteq SH(A)$ for all strict dynamic scheduling strategies Σ. That is, strict Σ-histories are always strict histories. We can also note that definition 10.12 and theorem 10.1 carry over to strict dynamic scheduling strategies.

11. Applications

The theory and the methods developed in this thesis will now be illustrated by some examples. Our aim is to illustrate the concepts. Therefore, we will first show some examples of synthesis of simple systems already known from the literature, so that the reader can see how the basic concepts of the theory are applied without being obscured by numerous details. Then we will conclude this section by an example of synthesis of a more complex system, showing the importance of the generality of the theory compared with earlier, more restricted approaches.

11.1 Forward solving of lower-triangular linear equation systems

The first model problem is forward solving of lower-triangular linear equation systems. We will synthesize two systolic arrays doing this. The first is a variation of the well-known array of Kung and Leiserson, [KLe80]. The second was given in [D84].

We will synthesize hardware solving the linear system of equations $Ax = b$, where $A = \{a_{ik}\}$ is an $n \times n$ lower triangular matrix. Especially, we want to find solutions with a high cell utilization when the system matrix A is band matrix.

First we determine the algebra $\mathcal{A}$ that will be used. It is chosen to be the one with the set of the real numbers as sort and binary operators $+, -, \cdot, /$ (plus, minus, times, divides). Note that we really cannot implement these operators in hardware, since real numbers in general require an infinite representation; any implementation must be an approximate one, using for instance floating-point numbers. We prefer, however, to perform the synthesis using the original algebra where the problem is posed, since this gives us some more freedom to perform algebraic transformation of expressions. ("+" on real numbers is for instance associative, but the corresponding operator on floating-point numbers is not. This associativity can be used to rewrite expressions in order to obtain a better dependence structure.)

The next step in the process is to specify the output functions $\{ p_i \mid 1 \leq i \leq n \}$ as functions of the free variables $\{ b_i \mid 1 \leq i \leq n \}$ and $\{ a_{ik} \mid 1 \leq i \leq n, 1 \leq k \leq i \}$:

Output specification.

$$
\begin{aligned}
p_1 &= \phi(b_1/a_{11}) \\
p_2 &= \phi((b_2 - a_{21}(b_1/a_{11}))/a_{22}) \\
p_3 &= \phi((b_3 - a_{32}(b_2 - a_{21}(b_1/a_{11}))/a_{22} - a_{31}(b_1/a_{11}))/a_{33}) \\
&\ \ \vdots
\end{aligned}
$$

$$p_i = \phi((b_i - \sum_{k=1}^{i-1} a_{ik}(\ldots))/a_{ii})$$

$$\vdots$$

The output functions are polynomials themselves, since they are defined by formal expressions. Thus, they are structurally data-independent in $\mathcal{A}$, and every output function $\phi(\mathbf{p}_i)$ is always computed by evaluating the same expression $\mathbf{p}_i$ regardless of indata.

Next, we determine a schedule of this output specification. A convenient start is to find a system of recurrence equations, or CCSA-set, that recursively defines the output functions. The following well-known recurrence equation will do the job:

CCSA-set M.

$1 \leq i \leq n$:

$$s_{i,1} \leftarrow b_i$$

$1 < i \leq n$, $1 \leq k < i$:

$$s_{i,k+1} \leftarrow s_{i,k} - a_{ik}x_k$$

$1 \leq k \leq n$:

$$x_k \leftarrow s_{k,k}/a_{kk}$$

Here we have introduced the new assigned variables $s_{i,k}$, $1 \leq i \leq n$, $1 \leq k \leq i$ and x_i, $1 \leq i \leq n$. It is easily verified that $\phi_{\mathbf{F}}(x_i) = p_i$ for all i, where $\mathbf{F}$ is the recursion scheme implicitly defined by the recurrence equation above.

The CCSA-set above directly defines a set of cell actions. Let us examine it a bit closer. In the set of computational events that this set of cell actions is derived from, every computational event is labelled with a unique expression, so for simplicity we identify computational events and expressions. $s_{i,1} \leftarrow b_i$ represents a block containing one expression, the input variable b_i. The cs-assignment $s_{i,k+1} \leftarrow s_{i,k} - a_{ik}x_k$ represents a block containing the input variable a_{ik}, one expression with multiplication as top operator and one with subtraction. $x_k \leftarrow s_{k,k}/a_{kk}$, finally, contains the input variable a_{kk} and one division. Thus, the complexity of each cell action is at most two operator applications. The situation is depicted in figure 11.1, where the corresponding computation network is drawn.

Note that the algebra $\mathcal{A}$ is partial, since division with zero is undefined. This leads to the condition that all diagonal elements in A must

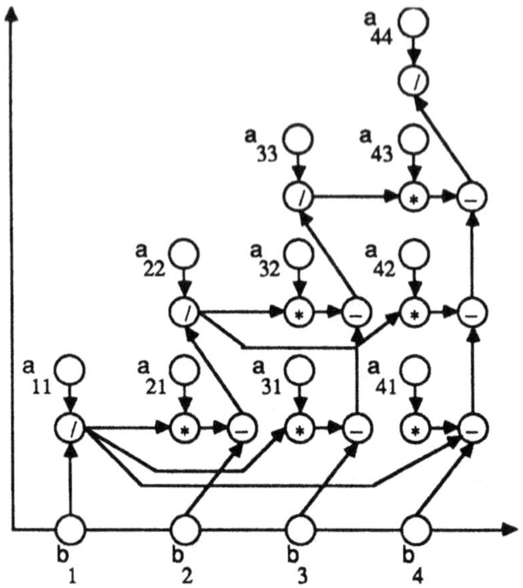

Figure 11.1: Data dependencies and partition of the underlying set of expressions, $n = 4$.

be nonzero. For a triangular matrix this is equivalent to the matrix being nonsingular.

By identifying variables in the domain and the range of different assignments, we can obtain the data dependence relation $\prec_M$ on the CCSA-set defined by the recurrence. The result is the following:

$$s_{1,1} \leftarrow b_1 \quad \prec_M \quad x_1 \leftarrow s_{1,1}/a_{11}$$

$1 < i \leq n,\ 1 \leq k < i$:

$$x_k \leftarrow s_{kk}/a_{kk} \quad \prec_M \quad s_{i,k+1} \leftarrow s_{i,k} - a_{ik}x_k$$

$1 < i \leq n$:

$$s_{i,1} \leftarrow b_i \quad \prec_M \quad s_{i,2} \leftarrow s_{i,1} - a_{i1}x_1$$

$2 < i \leq n,\ 1 < k < i$:

$$s_{i,k} \leftarrow s_{i,k-1} - a_{ik-1}x_{k-1} \quad \prec_M \quad s_{i,k+1} \leftarrow s_{i,k} - a_{ik}x_k$$

In figure 11.2 the data dependence relation $\prec_M$ on the CCSA-set M is shown. Let us now specify an indexing function G from the CCSA-set to index space. As index space we choose Z^2:

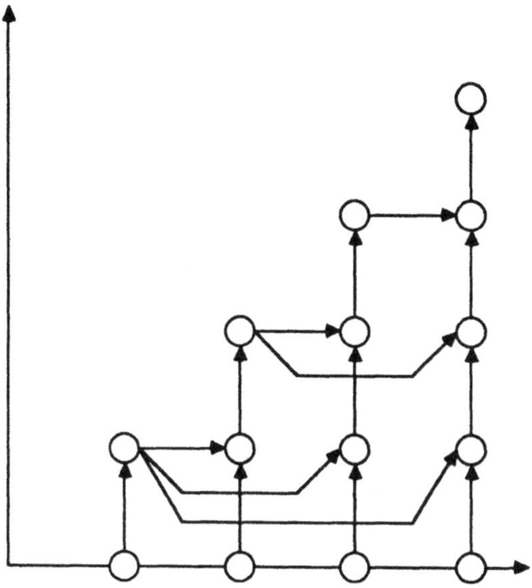

Figure 11.2: Data dependencies of the CCSA-set M, $n = 4$.

Indexing function G:

$1 \leq i \leq n$:

$$s_{i,1} \leftarrow b_i \quad \mapsto \quad \binom{i}{0}$$

$1 < i \leq n,\ 1 \leq k < i$:

$$s_{i,k+1} \leftarrow s_{i,k} - a_{ik}x_k \quad \mapsto \quad \binom{i}{k}$$

$1 \leq k \leq n$:

$$x_k \leftarrow s_{k,k}/a_{kk} \quad \mapsto \quad \binom{k}{k}$$

It is easily seen that G is 1-1 and thus it is a correct indexing function from M to the index set $G(M)$. This choice of G gives rise to the following set of data dependence vectors:

For each $s_{i,k+1}$:

$$\binom{i}{k+1} - \binom{i}{k} = \binom{0}{1}$$

For each x_k $(1 < i \leq n,\ 1 \leq k < i)$:

$$\begin{pmatrix} i-k \\ 0 \end{pmatrix}$$

Note that $i - k > 0$ always for the data dependence vectors given by any x_k.

Next step is to map the index vectors to a discrete space-time S. When this is done we have a complete schedule for the output functions p_i. We choose $S = N \times Z$. Thus, the embedding vector space is Z^2. The following linear mapping $L: Z^2 \to Z^2$ gives a slight variation of the systolic array of Kung and Leiserson, [KLe80]:

Index mapping L:

$$\begin{pmatrix} t \\ y \end{pmatrix} = \begin{pmatrix} 1 & 1 \\ 1 & -1 \end{pmatrix} \begin{pmatrix} i \\ k \end{pmatrix}$$

What about correctness? The matrix is nonsingular. It remains to check the time component of the mapped data dependence vectors, i.e. the design vectors of the mapped precedence relation. They turn out to be the following:

For each $s_{i,k+1}$:

$$\begin{pmatrix} 1 & 1 \\ 1 & -1 \end{pmatrix} \begin{pmatrix} 0 \\ 1 \end{pmatrix} = \begin{pmatrix} 1 \\ -1 \end{pmatrix}$$

For each x_k $(1 < i \leq n,\ 1 \leq k < i)$:

$$\begin{pmatrix} 1 & 1 \\ 1 & -1 \end{pmatrix} \begin{pmatrix} i-k \\ 0 \end{pmatrix} = \begin{pmatrix} i-k \\ i-k \end{pmatrix}$$

In figure 11.3 a space-time diagram is drawn, showing the position of each mapped assignment and the mapped precedence relation $\prec_{MG \circ L}$. Each arc in the graph of $\prec_{MG \circ L}$ can be seen as a design vector drawn between the mapped cs-assignments giving rise to it. The time component of each design vector is strictly greater than zero. This implies that the mapped precedence relation is ripple-free, which is desirable since then no zero-delay lines between cells are needed. Such lines would otherwise be likely to require a longer clock cycle; if two cells are connected by a zero-delay line, then the signals must ripple through both of them and the outputs become stable before the next clock tick arrives.

A design vector gives the distance and direction in space-time that the corresponding operand must be transmitted over. In the case of s-operands, the design vector $(1 \quad -1)$ provides the following information:

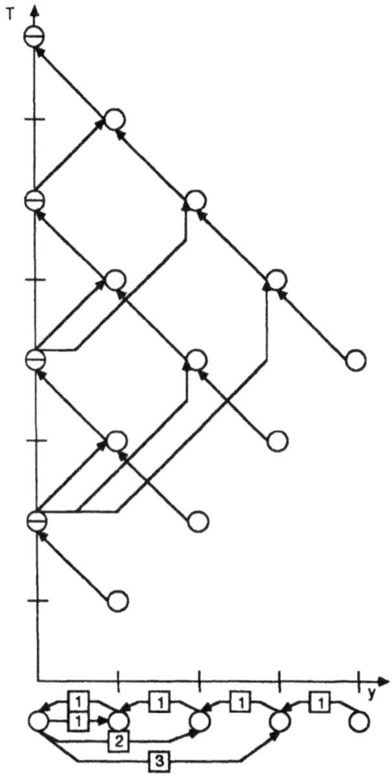

Figure 11.3: Space-time diagram with mapped precedence relation $\prec_{MG \circ L}$, $n = 4$. Fixed hardware projection.

the resulting operands are transmitted every time-step to the left neighbour of the cell where the computation is performed (if we let the space-axis have the positive direction to the right). So in this case we really have *nearest-neighbour communications*, which is a desirable property for systems like systolic arrays.

Let us now consider the x-operands. Every x_k has design vectors $(i-k \quad i-k)$, for $i = k+1, \ldots, n$. This means that x_k has fanout $n-k-1$. For $i-k > 1$, x_k must be sent $i-k$ length units with a delay of $i-k$ time units. The fixed hardware projection of the mapped precedence relation is shown in figure 11.3. This is a design where every x-operand is transmitted from its origin and then moves to its destinations following different wires, without being used more than once along each path. This is of course not desirable. The number of wires from the cells where x-operands are computed grows as n, and the communication pattern is not nearest-neighbour.

The solution is of course to reduce the fanout by transforming the CCSA-set according to definition 8.8 and theorem 8.2. The mapped assignments to which each x_k is sent are already totally ordered in time, which gives a causal x_k-enumeration. If we transform the CCSA-set according to this enumeration, then we obtain the following CCSA-set:

Transformed CCSA-set M'.

$1 \leq i \leq n$:

$$s_{i,1} \leftarrow b_i$$

$1 < i \leq n,\ 1 \leq k < i$:

$$s_{i,k+1} \leftarrow s_{i,k} - a_{ik}x_{k,i}$$
$$x_{k,i+1} \leftarrow x_{k,i}$$

$1 \leq k \leq n$:

$$x_{k,k+1} \leftarrow s_{k,k}/a_{kk}$$

M' contains these additional data dependencies:

$1 \leq k < n$:

$$x_{k,k+1} \leftarrow s_{k,k}/a_{kk} \quad \prec_{M'} \quad \begin{array}{l} s_{k+1,k+1} \leftarrow s_{k+1,k} - a_{k+1k}x_{k,k+1} \\ x_{k,k+2} \leftarrow x_{k,k+1} \end{array}$$

$1 < i < n,\ 1 \leq k < i$:

$$\begin{array}{l} s_{i+1,k+1} \leftarrow s_{i+1,k} - a_{i+1k}x_{k,i+1} \\ x_{k,i+2} \leftarrow x_{k,i+1} \end{array} \quad \prec_{M'} \quad \begin{array}{l} s_{i,k+1} \leftarrow s_{i,k} - a_{ik}x_{k,i} \\ x_{k,i+1} \leftarrow x_{k,i} \end{array}$$

With the same indexing function G as before this gives rise to the data dependence vector

$$\begin{pmatrix} i+1 \\ k \end{pmatrix} - \begin{pmatrix} i \\ k \end{pmatrix} = \begin{pmatrix} 1 \\ 0 \end{pmatrix}$$

for all $x_{k,i}$. The resulting dependence structure for the transformed CCSA-set is shown in figure 11.4.

With the same linear mapping L from the transformed CCSA-set to index space as before, every $x_{k,i}$ will have the design vector

$$\begin{pmatrix} 1 & 1 \\ 1 & -1 \end{pmatrix} \begin{pmatrix} 1 \\ 0 \end{pmatrix} = \begin{pmatrix} 1 \\ 1 \end{pmatrix}.$$

This means that every x-operand moves one length unit, to the right neighbour of its origin, during one time unit. Thus, we have localized

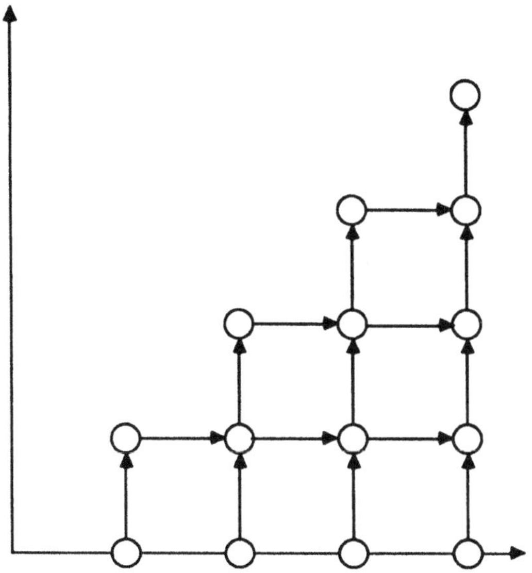

Figure 11.4: Data dependencies of the transformed CCSA-set, $n = 4$.

also the communication pattern of these operands. This can be seen in the fixed hardware projection. Let us summarize the resulting schedule:

CCSA-set M', indexing function:

$1 \leq i \leq n$:

$$s_{i,1} \leftarrow b_i \quad \mapsto \quad \begin{pmatrix} i \\ 0 \end{pmatrix}$$

$1 < i \leq n$, $1 \leq k < i$:

$$\begin{matrix} s_{i,k+1} \leftarrow s_{i,k} - a_{ik}x_{k,i} \\ x_{k,i+1} \leftarrow x_{k,i} \end{matrix} \quad \mapsto \quad \begin{pmatrix} i \\ k \end{pmatrix}$$

$1 \leq k \leq n$:

$$x_{k,k+1} \leftarrow s_{k,k}/a_{kk} \quad \mapsto \quad \begin{pmatrix} k \\ k \end{pmatrix}$$

Mapping from index space to space-time:

$$\begin{pmatrix} t \\ y \end{pmatrix} = \begin{pmatrix} 1 & 1 \\ 1 & -1 \end{pmatrix} \begin{pmatrix} i \\ k \end{pmatrix}$$

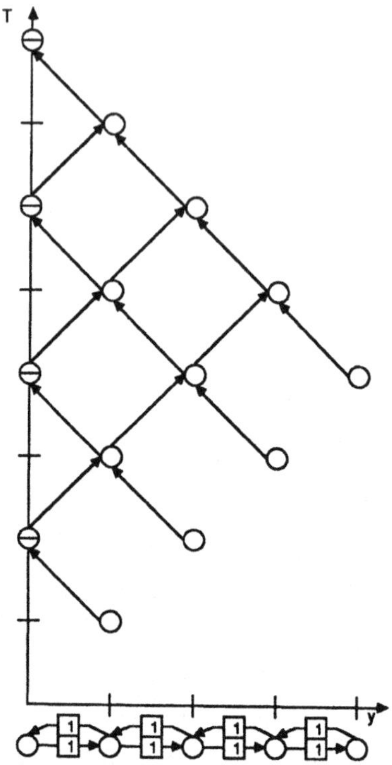

Figure 11.5: Space-time diagram of transformed CCSA-set with mapped precedence relation $\prec_{M'G\circ L}$, $n = 4$. Fixed hardware projection.

Let us now check the solution to see if it is a good one. First we can state that the correct mapping $L \circ G$, given by the indexing function G and the linear index mapping L above, also is correct with respect to a substitution field that has the fixed 1-i/o property. Consider the following substitution field Φ in the space Z:

$$\sigma_0^\Phi = \{x \leftarrow s/a\} \qquad ot_0^\Phi(x) = \{\langle 1,1,x\rangle\} \qquad \begin{array}{l} it_0^\Phi(s) = \langle 1,1,s\rangle \\ it_0^\Phi(a) = \langle 0,0,a\rangle \end{array}$$

$1 \leq y < n$:

$$\sigma_y^\Phi = \{s \leftarrow s - ax, x \leftarrow x\} \qquad \begin{array}{l} ot_y^\Phi(x) = \{\langle 1, y+1, x\rangle\} \\ ot_y^\Phi(s) = \{\langle 1, y-1, s\rangle\} \end{array}$$

$$it_y^\Phi(s) = \langle 1, y+1, s\rangle, \qquad it_y^\Phi(a) = \langle 0, y, a\rangle, \qquad it_y^\Phi(x) = \langle 1, y-1, x\rangle$$

$$\sigma_n^\Phi = \{s \leftarrow b\} \qquad ot_n^\Phi(s) = \{\langle 1, n-1, s\rangle\} \qquad it_n^\Phi(b) = \langle 0, n, b\rangle$$

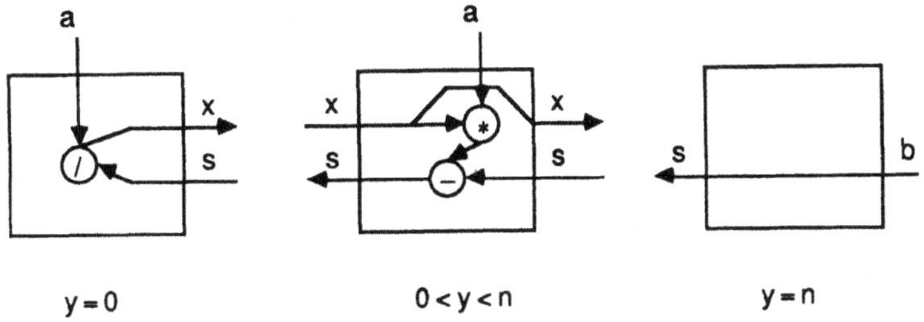

Figure 11.6: Substitution field Φ in space.

$\hat{\Phi}$, the substitution field in $N \times Z$ generated by Φ, has the fixed 1-i/o property. It is not hard to verify that $L \circ G$ is a correct mapping with respect to $\hat{\Phi}$.

Is the solution suitable when the lower-triangular system matrix A is band matrix? The answer is yes, and the reason is the following: for a band matrix all diagonals outside the band contain zeros only. Thus, the corresponding assignments need not be carried out. The matrix elements a_{ik} on a diagonal are characterized by that $i - k$ is constant. According to the indexing function G, a_{ik} is used by the cs-assignment with index vector $(i \quad k)$. This index vector is mapped to the space coordinate $i - k$ by the index mapping L. Therefore, the elements on a given diagonal in A are always processed in the same cell. It follows that we only have to provide as many cells as there are non-zero diagonals.

Another consequence is that all cs-assignments with index vector $(k \quad k)$ are mapped to the space coordinate 0. These cs-assignments are exactly those where a division is performed. Thus, only this cell needs to be capable of division, as can be seen by the substitution field Φ. Had the $(k \quad k)$-cs-assignments been mapped to other space points than 0, then these cells would also have needed to be capable of division, as well as the other operations they would have to perform. This would lead to a more complicated substitution field in space, demanding more complex cells. Alternatively the fixed 1-i/o property could be sacrificed which means that programmable hardware is used.

What about cell utilization? If we consider system matrices that are band matrices only, then the systolic array described above has cell utilization 1/2 in the limit case, when the ratio between the dimension of the matrix and the bandwidth goes to infinity. This is due to the fact that the determinant of

$$\begin{pmatrix} 1 & 1 \\ 1 & -1 \end{pmatrix}$$

is -2. Thus, the area of the set of index vectors is doubled when mapped to space-time. This expansion creates holes in space-time that limits the maximal cell utilization.

Is there any easy way to increase the cell utilization without sacrificing properties such as the fixed 1-i/o property, fitness for system matrices with band structure and no ripple? The demand that the elements of a certain diagonal are to be processed at a single cell leads to the condition on the space row of the transformation matrix L that $l_{22} = -l_{21}$. Then there are essentially two solutions (except for trivial reflections of the space-coordinate) that have a determinant of ± 1 and do not violate the condition, that all index vectors are to be mapped to integer lattice points in the space-time:

$$\begin{pmatrix} 0 & 1 \\ 1 & -1 \end{pmatrix} \quad \text{or} \quad \begin{pmatrix} 1 & 0 \\ 1 & -1 \end{pmatrix}.$$

These linear index mappings are correct and they yield an asymptotic cell utilization of 1 for system matrices with band structure, but unfortunalely they introduce ripple. The design vectors for either x- or s-operands will have a time component of zero. Since every $x_{k,i}$ except for $x_{i-1,i}$ depends on $x_{k,i-1}$, and correspondingly for the s-operands, it follows that either every x-operand or every s-operand must ripple through all cells. If the clock cycle has to be more than doubled in order to give this time to happen, then nothing is gained by the increased cell utilization.

If we are to solve *two* systems of equations, then we may increase the cell utilization by the well-known technique of *interleaving*, that allows the systems to be solved simultaneously. Interleaving is the procedure of "filling in the holes" of one space-time schedule with the schedule of the other task to be performed. If we have two systems of equations of the same size n, then this can be done simply by translating the mapped cs-assignments of one of the recurrences one step upwards in time compared with the other. (This is easily seen in the space-time diagram of figure 11.5.)

This interleaved system can be described formally in the following way. The sets of index vectors of the two CCSA-sets must be distinguished in some way. A convenient way is to add an extra dimension to the index space and transform the old index vectors, according to the following:

$$(i \quad k) \mapsto (i \quad k \quad 0)$$

for the first system of equations and

$$(i \quad k) \mapsto (i \quad k \quad 1)$$

for the second. This yields a new set of index vectors in Z^3 for the combined problem. We can now extend the previous mapping $L: Z^2 \rightarrow Z^2$ to the following mapping from Z^3 to Z^2:

$$\begin{pmatrix} t \\ y \end{pmatrix} = \begin{pmatrix} 1 & 1 & 1 \\ 1 & -1 & 0 \end{pmatrix} \begin{pmatrix} i \\ k \\ l \end{pmatrix}.$$

This mapping will accomplish exactly the displacement in time mentioned above. Note that the index mapping is the restriction of this function to the set of index vectors; it is 1-1 even though the total function is not. It is not possible to interleave more than two systems of equations on the array: if we try to extend the set of index vectors along the l-axis, then the index mapping will not be 1-1 any more.

The question remains if it is possible to increase the cell utilization when there is only one system of equations to solve. Apparently it cannot be done using the indexing function G and a linear mapping from the set of index vectors $G(M)$. Could it then be possible to find some nonlinear, possibly only weakly correct mapping, that gives a high cell utilization without spoiling the desirable properties mentioned above?

The answer is yes. A solution is sketched in [D84]. There, a new design with asymptotic cell utilization 1 for system matrices with band structure, still having the desired properties above, is derived from the Kung-Leiserson design. It is done using an informal technique of manipulating delay elements, thereby altering the space-time structure.

In the formal framework developed here, a transformation between different designs is, as earlier mentioned, best seen as a *space-time transformation.* Here we will give the space-time transformation that yields the new design.

In order to gain maximal freedom with respect to data dependencies, we start with the original CCSA-set for the Kung-Leiserson design, rather than the final one where the fanout of the x-operands is reduced. Consider therefore the space-time diagram in figure 11.3. We want to cluster the points in space-time in a way not violating the mapped data dependencies, so that the resulting structure can be compacted, thereby eliminating "holes" in space-time where cells are idle. It turns out to be advantageous to cluster the space-time points pairwise, in the "north-west-south-east" direction, in the following way: for all $(\, t \quad y\,)$ such that t, y are even, map $(\, t-1 \quad y+1\,)$ onto $(\, t \quad y\,)$. The resulting structure can be compacted by a simple scaling of a factor $1/2$. The steps are shown in figure 11.7.

The following space-time transformation A from our two-dimensional space-time back to itself will do both clustering and scaling:

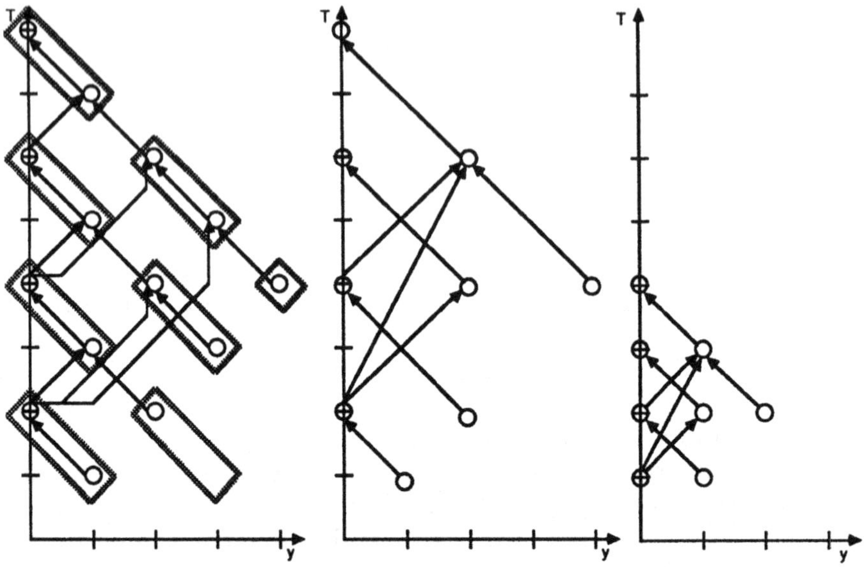

Figure 11.7: Clustering and compacting of communication structure, $n = 4$.

Space-time transformation A:

$$\begin{pmatrix} t' \\ y' \end{pmatrix} = \begin{pmatrix} \lceil t/2 \rceil \\ \lfloor y/2 \rfloor \end{pmatrix}.$$

It is easily seen in figure 11.7 that the precedence relation, given by the function A and the previous mapped precedence relation $\prec_{MG\circ L}$, really is a communication ordering. (For a more thorough investigation of this: see [Li85].) Thus, A is a $\prec_{MG\circ L}$-allowed space-time transformation. By theorem 7.3 follows that $A \circ L \circ G$ is a weakly correct mapping $M \to S$.

Let us describe the resulting schedule in the same manner as the description of the Kung-Leiserson schedule, using M, G and $A \circ L$:

CCSA-set M, indexing function G:

$1 \leq i \leq n$:

$$s_{i,1} \leftarrow b_i \quad \mapsto \quad \begin{pmatrix} i \\ 0 \end{pmatrix}$$

$1 < i \leq n$, $1 \leq k < i$:

$$s_{i,k+1} \leftarrow s_{i,k} - a_{ik}x_k \quad \mapsto \quad \begin{pmatrix} i \\ k \end{pmatrix}$$

$1 \leq k \leq n$:

$$x_k \leftarrow s_{k,k}/a_{kk} \quad \mapsto \quad \begin{pmatrix} k \\ k \end{pmatrix}$$

Mapping $A \circ L$ from index space to space-time:

$$\begin{pmatrix} t \\ y \end{pmatrix} = \begin{pmatrix} \lceil (i+k)/2 \rceil \\ \lfloor (i-k)/2 \rfloor \end{pmatrix}$$

By theorem 7.1 (which carries over to CCSA-sets from sets of cell actions) we can give an alternative description of the design by constructing a new CCSA-set M'', where the cs-assignments from the old CCSA-set M are merged according to the partition induced by the weakly correct mapping. From M'' there exists a correct mapping, that gives the same distribution in space-time of assignments as the weakly correct mapping from M. This way of describing the system may be preferred, since it gives a desirable separation between cs-assignments, described by the CCSA-set, and the distribution in space-time of these cs-assignments, described by the correct mapping.

Let us determine the partition of index vectors induced by the weakly correct index mapping $A \circ L$. We then merge the cs-assignments mapped to the same partition by the indexing function G. This is an alternative way to accomplish the merging given by theorem 7.1.

Consider an arbitrary index vector $(\, i \quad k\,)$ such that $i + k$ is even. For such an index vector,

$$A \circ L \begin{pmatrix} i \\ k \end{pmatrix} = \begin{pmatrix} \lceil (i+k)/2 \rceil \\ \lfloor (i-k)/2 \rfloor \end{pmatrix} = \begin{pmatrix} \lceil (i+k-1)/2 \rceil \\ \lfloor (i-k+1)/2 \rfloor \end{pmatrix} =$$

$$\begin{pmatrix} \lceil (i+(k-1))/2 \rceil \\ \lfloor (i-(k-1))/2 \rfloor \end{pmatrix} = A \circ L \begin{pmatrix} i \\ k-1 \end{pmatrix}.$$

Correspondingly, for any index vector $(\, i \quad k\,)$ such that $i + k$ is odd, it holds that

$$A \circ L \begin{pmatrix} i \\ k \end{pmatrix} = A \circ L \begin{pmatrix} i \\ k+1 \end{pmatrix}.$$

This covers all possible cases. Thus, the partition of the set of index vectors $G(M)$ induced by $A \circ L$ looks like follows: for all index vectors $(\, i \quad k\,)$ such that $i + k$ is even, if $(\, i \quad k-1\,) \in G(M)$ then

$$[(\, i \quad k\,)]\varepsilon_{A \circ L} = \{(\, i \quad k\,), (\, i \quad k-1\,)\}$$

else

$$[(\, i \quad k\,)]\varepsilon_{A \circ L} = \{(\, i \quad k\,)\}.$$

For all index vectors $(\,i \quad k\,)$ such that $i+k$ is odd, if $(\,i \quad k+1\,) \in G(M)$ then

$$[(\,i \quad k\,)]\varepsilon_{A\circ L} = \{(\,i \quad k\,),(\,i \quad k+1\,)\}$$

else

$$[(\,i \quad k\,)]\varepsilon_{A\circ L} = \{(\,i \quad k\,)\}.$$

The set of merged cs-assignments is given below. In order to simplify the notation we introduce the two functions f_o, f_e:

$$f_o(n) = 2\lceil n/2 \rceil - 1 = \begin{cases} n & \text{if } n \text{ is odd} \\ n-1 & \text{if } n \text{ is even} \end{cases}$$

$$f_e(n) = 2\lfloor n/2 \rfloor = \begin{cases} n & \text{if } n \text{ is even} \\ n-1 & \text{if } n \text{ is odd.} \end{cases}$$

We immediately transform the merged cs-assignments in M'' to equivalent pure ones. In the first assignment,

$$x_1 \leftarrow s_{1,1}/a_{11} \qquad (\,1 \quad 1\,)$$
$$s_{1,1} \leftarrow b_1 \qquad (\,1 \quad 0\,)$$

b_1 is substituted for $s_{1,1}$. Since $s_{1,1}$ neither is used by any other cs-assignment, nor defines any output function the second assignment can be dropped. This yields the new cs-assignment

$$x_1 \leftarrow b_1/a_{11}.$$

For $i = 2, 4, \ldots, f_e(n)$, M'' contains the assignments

$$s_{i,1} \leftarrow b_i.$$

These are mapped to distinct points in space and they are therefore not merged with any other cs-assignments. For $i = 3, 5, \ldots, f_o(n)$, M'' contains

$$s_{i,2} \leftarrow s_{i,1} - a_{i1}x_1 \qquad (\,i \quad 1\,)$$
$$s_{i,1} \leftarrow b_i. \qquad (\,i \quad 0\,)$$

In the same way as above they give the new cs-assignments

$$s_{i,2} \leftarrow b_i - a_{i1}x_1.$$

For $i = 5, 7, \ldots, f_o(n)$, $k = 3, 5, \ldots, i-2$ and $i = 4, 6, \ldots, f_e(n)$, $k = 2, 4, \ldots, i-2$,

$$s_{i,k+1} \leftarrow s_{i,k} - a_{ik}x_k \qquad (\,i \quad k\,)$$

$$s_{i,k} \leftarrow s_{i,k-1} - a_{ik-1}x_{k-1} \qquad (i \quad k-1)$$

give the new cs-assignments

$$s_{i,k+1} \leftarrow s_{i,k-1} - a_{ik-1}x_{k-1} - a_{ik}x_k.$$

Finally, for $k = 2, 3, \ldots, n$,

$$x_k \leftarrow s_{k,k}/a_{kk} \qquad (k \quad k)$$
$$s_{k,k} \leftarrow s_{k,k-1} - a_{kk-1}x_{k-1} \qquad (k \quad k-1)$$

yield pure cs-assignments

$$x_k \leftarrow (s_{k,k-1} - a_{kk-1}x_{k-1})/a_{kk}.$$

We now specify the indexing function G'' from the CCSA-set M'' above. For every new cs-assignment, we choose one of the index vectors of the merged cs-assignments forming it. This makes it possible to derive the correct mapping from CCSA-set to space-time, by simply composing the new indexing function with the old mapping $A \circ L$ from index space to space-time:

$$x_1 \leftarrow b_1/a_{11} \quad \mapsto \quad \begin{pmatrix} 1 \\ 1 \end{pmatrix}$$

$i = 2, 4, \ldots, f_e(n)$:

$$s_{i,1} \leftarrow b_i \quad \mapsto \quad \begin{pmatrix} i \\ 0 \end{pmatrix}$$

$i = 3, 5, \ldots, f_o(n)$:

$$s_{i,2} \leftarrow b_i - a_{i1}x_1 \quad \mapsto \quad \begin{pmatrix} i \\ 1 \end{pmatrix}$$

$i = 5, 7, \ldots, f_o(n)$, $k = 3, 5, \ldots, i-2$ and
$i = 4, 6, \ldots, f_e(n)$, $k = 2, 4, \ldots, i-2$:

$$s_{i,k+1} \leftarrow s_{i,k-1} - a_{ik-1}x_{k-1} - a_{ik}x_k \quad \mapsto \quad \begin{pmatrix} i \\ k \end{pmatrix}$$

$k = 2, 3, \ldots, n$:

$$x_k \leftarrow (s_{k,k-1} - a_{kk-1}x_{k-1})/a_{kk} \quad \mapsto \quad \begin{pmatrix} k \\ k \end{pmatrix}$$

Note that all index vectors are chosen so that the sum of the components is even. For all i, k such that $i+k$ is even holds that $\lceil (i+k)/2 \rceil = (i+k)/2$

and $\lfloor (i-k)/2 \rfloor = (i-k)/2$. Thus, for all index vectors $(i \quad k)$ in the new set of index vectors,

$$A \circ L \begin{pmatrix} i \\ k \end{pmatrix} = \begin{pmatrix} \lceil (i+k)/2 \rceil \\ \lfloor (i-k)/2 \rfloor \end{pmatrix} = \begin{pmatrix} i/2 + k/2 \\ i/2 - k/2 \end{pmatrix} = \begin{pmatrix} 1/2 & 1/2 \\ 1/2 & -1/2 \end{pmatrix} \cdot \begin{pmatrix} i \\ k \end{pmatrix}.$$

An interesting point is that the correct index mapping $A \circ L$ is linear. This reflects how the regularity of the new recurrences, given by the CCSA-set M'', directly gives the regularity of the space-time pattern and thus the regularity of the resulting hardware too. Note also that the matrix elements are fractions of integers. A usual constraint, [Mo82,MiWi84], is to consider matrices with integer elements only. While this restriction ensures that the image of an index vector always is an integer lattice point in space-time, it sometimes prunes away interesting solutions, as the one above. In our example no index vectors are mapped to non-integer coordinates, because the index mapping is the *restriction* of $A \circ L$ to the set of index vectors $G''(M'')$. The index vectors in $G''(M'')$ are located so that their images always are integer lattice points in space-time.

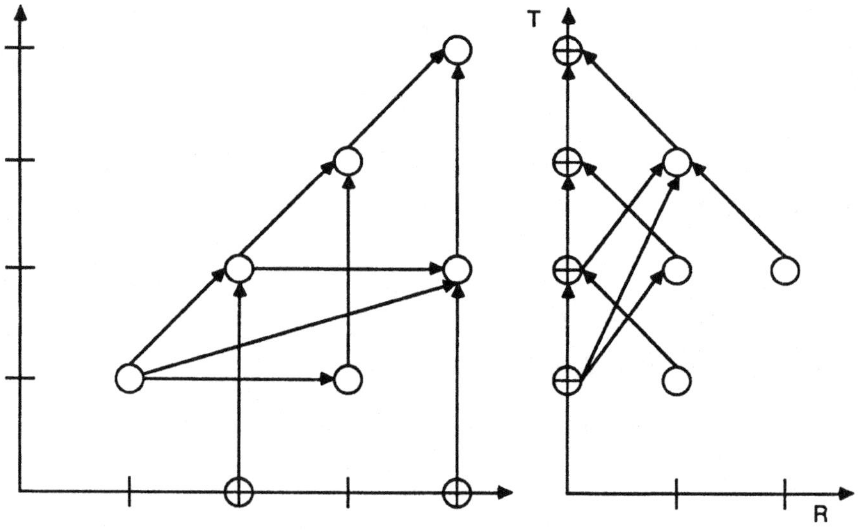

Figure 11.8: Data dependencies of M''. Space-time diagram with mapped precedence relation $\prec_{M'' A \circ L \circ G''}$, $n = 4$.

Since the mapping is linear, it is possible to analyze the communication requirements using data dependence vectors in index space. Figure 11.8 shows M'' with data dependencies and the space-time diagram for the mapped precedence relation. Note that we have to reduce the fanout of each x_k in the same way as for the Kung-Leiserson design. This can be done by introducing fresh variables $x_{k,i}$. The result is the following CCSA-set:

Transformed CCSA-set M''' with reduced fanout:

$$x_{1,3} \leftarrow b_1/a_{11}$$

$i = 3, 5, \ldots, f_o(n)$:

$$s_{i,2} \leftarrow b_i - a_{i1}x_{1,i}$$

$i = 5, 7, \ldots, f_o(n)$, $k = 3, 5, \ldots, i-2$ and
$i = 4, 6, \ldots, f_e(n)$, $k = 2, 4, \ldots, i-2$:

$$s_{i,k+1} \leftarrow s_{i,k-1} - a_{ik-1}x_{k-1,i+1} - a_{ik}x_{k,i}$$

$$x_{k,i+2} \leftarrow x_{k,i}$$

$k = 2, 3, \ldots, n$:

$$x_{k,k+2} \leftarrow (s_{k,k-1} - a_{kk-1}x_{k-1,i+1})/a_{kk}$$

This gives the following ddv's for each x-operand:

$$\begin{pmatrix}1\\1\end{pmatrix}, \begin{pmatrix}2\\0\end{pmatrix}$$

If the resulting data dependence vectors are transformed by the linear mapping found above, then the design vectors turn out to be $(1 \quad -1)$ for every $s_{i,k}$ and $(1 \quad 0)$, $(1 \quad 1)$ for every $x_{k,i}$.

Thus, s-operands moves in the negative y-direction, one length unit per time unit. For x-operands we actually have a fanout of two, since we do not utilize the x_k-enumeration fully. (The reason for this is that we want to avoid zero-delay lines.) The design vector $(1 \quad 0)$ means that the operand is used at the same cell next time. It must therefore be stored in an internal register. The design vector $(1 \quad 1)$, on the other hand, implies that the operand also must be transferred one length unit to the right, where it is to be used next time. Thus, the fanout does not cause any non-local communication. A space-time diagram is shown in figure 11.9.

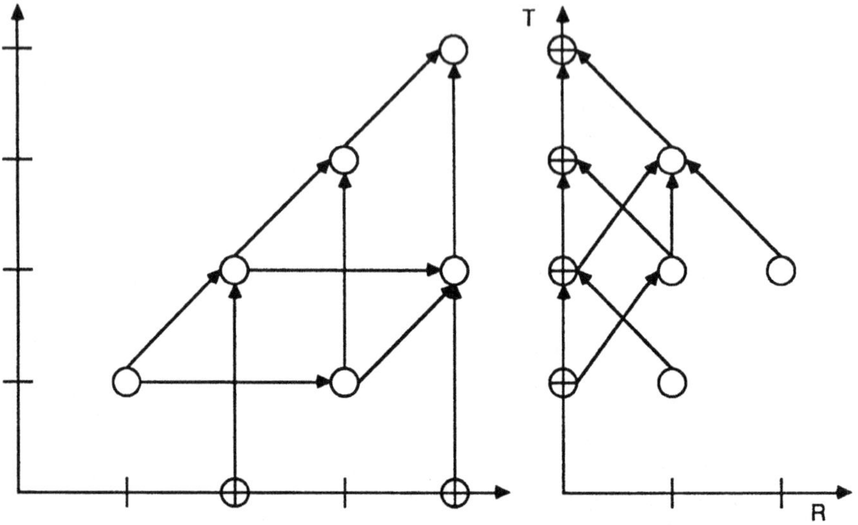

Figure 11.9: Space-time diagram with mapped precedence relation $\prec_{M''' A \circ L \circ G}$, $n = 4$.

Also this schedule can be supported by fixed hardware. Consider the following substitution field Φ' in space:

$$\sigma_0^{\Phi'} = \{x \leftarrow (s - ax)/a'\} \qquad ot_0^{\Phi'}(x) = \{\langle 1,0,x\rangle, \langle 1,1,x\rangle\}$$

$$it_0^{\Phi'}(s) = \langle 1,1,s\rangle \qquad it_0^{\Phi'}(a) = \langle 0,0,a\rangle$$

$$it_0^{\Phi'}(x) = \langle 1,0,x\rangle \qquad it_0^{\Phi'}(a') = \langle 0,0,a'\rangle$$

$1 \leq y < \lceil n/2 \rceil$:

$$\sigma_y^{\Phi'} = \{s \leftarrow s - ax - a'x', x \leftarrow x'\} \qquad \begin{array}{l} ot_y^{\Phi'}(s) = \{\langle 1, y-1, s\rangle\} \\ ot_y^{\Phi'}(x) = \{\langle 1, y+1, x\rangle, \langle 1, y, x\rangle\} \end{array}$$

$$it_y^{\Phi'}(s) = \langle 1, y+1, s\rangle \qquad it_y^{\Phi'}(a) = \langle 0, y, a\rangle \qquad it_y^{\Phi'}(x) = \langle 1, y, x\rangle$$

$$it_y^{\Phi'}(a') = \langle 0, y, a'\rangle \qquad it_y^{\Phi'}(x') = \langle 1, y-1, x'\rangle$$

$$\sigma_{\lceil n/2 \rceil}^{\Phi'} = \{s \leftarrow b - ax\} \qquad ot_{\lceil n/2 \rceil}^{\Phi'}(s) = \{\langle 1, \lceil n/2 \rceil - 1, s\rangle\}$$

$$it^{\Phi'}_{\lceil n/2\rceil}(b) = \langle 0, \lceil n/2\rceil, b\rangle \qquad it^{\Phi'}_{\lceil n/2\rceil}(a) = \langle 0, \lceil n/2\rceil, a\rangle$$
$$it^{\Phi'}_{\lceil n/2\rceil}(x) = \langle 1, \lceil n/2\rceil - 1, x\rangle$$

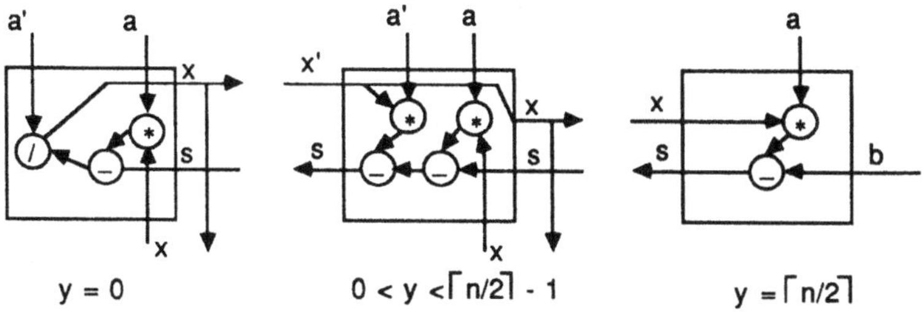

Figure 11.10: Substitution field Φ' in space.

It is not hard to verify that $A \circ L \circ G''$ is a correct mapping with respect to $\hat{\Phi}'$. Note that the substitutions $\sigma_y^{\hat{\Phi}'}$ are more complex than the substitutions $\sigma_y^{\hat{\Phi}}$. This balances the reduction of the number of cells from n to $\lceil n/2\rceil$. Thus really nothing is gained in terms of hardware, which easily could be believed if only the numbers of cells were compared. The gain is increased utilization of the hardware. Let us finally summarize our description of this schedule:

CCSA-set M''', index function G'':

$$x_{1,3} \leftarrow b_1/a_{11} \quad \mapsto \quad \begin{pmatrix} 1 \\ 1 \end{pmatrix}$$

$i = 2, 4, \ldots, f_e(n)$:

$$s_{i,1} \leftarrow b_i \quad \mapsto \quad \begin{pmatrix} i \\ 0 \end{pmatrix}$$

$i = 3, 5, \ldots, f_o(n)$:

$$s_{i,2} \leftarrow b_i - a_{i1}x_{1,i} \quad \mapsto \quad \begin{pmatrix} i \\ 1 \end{pmatrix}$$

$i = 5, 7, \ldots, f_o(n)$, $k = 3, 5, \ldots, i-2$ and
$i = 4, 6, \ldots, f_e(n)$, $k = 2, 4, \ldots, i-2$:

$$\left.\begin{array}{l} s_{i,k+1} \leftarrow s_{i,k-1} - a_{ik-1}x_{k-1,i+1} - a_{ik}x_{k,i} \\ x_{k,i+2} \leftarrow x_{k,i} \end{array}\right. \quad \mapsto \quad \begin{pmatrix} i \\ k \end{pmatrix}$$

$k = 2, 3, \ldots, n$:

$$x_{k,k+2} \leftarrow (s_{k,k-1} - a_{kk-1}x_{k-1})/a_{kk} \quad \mapsto \quad \begin{pmatrix} k \\ k \end{pmatrix}$$

Mapping $A \circ L$ from set of index vectors to space-time:

$$\begin{pmatrix} t \\ y \end{pmatrix} = \begin{pmatrix} 1/2 & 1/2 \\ 1/2 & -1/2 \end{pmatrix} \begin{pmatrix} i \\ k \end{pmatrix}$$

11.2 Matrix multiplication

In this section we will consider systolic implementations of matrix multiplication. Many such implementations have been proposed, such as straight-forward two-dimensional arrays, [KLe80,WD81,Kat82,CM83, U84], a linear array, [RFS82B], or a fault-tolerant array, [VRF83]. It is also very popular to describe, synthesize and verify simple arrays for matrix multiplication, in order to demonstrate different theories and methodologies for reasoning about parallel hardware, [MR82,C83,C86C, Q83,Li83,CS83,MiWi84,LW85]. A classification of regular systolic networks that covers most two-dimensional arrays above is found in [Le84].

Almost all systolic arrays for matrix multiplication implement the following recurrence equation or closely related ones (see, however, [PV80] for a recursive design based on matrix partitioning where this is not true):

Recurrence equation M for computing $C = A \times B$:

$1 \leq i \leq m$, $1 \leq j \leq n$:

$$c_{ij1} \leftarrow 0$$

$1 \leq i \leq m$, $1 \leq j \leq n$, $1 \leq k \leq p$:

$$c_{ijk+1} \leftarrow c_{ijk} + a_{ik}b_{kj}$$

This recurrence equation defines the elements $c_{ij} = c_{ijn+1}$ in the $m \times n$-matrix C as polynomials in the matrix elements of the $m \times p$-matrix A and the $p \times n$-matrix B.

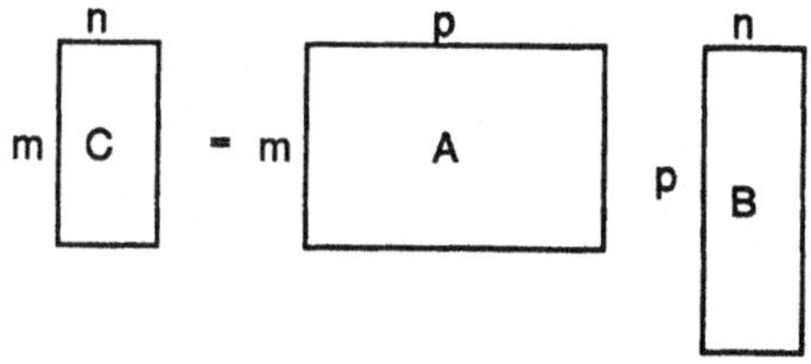

Figure 11.11: $C = A \times B$.

Here we will define an indexing G of the recurrence M and synthesize a number of systolic designs for it, by linear index mappings from $G(M)$ to a three-dimensional space-time. The reason why we consider such a well-analyzed field as synthesis of systolic arrays for matrix multiplication is, that it offers an unusual possibility to demonstrate how different constraints on the resulting array can be translated into constraints on the matrix of the linear index mapping.

Consider the indexing function G from M to Z^3, defined by the following:

Indexing function $G: M \to Z^3$:

$1 \leq i \leq m$, $1 \leq j \leq n$:

$$c_{ij1} \leftarrow 0 \quad \mapsto \quad \begin{pmatrix} i \\ j \\ 0 \end{pmatrix}$$

$1 \leq i \leq m$, $1 \leq j \leq n$, $1 \leq k \leq p$:

$$c_{ijk+1} \leftarrow c_{ijk} + a_{ik}b_{kj} \quad \mapsto \quad \begin{pmatrix} i \\ j \\ k \end{pmatrix}$$

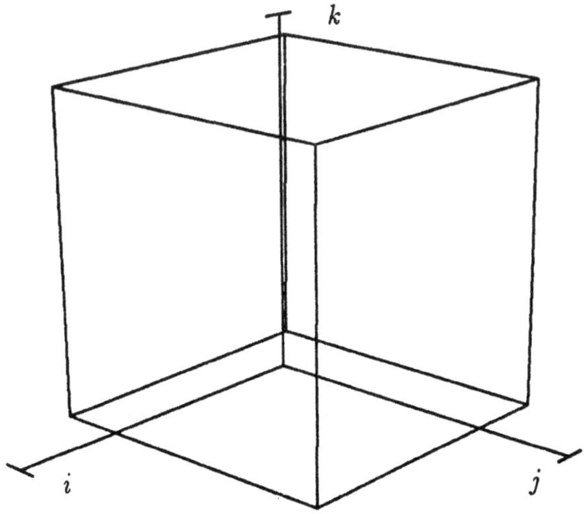

Figure 11.12: $G(M)$, $m = n = p$.

This defines $G(M)$ as a subset of Z^3. We can think of it as a continuous volume, a geometrical object that can be rotated, stretched and compressed by linear transformations. Thus, geometrical intuition can be used as a guidance to find suitable linear index mappings. We can also compute the space-time hull from the mapped boundaries of the object. This provides a simple way to compute execution time and cell utilization.

Let us now consider the transformation matrix L. Our task is to determine the elements of L.

$$L = \begin{pmatrix} l_{11} & l_{12} & l_{13} \\ l_{21} & l_{22} & l_{23} \\ l_{31} & l_{32} & l_{33} \end{pmatrix}$$

Different choices of elements will lead to different designs. What restrictions are there?

The first restriction is that we will only consider *correct* index mappings, to a three-dimensional space-time. Therefore, because of the shape of $G(M)$, it follows that L must be non-singular when $m > 1$, $n > 1$ and $p > 1$. Second, there is the causality criterion that the time component of the mapped data dependence vectors must be nonnegative. The data dependence, given by c-operands transferred between different iterations, yields the following data dependence vector:

$$\begin{pmatrix} 0 \\ 0 \\ 1 \end{pmatrix}$$

Thus, it must hold that $l_{13} \geq 0$, or, if we want the implementation to be ripple-free, that $l_{13} > 0$. Here we will restrict the attention to ripple-free implementations only. We also want to keep the computational density high. Therefore, by the argument of the determinant being a scale factor of the mapping, matrices with a small determinant should be looked for.

An optimal design with respect to these restrictions is given by the following index mapping L_1:

$$\begin{pmatrix} t \\ x \\ y \end{pmatrix} = \begin{pmatrix} 0 & 0 & 1 \\ 1 & 0 & 0 \\ 0 & 1 & 0 \end{pmatrix} \begin{pmatrix} i \\ j \\ k \end{pmatrix}$$

This array was presented in [CM83]. What are the properties of this design? The determinant of the transformation matrix is 1. k is mapped directly to t, i directly to x and j to y. It is thus immediately seen that we need mn cells and that the execution time (disregarding the "input" assignments $c_{ij1} \leftarrow 0$) is p. Therefore, the cell utilization is 1 which is maximal.

Let us now see how the elements in the matrices move. Every c_{ijk} is computed at $(k-1, i, j)$ and used at (k, i, j). Since the spatial component is independent of k, this means that every c_{ij} is fixed at the point $(i \quad j)$ in space, where it is updated in every time step. An alternative way to find this behaviour is to consider the data dependence vector $(0 \quad 0 \quad 1)$ caused by the variables c_{ijk}. This vector is transformed into the design vector $(1 \quad 0 \quad 0)$, which indicates that every c_{ijk} is stored one time step in an internal register of the cell, between its creation and its use.

Consider now the element a_{ik} of A for some i and k. a_{ik} is used by every cs-assignment with index vector $(i \quad j \quad k)$, for $j = 1, \ldots, n$. But the time does not vary with j, since $l_{12} = 0$. Thus, a_{ik} is used simultaneously by all these cs-assignments. This means that the value of a_{ik} has to be *broadcast* to all these cs-assignments. Every b_{kj} is also broadcast in the same way. This broadcast is likely to require a slowdown of the clock. Therefore, this first solution is not the winner it seemed to be.

The remedy for the broadcast is to skew in time the cs-assignments using the same inputs. This happens if the time for a cs-assignment becomes dependent of i and j also, that is: if l_{11} and l_{12} are non-zero. If both are set to one, then the following index mapping L_2 is obtained:

$$\begin{pmatrix} t \\ x \\ y \end{pmatrix} = \begin{pmatrix} 1 & 1 & 1 \\ 1 & 0 & 0 \\ 0 & 1 & 0 \end{pmatrix} \begin{pmatrix} i \\ j \\ k \end{pmatrix}$$

This systolic array architecture was proposed in [Kat82]. The determinant of L_2 is one, which indicates a high cell utilization. The design still uses mn cells, since the space part of L_2 is identical with the space part of L_1. Execution time is now $m + n + p$, which can be seen by checking the time coordinates of the mapped corners of the set of index vectors. Thus, the cell utilization is (disregarding all $c_{ij1} \leftarrow 0$)

$$mnp/(mn(m+n+p)) = p/(m+n+p).$$

So when p is much greater than $m+n$, the cell utilization is close to one. In the limit case, when infinite matrices are multiplied, it is equal to one. This design is indeed suitable for multiplication of such matrices.

What about the movements of c_{ij}, a_{ik}, b_{kj} now? The elements of C are fixed in the array as before, since the k-column of the transformation

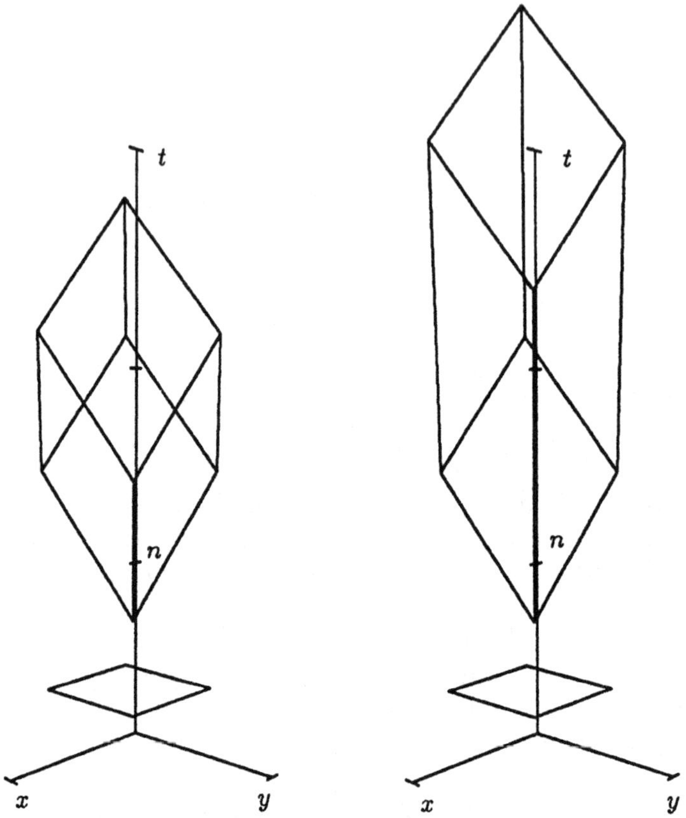

Figure 11.13: Transformed set of index vectors for L_2. (a) with $p = m = n$, (b) with $m = n$ and $p = 2n$.

matrix is the same. For any i and k, a_{ik} will now be used at the space-time points

$$\begin{pmatrix} t \\ x \\ y \end{pmatrix} = \begin{pmatrix} i+j+k \\ i \\ j \end{pmatrix}.$$

Thus, $j = t - i - k$ and we can parameterize the space coordinates x, y as functions of t for the usages of a_{ik}. The result is $x = i$ and $y = t - i - k$. Thus a_{ik} moves in the positive y-direction. In the same manner, by substituting $t - j - k$ for i, it is found that every b_{kj} moves in the positive x-direction.

The parameterization above can also be used, if we want to derive the orientations of the matrices A and B at given times. The rows of A are given by a fixed i and varying k and the columns are given by a fixed k and varying i. So at any given time, the rows of A are oriented in the

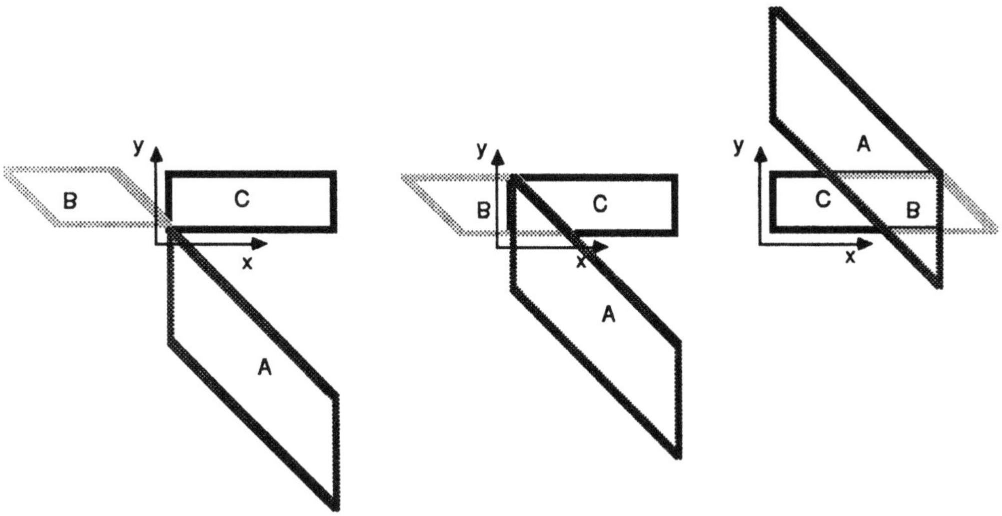

Figure 11.14: "Snapshots" of the L_2-array.

spatial direction $(x \quad y) = (0 \quad -1)$ and the columns in the direction $(1 \quad -1)$. In the same way the row and column orientation for B can be found: it turns out to be $(-1 \quad 0)$ for the columns and $(-1 \quad 1)$ for the rows. Thus, we have formally derived a basis for the snapshots in figure 11.14.

Still nothing has been mentioned about *how* the elements in A and B are going to be transferred to their places of usage; since they only appear as free variables without a source, the cs-assignment description does not contain any information about where they are to come from. The natural way to distribute such a value is to supply it to the first assignment using it and then let it be sent from that assignment to the next one using it. We can derive this behaviour formally in the following way: we introduce a single assignment where the free variable (matrix element) is assigned to a bound variable, that in turn is substituted for all old occurences of the free one. Then fanout reduction can be applied, according to the total time ordering imposed on the events of usage by L_2. The result is the following modified recurrence equation:

Modified recurrence equation M', indexing function:

$1 \leq i \leq m$, $1 \leq j \leq n$:

$$c_{ij1} \leftarrow 0 \qquad \mapsto \qquad \begin{pmatrix} i \\ j \\ 0 \end{pmatrix}$$

$1 \leq i \leq m$, $1 \leq k \leq p$:

$$a_{ik1} \leftarrow a_{ik} \qquad \mapsto \qquad \begin{pmatrix} i \\ 0 \\ k \end{pmatrix}$$

$1 \leq j \leq n$, $1 \leq k \leq p$:

$$b_{kj1} \leftarrow b_{kj} \qquad \mapsto \qquad \begin{pmatrix} 0 \\ j \\ k \end{pmatrix}$$

$1 \leq i \leq m$, $1 \leq j \leq n$, $1 \leq k \leq p$:

$$\begin{array}{l} c_{ijk+1} \leftarrow c_{ijk} + a_{ikj} b_{kji} \\ a_{ikj+1} \leftarrow a_{ikj} \\ b_{kji+1} \leftarrow b_{kji} \end{array} \qquad \mapsto \qquad \begin{pmatrix} i \\ j \\ k \end{pmatrix}$$

This modified CCSA-set, together with the indexing function, gives the following new data dependence vectors:

$$a\text{-operands:} \quad \begin{pmatrix} 0 \\ 1 \\ 0 \end{pmatrix} \qquad b\text{-operands:} \quad \begin{pmatrix} 1 \\ 0 \\ 0 \end{pmatrix}$$

The design vectors are the following:

$$\begin{pmatrix} 1 & 1 & 1 \\ 1 & 0 & 0 \\ 0 & 1 & 0 \end{pmatrix} \begin{pmatrix} 0 \\ 1 \\ 0 \end{pmatrix} = \begin{pmatrix} 1 \\ 0 \\ 1 \end{pmatrix}$$

for the a-operands and

$$\begin{pmatrix} 1 & 1 & 1 \\ 1 & 0 & 0 \\ 0 & 1 & 0 \end{pmatrix} \begin{pmatrix} 1 \\ 0 \\ 0 \end{pmatrix} = \begin{pmatrix} 1 \\ 1 \\ 0 \end{pmatrix}$$

for the b-operands. Thus, the transfer of every a-operand is supported by a line with unit delay in the positive y-direction and the transfer of every b-operand with a line in the positive x-direction.

Note that any cs-assignment in M' (except for the "input" assignments) can be matched against the substitution

$$\{c \leftarrow c + ab, a \leftarrow a, b \leftarrow b\}$$

that describes the action of an inner-product cell, [KLe80], fully.

There are two linear index mappings that give arrays very similar to the one given by L_2:

$$L_3, \text{ "fixed } A\text{"}: \quad \begin{pmatrix} 1 & 1 & 1 \\ 1 & 0 & 0 \\ 0 & 0 & 1 \end{pmatrix} \qquad L_4, \text{ "fixed } B\text{"}: \quad \begin{pmatrix} 1 & 1 & 1 \\ 0 & 1 & 0 \\ 0 & 0 & 1 \end{pmatrix}.$$

In the L_3-array, the elements of A are fixed. It needs mp cells. Execution time is still $m+n+p$, so the cell utilization is $mnp/(mp(m+n+p)) = n/(m+n+p)$, which gets close to 1 when n grows very big compared with $m+p$. This array is therefore suitable if a moderately big A-matrix is to be multiplied with a very "long" matrix B. In the L_4-array the elements of B are fixed, it has np cells and cell utilization $m/(m+n+p)$ and is therefore good for multiplying a "tall" A with a comparatively small B. Their connection patterns can easily be derived in the same way as for L_2.

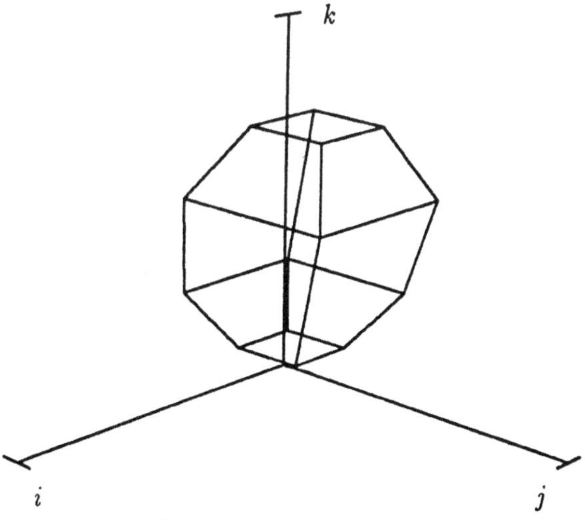

Figure 11.15: Set of index vectors for multiplication of two 5×5 band matrices, each with one nonzero sub- and superdiagonal.

Let us now turn to the problem of multiplying two $n \times n$ band matrices. It is desirable to find a schedule that takes advantage of the knowledge that iterations where either the a- or b-operand is zero need not be carried out. Using this knowledge, the number of cells can be reduced to the magnitude of the bandwidth of the resulting matrix.

The idea is as follows: the schedule will take advantage of the band structure if the elements in A and B move in the same direction as the respective diagonals. If so, there only need to be cells where non-zero diagonals from A and B intersect. Everywhere else zero-diagonals will intersect, and the cs-assignments scheduled there need not be carried out.

Assume that we use the recurrence equation M'. This implies that the transformation matrix must be of the form

$$\begin{pmatrix} 1 & 1 & 1 \\ l_{21} & l_{22} & l_{23} \\ l_{31} & l_{32} & l_{33} \end{pmatrix}.$$

Consider any element a_{ik} in A. All other elements $a_{i+\delta k+\delta}$ are on the same diagonal. Regardless of j, the direction of this diagonal in space must be the same as the direction of movement of a_{ik}. The direction of the diagonal in space is

$$\begin{pmatrix} l_{21} & l_{22} & l_{23} \\ l_{31} & l_{32} & l_{33} \end{pmatrix} \begin{pmatrix} i+\delta-i \\ j-j' \\ k+\delta-k \end{pmatrix} = \begin{pmatrix} l_{21}\delta + l_{22}(j-j') + l_{23}\delta \\ l_{31}\delta + l_{32}(j-j') + l_{33}\delta \end{pmatrix}$$

where δ is chosen to be non-zero. Similarly, the direction of movement of a_{ik} is the direction of the design vector of the a-operands, or

$$\begin{pmatrix} l_{21} & l_{22} & l_{23} \\ l_{31} & l_{32} & l_{33} \end{pmatrix} \begin{pmatrix} 0 \\ 1 \\ 0 \end{pmatrix} = \begin{pmatrix} l_{22} \\ l_{32} \end{pmatrix}.$$

If these directions are to be the same, it must hold that

$$\begin{pmatrix} l_{21}\delta + l_{22}(j-j') + l_{23}\delta \\ l_{31}\delta + l_{32}(j-j') + l_{33}\delta \end{pmatrix} = c_a' \begin{pmatrix} l_{22} \\ l_{32} \end{pmatrix}$$

or

$$\begin{pmatrix} l_{21} + l_{23} \\ l_{31} + l_{33} \end{pmatrix} = \frac{c_a' - (j-j')}{\delta} \begin{pmatrix} l_{22} \\ l_{32} \end{pmatrix} = c_a \begin{pmatrix} l_{22} \\ l_{32} \end{pmatrix}.$$

A similar reasoning for the diagonals of B gives the following equation:

$$\begin{pmatrix} l_{22} + l_{23} \\ l_{32} + l_{33} \end{pmatrix} = c_b \begin{pmatrix} l_{21} \\ l_{31} \end{pmatrix}.$$

See also [Li83]. These equations are essentially the same as the corresponding constraints in [LW85]. It can be shown (straightforward but tedious) that the only values of c_a and c_b that do not give a singular transformation matrix are $c_a = c_b = -1$. The following equation for the spatial part of the matrix follows:

$$l_{21} + l_{22} + l_{23} = 0$$
$$l_{31} + l_{32} + l_{33} = 0$$

If we want the cells to communicate with their neighbours only, then the values of these elements are limited to 0, 1, −1. (Otherwise some of the data dependence vectors $(0 \quad 0 \quad 1)$, $(0 \quad 1 \quad 0)$, $(1 \quad 0 \quad 0)$ would give a design vector implying non-local data transfers.) Furthermore, we still have the condition that the transformation matrix must be non-singular.

An index mapping fulfilling the conditions above is the following, L_5:

$$\begin{pmatrix} t \\ x \\ y \end{pmatrix} = \begin{pmatrix} 1 & 1 & 1 \\ -1 & 0 & 1 \\ 0 & -1 & 1 \end{pmatrix} \begin{pmatrix} i \\ j \\ k \end{pmatrix}$$

If we map the data dependence vectors $(0 \quad 0 \quad 1), (0 \quad 1 \quad 0), (1 \quad 0 \quad 0)$ given by the c-, a- and b-operands, then we immediately obtain the following design vectors:

$$c\text{-operands:} \quad \begin{pmatrix} 1 \\ 1 \\ 1 \end{pmatrix} \qquad a\text{-operands:} \quad \begin{pmatrix} 1 \\ 0 \\ -1 \end{pmatrix} \qquad b\text{-operands:} \quad \begin{pmatrix} 1 \\ -1 \\ 0 \end{pmatrix}$$

L_5 actually gives the well-known systolic array for multiplication of band matrices of Kung and Leiserson, [KLe80]. In figure 11.16 it can be seen why this index mapping gives a design that is good for band matrices. Cells need to be provided for in the projection to space of the mapped set of index vectors. If we build an array for multiplying dense $n \times n$-matrices using the Kung-Leiserson design, then $O(n^2)$ cells are needed, and the outermost cells will be very poorly utilized. For band matrices with bandwidths b_A, b_B the number of processors will be $O(b_A b_B)$. The execution time is in both cases $3n - 2$. The determinant of L_5 is three which indicates a cell utilization of $1/3$. It is indeed well-known that the Kung-Leiserson array approaches this cell utilization when the bands are very long. This low cell utilization implies that there are holes in the mapped set of index vectors in figure 11.16.

Exactly as for the Kung-Leiserson design of a forward-solving array considered in the previous example, it is possible to increase the cell utilization by interleaving. If we are to perform three matrix multiplications

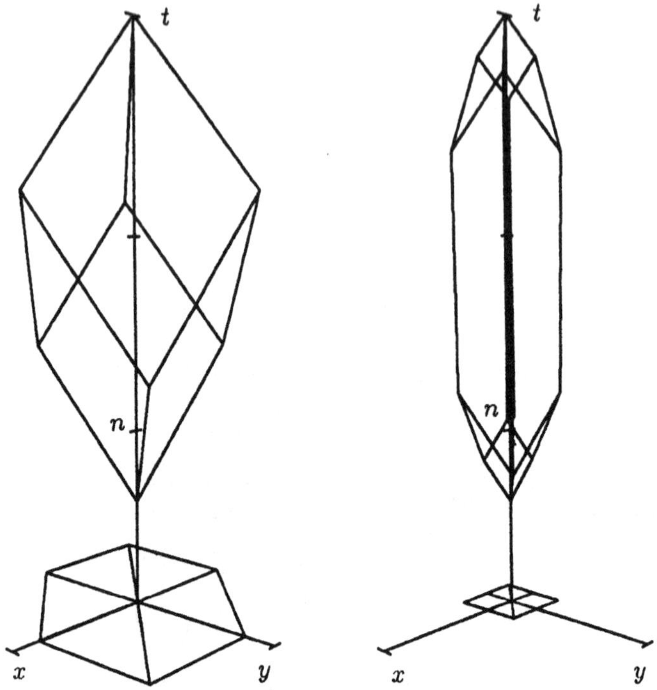

Figure 11.16: Mapped sets of index vectors and projection on space, for multiplication of (a) 5 × 5 dense matrices and (b) 5 × 5 band matrices (one sub-, one superdiagonal) under L_5.

$A_l \times B_l$, $l = 0, 1, 2$, of the same size, then these can be interleaved on the array in the following way:

1. For matrix multiplication no. l, change all index vectors $(\,i \quad j \quad k\,)$ to $(\,i \quad j \quad k \quad l\,)$.
2. Form the new set of index vectors as the union of the three extended index sets.
3. Map the new set of index vectors to three-dimensional space-time by the following mapping:

$$\begin{pmatrix} t \\ x \\ y \end{pmatrix} = \begin{pmatrix} 1 & 1 & 1 & 1 \\ -1 & 0 & 1 & 0 \\ 0 & -1 & 1 & 0 \end{pmatrix} \begin{pmatrix} i \\ j \\ k \\ l \end{pmatrix}$$

The last column of the transformation matrix will cause a displacement of l in time for multiplication l.

The question naturally arises, if there is a systolic array for band matrix multiplication that has cell utilization one in the limit. Such arrays are known and we will derive one of them. It is not hard to see that the restrictions posed on the spatial part of

$$\begin{pmatrix} 1 & 1 & 1 \\ l_{21} & l_{22} & l_{23} \\ l_{31} & l_{32} & l_{33} \end{pmatrix}$$

forces the determinant to be either 3 or -3, which will keep the cell utilization down. The remedy is to alter the time row; if one or two elements are -1, then the determinant will become 1 or -1, respectively. l_{13} cannot be negated unless another recurrence than M is used, since a negative value of l_{13} would violate the time causality with respect to the c-operands. The following index mapping, L_6, preserves causality with respect to the dependencies of M:

$$\begin{pmatrix} t \\ x \\ y \end{pmatrix} = \begin{pmatrix} 1 & -1 & 1 \\ -1 & 0 & 1 \\ 0 & -1 & 1 \end{pmatrix} \begin{pmatrix} i \\ j \\ k \end{pmatrix}$$

This array was first found by Schreiber, [Sch83], and it was described in [Li83] using L_6.

Note that the index mapping L_6 imposes a different time ordering than before on the usage of the elements of A, since now j *decreases* when the time increases. Therefore, the modified recurrence M' cannot be used, since the fanout reduction of the a-operands there is based on a time ordering where j increases with increasing time. Instead we can derive a new recurrence equation M'' from M (with $m = p = n$) and the new time ordering:

Modified recurrence equation M'', indexing function:

$1 \leq i, j \leq n$:

$$c_{ij1} \leftarrow 0 \quad \mapsto \quad \begin{pmatrix} i \\ j \\ 0 \end{pmatrix}$$

$1 \leq i, k \leq n$:

$$a_{ikn} \leftarrow a_{ik} \quad \mapsto \quad \begin{pmatrix} i \\ n+1 \\ k \end{pmatrix}$$

$1 \leq j,k \leq n$:

$$b_{kj1} \leftarrow b_{kj} \quad \mapsto \quad \begin{pmatrix} 0 \\ j \\ k \end{pmatrix}$$

$1 \leq i,j,k \leq n$:

$$\begin{array}{l} c_{ijk+1} \leftarrow c_{ijk} + a_{ikj} b_{kji} \\ a_{ikj-1} \leftarrow a_{ikj} \\ b_{kji+1} \leftarrow b_{kji} \end{array} \quad \mapsto \quad \begin{pmatrix} i \\ j \\ k \end{pmatrix}$$

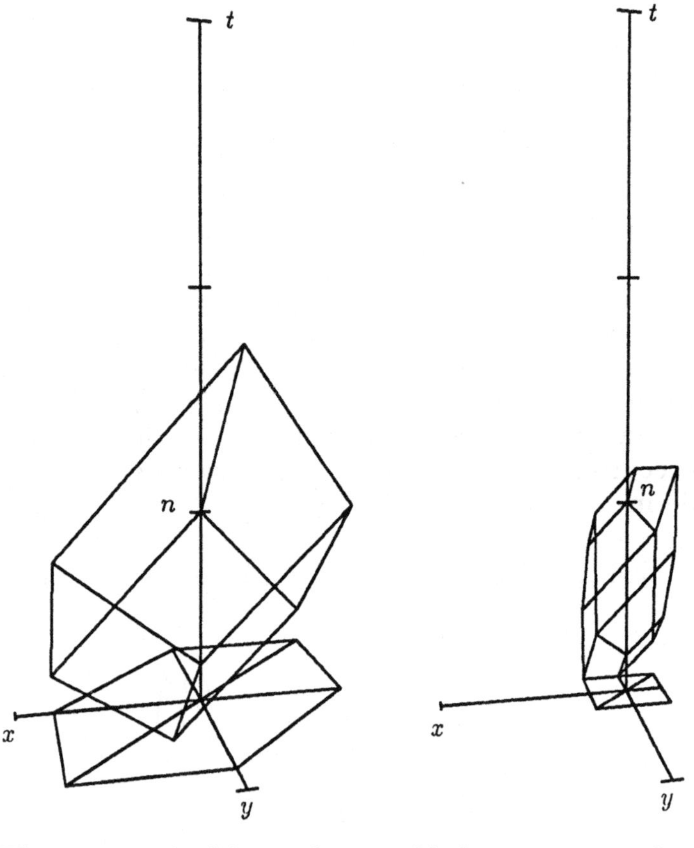

Figure 11.17: Mapped sets of index vectors and projection on space, for multiplication of (a) 5×5 dense matrices and (b) 5×5 band matrices (one sub-, one superdiagonal) under L_6.

Now the data dependence vector for a-operands is $(\,0 \quad -1 \quad 0\,)$, so the design vector for these operands becomes

$$\begin{pmatrix} 1 & -1 & 1 \\ -1 & 0 & 1 \\ 0 & -1 & 1 \end{pmatrix} \begin{pmatrix} 0 \\ -1 \\ 0 \end{pmatrix} = \begin{pmatrix} 1 \\ 0 \\ 1 \end{pmatrix}$$

which has a positive time component.

The condition on the direction of the diagonals in space to be parallel with the data flow was derived under the assumption that the time part of it was $(\,1 \quad 1 \quad 1\,)$. The direction of the diagonals of A will, however, still be parallel with the direction of the movement of the elements under the schedule given by L_6. The reason is that the new time ordering causes a simple reversal of the direction of the flow of a-elements, since the spatial part of L_6 is the same as the one of L_5 and the new design vector for a-operands is the negated old one. Thus, L_6 is also suitable for multiplication of band matrices. The determinant of L_6 is 1, and the cell utilization is indeed close to 1 when the bands are long.

Note that the array above is *not* the same as the one of Weiser and Davis, [WD81,LW85]. That array is also suitable for band matrices and it has a cell utilization close to 1 for such matrices, but it does not implement the recurrence equation M. Instead it is based on the following recurrence:

Recurrence equation M''' for computing $C = A \times B$, indexing function:

$1 \le i \le m$, $1 \le j \le n$:

$$c_{ijn} \leftarrow 0 \quad \mapsto \quad \begin{pmatrix} i \\ j \\ n+1 \end{pmatrix}$$

$1 \le i \le m$, $1 \le j \le n$, $1 \le k \le p$:

$$c_{ijk-1} \leftarrow c_{ijk} + a_{ik}b_{kj} \quad \mapsto \quad \begin{pmatrix} i \\ j \\ k \end{pmatrix}$$

Under this recurrence and indexing function, every c-operand has a data dependence vector $(\,0 \quad 0 \quad -1\,)$. So the element l_{13} of the transformation matrix must now be negative instead of positive. The following index function L_7 yields the Weiser-Davis array:

$$\begin{pmatrix} t \\ x \\ y \end{pmatrix} = \begin{pmatrix} 1 & 1 & -1 \\ -1 & 0 & 1 \\ 0 & -1 & 1 \end{pmatrix} \begin{pmatrix} i \\ j \\ k \end{pmatrix}$$

From the transformation matrix it is easily verified that this array has the same good properties as the array given by L_6.

11.3 *LU*-decomposition

LU-decomposition is also a nice model problem for synthesis of systolic arrays. Some other authors have used it to exemplify the method of linear mappings, [Mo82,C85B,Raj86]. Here we will derive a design closely related to the well-known array of Kung and Leiserson, [KLe80]. The method of linear index mappings will be used.

LU-decomposition is the unique decomposition of a non-singular* square matrix A into a lower-triangular matrix L, where all elements on the main diagonal are 1, and an upper-triangular matrix U, such that $A = LU$. The following equations provides an implicit output specification of the elements l_{ik}, u_{kj} in L and U as functions of the elements in A:

$$1 \leq i,j \leq n: \quad a_{ij} = \sum_{k=1}^{\min(i,j)} l_{ik}u_{kj} \qquad 1 \leq k \leq n: \quad l_{kk} = 1$$

The recurrence equation below, [KLe80], computes elements l_{ik}, u_{kj} that satisfy the equations above:

Recurrence for *LU*-decomposition:

$1 \leq i,j \leq n$:

$$a_{ij,1} \leftarrow a_{ij}$$

$1 < i,j \leq n$, $1 \leq k < \min(i,j)$:

$$a_{ij,k+1} \leftarrow a_{ij,k} - l_{ik}u_{kj}$$

$1 \leq k < n$, $k < i \leq n$:

$$l_{ik} \leftarrow a_{ik,k}/u_{kk}$$

$1 < j \leq n$, $1 \leq k \leq j$:

$$u_{kj} \leftarrow a_{kj,k}$$

Note that theorems 6.5 and 6.1 imply that the output functions of all l_{ik}, u_{kj} are polynomials in the free variables a_{ij}. Thus, the implicit output specification above is structurally data-independent.

This recurrence could be used as a starting point for synthesis. But before doing so the following observation should be made: the recurrence above requires $n(n-1)/2$ divisions. By a simple rewriting, these divisions

---

* Strictly speaking, all matrices A_{kk} formed as the intersection of the k first rows and columns of A should be non-singular, [DBA74].

can be replaced by n inversions and $n(n-1)/2$ multiplications. At first sight this may seem inferior, but the multiplications are probably so much easier to realize in hardware that the rewriting will be profitable. As will be seen further on, it is possible to schedule the rewritten recurrence in space-time so that the number of cells performing inversions is kept at an absolute minimum.

Rewritten recurrence M for LU-decomposition, indexing function:

$1 \le i, j \le n$:

$$a_{ij,1} \leftarrow a_{ij} \qquad \mapsto \qquad (i \quad j \quad 0)$$

$1 < i, j \le n$, $1 \le k < \min(i, j)$:

$$a_{ij,k+1} \leftarrow a_{ij,k} - l_{ik}u_{kj} \qquad \mapsto \qquad (i \quad j \quad k)$$

$1 \le k < n$, $k < i \le n$:

$$l_{ik} \leftarrow a_{ik,k}m_k \qquad \mapsto \qquad (i \quad k \quad k)$$

$1 < j \le n$, $1 \le k < j$:

$$u_{kj} = a_{kj,k} \qquad \mapsto \qquad (k \quad j \quad k)$$

$1 \le k \le n$:

$$\left.\begin{array}{l} u_{kk} \leftarrow a_{kk,k} \\ m_k \leftarrow 1/a_{kk,k} \end{array}\right. \qquad \mapsto \qquad (k \quad k \quad k)$$

A natural indexing function is given above. Z^3 is chosen as index space. It is therefore natural to try to find two-dimensional arrays by linear mappings to a three-dimensional space-time. An analysis of the data dependencies of this recurrence yields the following data dependence vectors for different variables:

$$a_{ik,j}: \begin{pmatrix} 0 \\ 0 \\ 1 \end{pmatrix} \qquad l_{ik}: \begin{pmatrix} 0 \\ j-k \\ 0 \end{pmatrix} \quad \text{for } j = k+1, \ldots, n$$

$$u_{kj}, m_k: \begin{pmatrix} i-k \\ 0 \\ 0 \end{pmatrix} \quad \text{for } i = k+1, \ldots, n$$

$j-k$ and $i-k$ above will always be positive. So in order to preserve causality and give a ripple-free implementation, all elements in the time row of the transformation matrix must be positive. Note also that l_{ik}, u_{kj}, m_k have a fanout of $n-k$.

The dependence pattern is very similar to the one of matrix multiplication. The a-, l- and u-operands have the same direction of their data

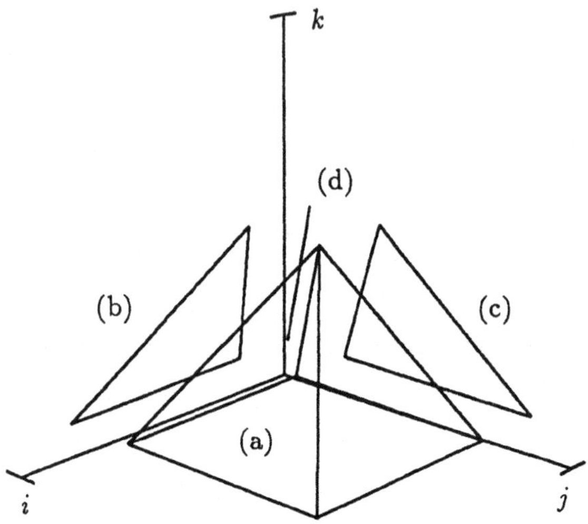

Figure 11.18: Set of index vectors for M: (a) iteration on a-operands, (b) assignment of l-operands, (c) assignment of u-operands and (d) inversion of diagonal elements.

dependence vectors as the c-, a- and b-operands for matrix multiplication. Therefore, a linear mapping from this set of index vectors will give a design that is good for LU-decomposition of band matrices exactly when it meets the condition to give a good design for multiplication of band matrices. L_5, for instance, will yield such a design:

$$\begin{pmatrix} t \\ x \\ y \end{pmatrix} = \begin{pmatrix} 1 & 1 & 1 \\ -1 & 0 & 1 \\ 0 & -1 & 1 \end{pmatrix} \begin{pmatrix} i \\ j \\ k \end{pmatrix}$$

The fanout of l_{ik}, u_{kj}, m_k can be reduced in the same way as for a- and b-operands in the matrix multiplication case, with the distinction that the variables here are already assigned. The result is a systolic array that is very similar to the array of Kung and Leiserson, with these good properties:

- Assignment of l-operands is always performed in a single row of processors. No other assignments are mapped to this row, so these cells can be specialized to perform an action of type $\{l \leftarrow am\}$.

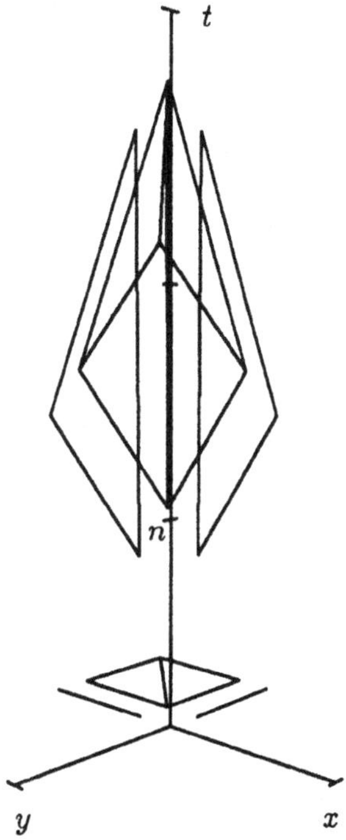

Figure 11.19: Mapped set of index vectors and space projection under L_5. Note that the different parts (a) – (d) of the set of index vectors are projected to distinct areas of the space.

- The same applies to assignment of u-operands. The cells in this row only need to perform an action of type $\{l \leftarrow a\}$, a trivial propagation of data.
- All assignments of a-operands is performed in an interior area. These cells will have to perform actions of type $\{a \leftarrow a - lu\}$.
- The assignments where m-operands are assigned inverted a-operands, finally, are all mapped to a single point in space. This cell alone needs to be able to perform inversion. Its action is of type $\{m \leftarrow 1/a, u \leftarrow a\}$.

Exactly as for the Kung and Leiserson array, this array has cell utilization 1/3 in the limit. The only difference between the arrays is that the Kung and Leiserson array was designed to utilize inner product cells (action of type $\{a \leftarrow a+lu\}$) as much as possible. This can be achieved by a simple rewriting of the recurrence equation M.

The striking similarity between the Kung and Leiserson arrays for matrix multiplication and LU-decomposition is totally explained by the analysis above. It is caused by the fact that the data dependence vectors are essentially the same for the two recurrences and that the same linear index mapping L_5 is used. The question naturally arises if it is possible to utilize the index mapping L_6,

$$\begin{pmatrix} t \\ x \\ y \end{pmatrix} = \begin{pmatrix} 1 & -1 & 1 \\ -1 & 0 & 1 \\ 0 & -1 & 1 \end{pmatrix} \begin{pmatrix} i \\ j \\ k \end{pmatrix}$$

to increase the cell utilization in the same manner as for multiplication of band matrices. The answer is no, and the reason is that L_6 will violate the causality for the l-operands, unless the recurrence M is rewritten or the indexing function is changed so that the l-operands have data dependence vectors of the form $(\,0 \quad -\delta \quad 0\,)$ (where $\delta > 0$), without altering the data dependence vectors of the other operands. In the case of matrix multiplication this was easily done, since the a-operands there are free variables; thus the mapping to space-time could be found first and the data dependencies introduced afterwards, by introducing an "input" assignment for them and then apply fanout reduction. But the l-operands are assigned variables, and the assignments where they are created have already been labelled with index vectors.

There are designs for LU-decomposition, though, that are based on linear mappings with an absolute value of the determinant of 1. The following, L_7, was proposed by Chen, [C85B]:

$$\begin{pmatrix} t \\ x \\ y \end{pmatrix} = \begin{pmatrix} 1 & 1 & 1 \\ 1 & 0 & 1 \\ 0 & 1 & 1 \end{pmatrix} \begin{pmatrix} i \\ j \\ k \end{pmatrix}$$

What properties will the resulting design have? Since the first row of L_7 is equal to the first row of L_5, the function from index vectors to time is the same. Both designs will have an execution time of $3n - 2$ (disregarding the "input" assignments of a-operands), where the first assignment processed is the one labelled with $(\,1 \quad 1 \quad 1\,)$, at $t = 3$, and the last one is $(\,n \quad n \quad n\,)$, at $t = 3n$. L_7 is not suitable for LU-decomposition of band matrices, since its spatial part does not satisfy the equation

$$l_{21} + l_{22} + l_{23} = 0$$
$$l_{31} + l_{32} + l_{33} = 0.$$

While the number of cells for the Kung-Leiserson design is proportional to the square of the bandwidth b and independent of the size n for a band matrix, the Chen design needs a number of cells that is proportional to n even for band matrices. In the limit, when infinite band matrices are decomposed, the Chen design needs an infinite number of cells, while

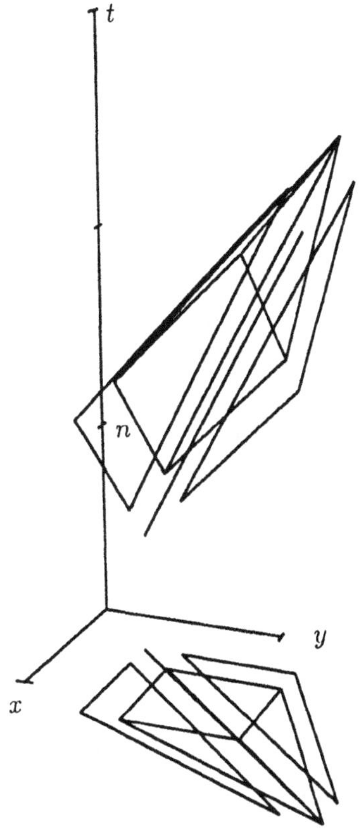

Figure 11.20: Mapped set of index vectors and space projection under L_7.

the number of cells for the Kung-Leiserson design still is proportional to b^2. For decomposition of full $n \times n$-matrices the Kung-Leiserson design needs a number of cells proportional to n^2, while the Chen design needs about $2n^2$ cells. The assignments performing inversions, with index vectors $(\,k \quad k \quad k\,)$, $1 \leq k \leq n$, are mapped to space coordinates $(\,2k \quad 2k\,)$ by L_7. So the Chen design requires n cells to be able to perform inversion, instead of one cell for the Kung-Leiserson array. The design thus seems to be inferior to the Kung-Leiserson design in all respects.

This example shows that it is necessary not to choose a too simple performance index. Apparently the scale factor provided by the determinant of the transformation matrix is not a sufficient measure. Other factors, such as that a "tilt" of the mapped set of index vectors in space-time, that will cause a "drift" of the computational activity in the array, are of equal importance. The design of Chen was selected to give an optimal schedule on a processor array with regular connections between processing elements in the spatial directions $(1\ \ 0)$, $(0\ \ 1)$, $(1\ \ 1)$. A conclusion to be drawn is that such a processor configuration is not a good one for LU-decomposition.

11.4 Networks for the fast Fourier transform

In this example we will show $\overset{stt}{\leftrightarrow}$-equivalence between two communication structures. The communication structures considered are the *perfect shuffle-structure* $\langle S_{ps}, \prec_{ps}\rangle$, describing a typical execution on the perfect shuffle network, [St71], and a communication structure $\langle S_j, \prec_j\rangle$ describing space-time communication in a network similar to a type of fast Fourier transform networks, [GW70,JWCD81,JC82,Sw86], that gives $O(\log n)$ space complexity and $O(n)$ time complexity (whereas the perfect shuffle gives the converse). We will also give a verification that a Fast Fourier Transform algorithm (FFT) can be implemented on these networks. As an intermediate vehicle we will use the *time-dependent binary cube-structure* $\langle S_{tbc}, \prec_{tbc}\rangle$.

The perfect shuffle-structure of order m is given by:

$S_{ps} = N \times \{0, \ldots, 2^m - 1\}$

$\prec_{ps}$: $1 \le t \le m+1$, $0 \le i < 2^m$:

$$\langle t-1, \lfloor i/2 \rfloor\rangle \prec_{ps} \langle t, i\rangle \text{ and } \langle t-1, \lfloor i/2 \rfloor + 2^{m-1}\rangle \prec_{ps} \langle t, i\rangle$$

The time-dependent binary m-cube structure is defined by:

$S_{tbc} = N \times \{0,1\}^m$

$\prec_{tbc}$: $1 \le t \le m+1$, $j_{m-t+1 \bmod m}, i_0, \ldots, i_{m-1} \in \{0,1\}$:

$$\langle t-1, i_{m-1}, \ldots, j_{m-t+1 \bmod m}, \ldots, i_0\rangle \prec_{tbc} \langle t, i_{m-1}, \ldots, i_{m-t+1 \bmod m}, \ldots, i_0\rangle$$

$\langle S_j, \prec_j\rangle$ is defined by:

$S_j = N \times N$

$\prec_j$: Let A_j be a bijection $S_{tbc} \rightarrow S_j$ such that the following holds:

$i_0, \ldots, i_{m-1} \in \{0,1\}$:

$$A_j(0, i_{m-1}, \ldots, i_0) = \langle \sum_{n=1}^{m-1} i_n 2^n + i_0 2^m, 0\rangle$$

$1 \le t \le m+1$:

$$A_j(t, i_{m-1}, \ldots, i_0) = \langle \sum_{n=0}^{m-1} i_n 2^n + 2^{m+1} - 2^{m+1-t}, t\rangle$$

(Such a function exists. The condition above specifies a bijective restriction A_j' of A_j to a finite subset S' of S_{tbc}. The cardinality of $S_{tbc} \setminus S'$, $|N|$, is the same as the cardinality of $S_j \setminus A'_j(S')$, so A'_j can be extended to a bijection $S_{tbc} \to S_j$.) Now $\prec_j$ is defined by $fi(\langle S_{tbc}, \prec_{tbc}\rangle, \langle S_j, \prec_j\rangle, A_j)$. By lemma 6.5, $\prec_j$ exists and it is easily shown to be a communication ordering. By definition A_j is $\prec_{tbc}$-allowed.

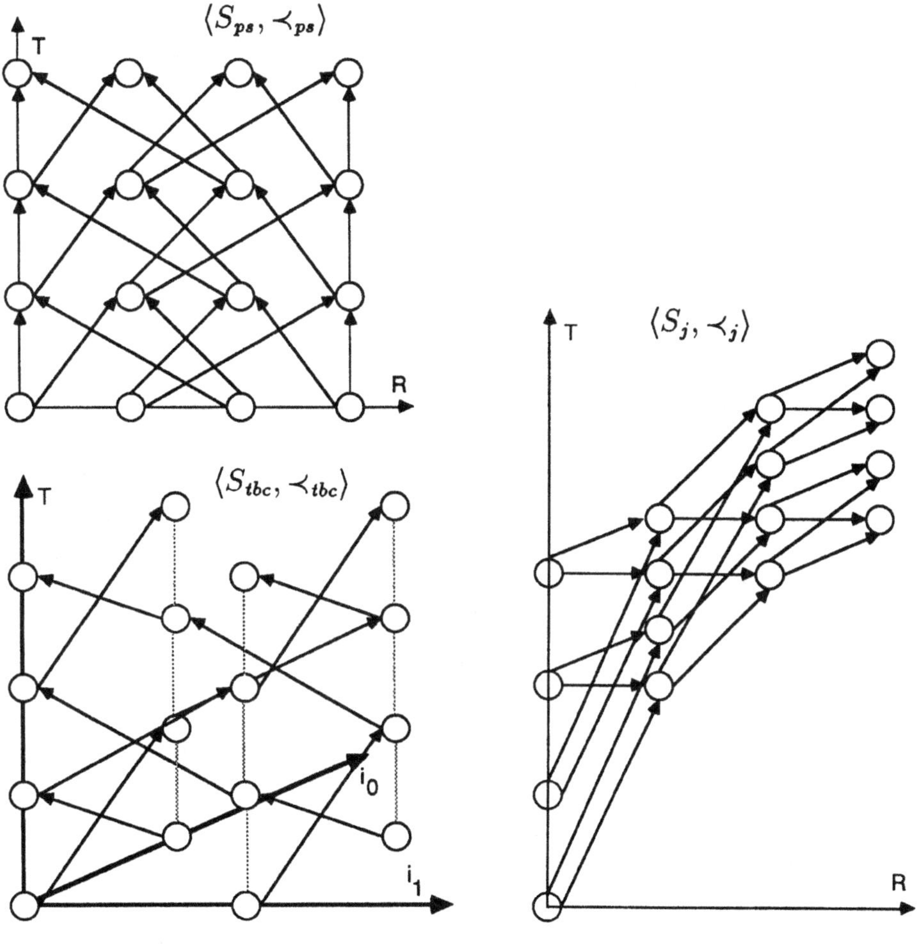

Figure 11.21: $\langle S_{tbc}, \prec_{tbc}\rangle$, $\langle S_{ps}, \prec_{ps}\rangle$, $\langle S_j, \prec_j\rangle$, $m = 2$.

We now show $\langle S_{ps}, \prec_{ps}\rangle \overset{stt}{\leftrightarrow} \langle S_j, \prec_j\rangle$:

$\langle S_j, \prec_j\rangle \overset{stt}{\leftrightarrow} \langle S_{tbc}, \prec_{tbc}\rangle$: Directly from proposition 7.17 and the definition of $\prec_j$.

$\langle S_{tbc}, \prec_{tbc}\rangle \overset{stt}{\leftrightarrow} \langle S_{ps}, \prec_{ps}\rangle$: Let A_{ps} be a bijection $S_{tbc} \rightarrow S_{ps}$ such that the following holds:

$i_0, \ldots, i_{m-1} \in \{0,1\}$, $0 \le t \le m+1$:

$$A_{ps}(t, i_{m-1}, \ldots, i_0) = \langle t, \sum_{n=0}^{m-1} i_{n-t+1 \bmod m} 2^n \rangle.$$

It is easy to verify that the restriction of A_{ps} specified above is 1-1. It can be shown that this restricted function can be extended to a bijection $S_{tbc} \rightarrow S_{ps}$ in the same way as for A_j. Next, define $\prec_{A_{ps}}$ on S_{ps} by $fi(\langle S_{tbc}, \prec_{tbc}\rangle, \langle S_{ps}, \prec_{A_{ps}}\rangle, A_{ps})$. Let us show that $\prec_{A_{ps}} = \prec_{ps}$:
From the definitions of $\prec_{tbc}$, A_{ps} follows that

$$\langle t-1, \sum_{n=0}^{m-2} i_{n-t+2 \bmod m} 2^n + j_{m-t+1 \bmod m} 2^{m-1}\rangle \prec_{A_{ps}} \langle t, \sum_{n=0}^{m-1} i_{n-t+1 \bmod m} 2^n \rangle$$

for all relevant values of t. Let $i = \sum_{n=0}^{m-1} i_{n-t+1 \bmod m} 2^n$. Then it holds that $\sum_{n=0}^{m-2} i_{n-t+2 \bmod m} 2^n = \lfloor i/2 \rfloor$ and $\langle t-1, \lfloor i/2 \rfloor + j_{m-t+1 \bmod m} 2^{m-1}\rangle \prec_{A_{ps}} \langle t, i\rangle$. It is now easy to check that really $\prec_{A_{ps}} = \prec_{ps}$. Thus, by proposition 7.17, $\langle S_{tbc}, \prec_{tbc}\rangle \overset{stt}{\leftrightarrow} \langle S_{ps}, \prec_{ps}\rangle$, and transitivity gives $\langle S_j, \prec_j\rangle \overset{stt}{\leftrightarrow} \langle S_{ps}, \prec_{ps}\rangle$.

So all sets of cell actions (cs-assignments) executable on any of the three structures are executable on the other two as well, and the mappings A_j, A_{ps} (and inverses) can be used to transform the schedules. Let us now schedule a FFT algorithm, computing the discrete Fourier transform $\sum_{k=0}^{N-1} a(k)\omega^{jk}$, $j = 0, \ldots, N-1$, $\omega = e^{2\pi i/N}$, $N = 2^m$, on the time-dependent binary $m-1$-cube:

Recurrence equation M (cf. [St71], j, k are expanded into binary representation):

$i_0, \ldots, i_{m-1} \in \{0,1\}$:

$$b_0(i_{m-1}, \ldots, i_0) \leftarrow a(\sum_{n=0}^{m-1} i_n 2^n)$$

$1 \le s \le m$:

$$b_s(i_{m-1}, \ldots, i_{m-s}, \ldots, i_0) \leftarrow$$

$$\sum_{k_{m-s} \in \{0,1\}} b_{s-1}(i_{m-1}, \ldots, k_{m-s}, \ldots, i_0)\omega^{(\sum_{n=0}^{s-1} i_{m-n-1} 2^n) k_{m-s} 2^{m-s}}$$

Result: $b_m(i_{m-1},\ldots,i_0) = \sum_{k=0}^{N-1} a(k)\omega^{jk}$, $j = \sum_{n=0}^{N} i_{N-n}2^n$ (binary reversed order)

The operations involved are addition and multiplication of complex numbers, and exponentiation of the N-th root of unity. An indexing function G from the CCSA-set M above to the index space $\{0,\ldots,m\} \times \{0,1\}^m$ can be defined in the following natural way: the assignments producing $b_s(i_{m-1},\ldots,i_0)$ are mapped to $(s, i_{m-1},\ldots,i_0)$ by G. This indexing function is 1-1.

The data dependence relation $\prec_M$ on M can be derived by identifying input and output variables of the assignments. This relation is mapped by G into the relation $\prec_{MG}$ on the index space. $\prec_{MG}$ has the following simple structure:

$i_0,\ldots,i_{m-1},k_{m-s} \in \{0,1\}$, $1 \leq s \leq m$:

$$(s-1, i_{m-1},\ldots,k_{m-s},\ldots,i_0) \prec_{MG} (s, i_{m-1},\ldots,i_{m-s},\ldots,i_0).$$

We map the index set $G(M)$ into $\langle S_{tbc}, \prec_{tbc}\rangle$ (of order m) by a weakly correct index mapping W as follows:

$i_0,\ldots,i_{m-1} \in \{0,1\}$:

$$W(0, i_{m-1},\ldots,i_0) = \langle 0, i_{m-2},\ldots,i_1,i_{m-1}\rangle$$

$i_0,\ldots,i_{m-1} \in \{0,1\}$, $s > 0$:

$$W(s, i_{m-1},\ldots,i_{m-s},\ldots,i_0) = \langle s, i_{m-1},\ldots,i_{m-s+1},i_{m-s-1},\ldots,i_0\rangle$$

The assignments in M are mapped pairwise by $W \circ G$ onto single points in S_{tbc}. Thus, we are running m-FFT on an $m-1$-cube. The reason is efficiency; the assignments mapped to the same point use the same indata and have subexpressions in common, so these subexpressions need not be evaluated twice, at different events.

Let us finally examine $\prec_{MW\circ G}$ on S_{tbc} in order to see that $\prec_{MW\circ G} \subseteq \prec_{tbc}$. $\prec_{MW\circ G}$ can be obtained from $\prec_{MG}$ and W, and the result is (renaming binary indices):

$i_0,\ldots,i_{m-2},k_{m-1} \in \{0,1\}$:

$$\langle 0, i_{m-2},\ldots,i_1,k_{m-1}\rangle \prec_{MW\circ G} \langle 1, i_{m-2},\ldots,i_0\rangle$$

$1 < s \leq m$, $i_0,\ldots,i_{m-2},j_{m-s} \in \{0,1\}$:

$$\langle s-1, i_{m-2},\ldots,j_{m-s},\ldots,i_0\rangle \prec_{MW\circ G} \langle s, i_{m-2},\ldots,i_{m-s},\ldots,i_0\rangle$$

It is easy to see that in fact $\prec_{MW\circ G} = \prec_{tbc}$. Therefore, this FFT algorithm is executable on all three space-time structures. $\langle S_j, \prec_j\rangle$ is especially interesting, since it has an advantage with respect to wiring; all connections are local in space. An implementation of a slightly different FFT algorithm on a similar network, described in terms of delay elements and functional units, is derived in [JC82]. FFT processors of this type have actually been built, for both radix-2 FFT [GW70] and radix-4 FFT [Sw86]. The example here shows that not only FFT, but *any* recursive partitioning-type algorithm that can be recursively described using binary indices and mapped to $\langle S_{tbc}, \prec_{tbc}\rangle$ as above can be scheduled on $\langle S_j, \prec_j\rangle$.

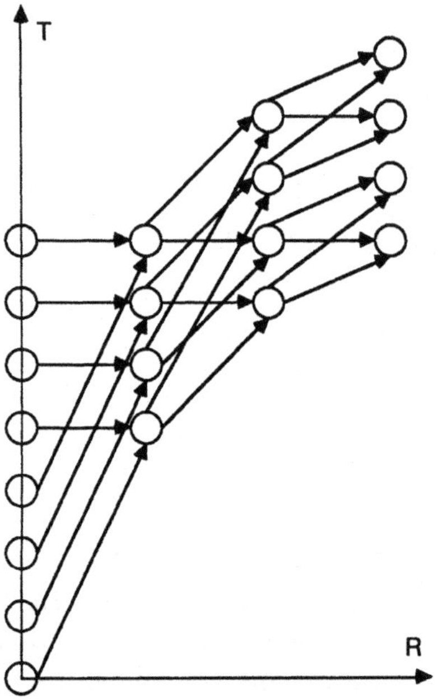

Figure 11.22: Space-time diagram for $\langle S, \prec_{MW' \circ G}\rangle$, $m = 3$.

Let us finally examine the FFT implementation on the communication structure $\langle S_j, \prec_j\rangle$ a bit closer. To be precise we will actually consider a slightly different communication ordering, that is somewhat easier to support by fixed hardware. We define an index mapping W' directly from the index set of M to S_j, according to the following:

$i_0, \ldots, i_{m-1} \in \{0, 1\}$:

$$W'(0, i_{m-1}, \ldots, i_0) = \langle \sum_{n=0}^{m-1} i_n 2^n, 0\rangle$$

$i_0, \ldots, i_{m-1} \in \{0,1\}$, $1 \le s \le m$:

$$W'(s, i_{m-1}, \ldots, i_0) = \langle \sum_{n=0}^{m-s-1} i_n 2^n + \sum_{n=m-s+1}^{m-1} i_n 2^{n-1} + 2^m - 2^{m-s}, s \rangle.$$

W' is identical to $A_{ps} \circ W$ except for $s = 0$. The index vectors $(0, i_{m-1}, \ldots, i_0)$ are mapped to $\langle \sum_{n=0}^{m-1} i_n 2^n, 0 \rangle$. These are the index vectors of the cs-assignments $b_0(0, i_{m-1}, \ldots, i_0) = a(i)$, where $i = \sum_{n=0}^{m-1} i_n 2^n$. Thus, W' gives a schedule where the coefficients $a(i)$ arrives as a time series at $x = 0$. The mapped precedence relation $\prec_{MW' \circ G}$ is easily seen to be a communication ordering. Figure 11.22 shows a space-time diagram for the resulting communication structure when $m = 3$.

When $\prec_{MW' \circ G}$ is determined, it is possible to find its fixed hardware projection $\Delta(\prec_{MW' \circ G})$. It consists of the following triples:

$$\langle 0, 1, 0 \rangle \qquad \langle 0, 1, 2^{m-1} \rangle$$

$1 < x \le m$:

$$\langle x-1, x, 0 \rangle \qquad \langle x-1, x, 2^{m-x} \rangle \qquad \langle x-1, x, 2^{m-x+1} \rangle$$

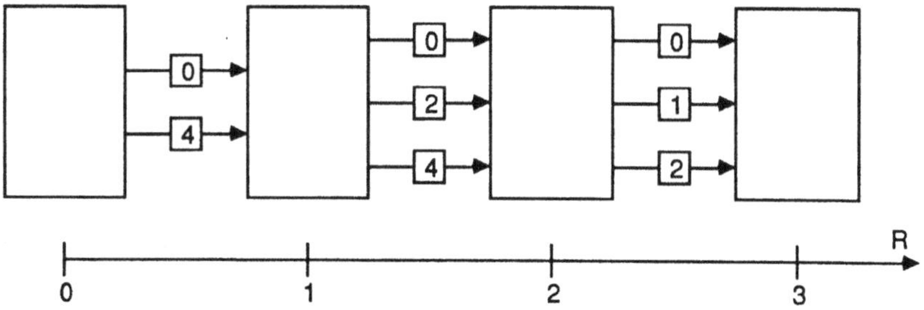

Figure 11.23: Graphical representation of $\Delta(\prec_{MW' \circ G})$ for $m = 3$.

The fixed hardware projection is a hardware description that concerns timing and spatial connections only. It does not describe what the nodes in the network are to perform, nor does it describe what is to be sent over the different lines at different events. In order to deal with these issues, a closer examination of the resulting cs-assignments in space-time and the communication between them must be made.

Let us first determine what the cells at different points in space must be able to do. The assignments producing $b_s(i_{m-1}, \ldots, i_{m-s+1}, 0, i_{m-s-1}, \ldots, i_0)$ and $b_s(i_{m-1}, \ldots, i_{m-s+1}, 1, i_{m-s-1}, \ldots, i_0)$ are, except for $s = 0$, mapped to the same point in space-time. The resulting merged assignment is, when $s > 1$,

$$b_s(i_{m-1},\ldots,i_{m-s+1},0,i_{m-s-1},\ldots,i_0) \leftarrow$$
$$\sum_{k_{m-s}\in\{0,1\}} b_{s-1}(i_{m-1},\ldots,k_{m-s},\ldots,i_0)\omega^{(\sum_{n=0}^{s-2} i_{m-n-1}2^n)k_{m-s}2^{m-s}}$$
$$b_s(i_{m-1},\ldots,i_{m-s+1},1,i_{m-s-1},\ldots,i_0) \leftarrow$$
$$\sum_{k_{m-s}\in\{0,1\}} b_{s-1}(i_{m-1},\ldots,k_{m-s},\ldots,i_0)\omega^{(2^{s-1}+\sum_{n=0}^{s-2} i_{m-n-1}2^n)k_{m-s}2^{m-s}}.$$

If we adopt the convention that $\sum_{n=0}^{-1} x_n = 0$ for any sequence $\{x_n\}_0^\infty$, then the merged assignment above is valid also for $s = 1$. Let us write the terms in the sums explicitly for $k_{m-s} = 0, 1$. The first sum becomes

$$b_{s-1}(i_{m-1},\ldots,i_{m-s+1},0,i_{m-s-1},\ldots,i_0)\omega^0+$$
$$b_{s-1}(i_{m-1},\ldots,i_{m-s+1},1,i_{m-s-1},\ldots,i_0)\omega^{(\sum_{n=0}^{s-2} i_{m-n-1}2^n)2^{m-s}} =$$
$$b_{s-1}(i_{m-1},\ldots,i_{m-s+1},0,i_{m-s-1},\ldots,i_0)+$$
$$b_{s-1}(i_{m-1},\ldots,i_{m-s+1},1,i_{m-s-1},\ldots,i_0)\omega^{2^{m-s}(\sum_{n=0}^{s-2} i_{m-n-1}2^n)}$$

and the second

$$b_{s-1}(i_{m-1},\ldots,i_{m-s+1},0,i_{m-s-1},\ldots,i_0)\omega^0+$$
$$b_{s-1}(i_{m-1},\ldots,i_{m-s+1},1,i_{m-s-1},\ldots,i_0)\omega^{(2^{s-1}+\sum_{n=0}^{s-2} i_{m-n-1}2^n)2^{m-s}} =$$
$$b_{s-1}(i_{m-1},\ldots,i_{m-s+1},0,i_{m-s-1},\ldots,i_0)+$$
$$b_{s-1}(i_{m-1},\ldots,i_{m-s+1},1,i_{m-s-1},\ldots,i_0)\omega^{2^{m-s}2^{s-1}+2^{m-s}(\sum_{n=0}^{s-2} i_{m-n-1}2^n)} =$$
$$(2^{m-s}2^{s-1} = 2^{m-1} = N/2,\ \omega^{N/2} = -1) =$$
$$b_{s-1}(i_{m-1},\ldots,i_{m-s+1},0,i_{m-s-1},\ldots,i_0)-$$
$$b_{s-1}(i_{m-1},\ldots,i_{m-s+1},1,i_{m-s-1},\ldots,i_0)\omega^{2^{m-s}(\sum_{n=0}^{s-2} i_{m-n-1}2^n)}.$$

Thus, the cs-assignment can be written

$$\begin{aligned}
&b_s(i_{m-1},\ldots,i_{m-s+1},0,i_{m-s-1},\ldots,i_0) \leftarrow \\
&\quad b_{s-1}(i_{m-1},\ldots,i_{m-s+1},0,i_{m-s-1},\ldots,i_0)+ \\
&\quad b_{s-1}(i_{m-1},\ldots,i_{m-s+1},1,i_{m-s-1},\ldots,i_0)\omega^{2^{m-s}(\sum_{n=0}^{s-2} i_{m-n-1}2^n)} \\
&b_s(i_{m-1},\ldots,i_{m-s+1},1,i_{m-s-1},\ldots,i_0) \leftarrow \\
&\quad b_{s-1}(i_{m-1},\ldots,i_{m-s+1},0,i_{m-s-1},\ldots,i_0)- \\
&\quad b_{s-1}(i_{m-1},\ldots,i_{m-s+1},1,i_{m-s-1},\ldots,i_0)\omega^{2^{m-s}(\sum_{n=0}^{s-2} i_{m-n-1}2^n)}.
\end{aligned}$$

Let us denote this assignment by a_{tx}, where $\langle t, x\rangle$ are the space-time coordinates given by W'. Consider the substitution

$$\sigma_x = \{b^{out}_{x,0} \leftarrow b^{in}_{x-1,0} + b^{in}_{x-1,1}w, b^{out}_{x,1} \leftarrow b^{in}_{x-1,0} - b^{in}_{x-1,1}w\}.$$

It is easily seen, that for all t, x under consideration holds that $\sigma_x \geq_\phi a_{tx}$. Thus, σ_x is a covering substitution for all x, and the cell at x can be designed to implement σ_x.

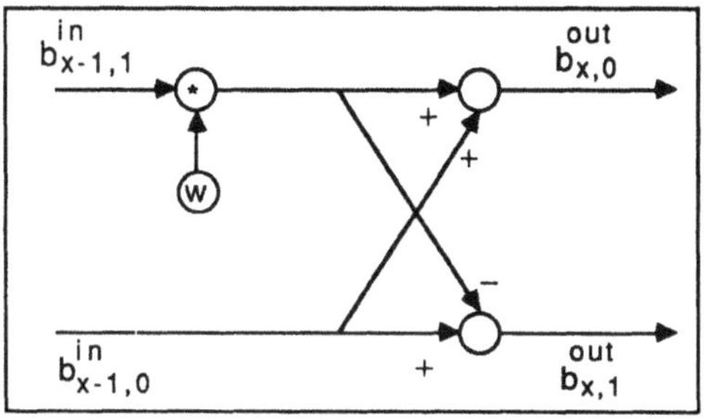

Figure 11.24: Inner structure of cell, $x > 1$

Note that the variable w matches the powers of ω. These are constants with respect to the input, but we have to provide them at the proper times and places as inputs to w. For transforms of fixed size and at a given cell x, this input becomes a function of time only, and it can be determined from a_{tx} and W'. The powers of ω that are to be used in a certain cell can therefore be calculated in advance, stored in local memory and then loaded into w as the execution goes on.

For $x = s = 1$ there is a simpler covering substitution than σ_x above. In this case $\omega^{2^{m-s}(\sum_{n=0}^{s-2} i_{m-n-1}2^n)} = \omega^{2^{m-s}(\sum_{n=0}^{-1} i_{m-n-1}2^n)} = \omega^0 = 1$, and thus a_{t1} can be simplified to

$$b_1(0, i_{m-2}, \ldots, i_0) \leftarrow b_0(0, i_{m-2}, \ldots, i_0) + b_0(1, i_{m-2}, \ldots, i_0)$$
$$b_1(1, i_{m-2}, \ldots, i_0) \leftarrow b_0(0, i_{m-2}, \ldots, i_0) - b_0(1, i_{m-2}, \ldots, i_0).$$

It follows that

$$\sigma_1 = \{b^{out}_{1,0} \leftarrow b^{in}_{0,0} + b^{in}_{0,1}, b^{out}_{1,1} \leftarrow b^{in}_{0,0} - b^{in}_{0,1}\}$$

is a covering substitution of $x = 1$.

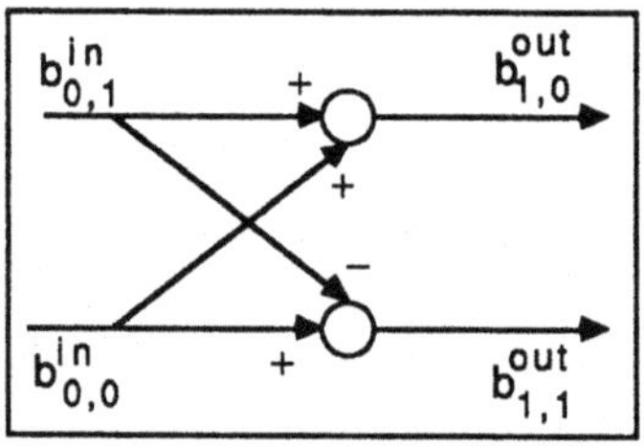

Figure 11.25: Inner structure of cell, $x = 1$

Let us now analyze the precise pattern of communication and derive a way to support it in hardware. How should the lines $b^{out}_{x,0}$, $b^{out}_{x,1}$, $b^{in}_{x,0}$, $b^{in}_{x,1}$ be connected?

First we consider the input to the cell at $x = 1$. This input arrives from the cell at $x = 0$. The assignments mapped to $x = 0$ are all of the form $\{b \leftarrow a\}$, where a is a free variable, so this cell is a simple input pad where the time series $a(k)$, $k = 1, \ldots, N-1$ arrives. For every a_{t1} mapped to $x = 1$, $b_0(0, i_{m-2}, \ldots, i_0)$ is matched against $b^{in}_{0,0}$ in σ_1 and $b_0(1, i_{m-2}, \ldots, i_0)$ is matched against $b^{in}_{0,1}$. $b_0(0, i_{m-2}, \ldots, i_0)$ is produced at $\langle \sum_{n=0}^{m-2} i_n 2^n, 0\rangle$, $b_0(1, i_{m-2}, \ldots, i_0)$ is produced at $\langle \sum_{n=0}^{m-2} i_n 2^n + 2^{m-1}, 0\rangle$ and they are both consumed at $\langle \sum_{n=0}^{m-2} i_n 2^n + 2^m - 2^{m-1}, 1\rangle = \langle \sum_{n=0}^{m-2} i_n 2^n + 2^{m-1}, 1\rangle$. The difference (design vector) for $b_0(0, i_{m-2}, \ldots, i_0)$ is $\langle 2^{m\ 1}, 1\rangle$ and for $b_0(1, i_{m-2}, \ldots, i_0)$ it is $\langle 0, 1\rangle$. Thus, $b^{in}_{0,0}$ can be hardwired to a line from the input pad at $x = 0$ with delay 2^{m-1} and $b^{in}_{0,1}$ can be hardwired to a line from $x = 0$ with zero delay.

Let us now consider the communication between the cells $x - 1$ and x, where $x > 1$. How should the output lines $b^{out}_{x-1,0}$, $b^{out}_{x-1,1}$ of σ_{x-1} and the input lines $b^{in}_{x-1,0}$, $b^{in}_{x-1,1}$ of σ_x be connected?

For every a_{t1} mapped to $x = s$, $b_{s-1}(i_{m-1}, \ldots, i_{m-s+1}, 0, i_{m-s-1}, \ldots, i_0)$ is matched against $b^{in}_{x-1,0}$ in σ_x and $b_{s-1}(i_{m-1}, \ldots, i_{m-s+1}, 1, i_{m-s-1}, \ldots, i_0)$

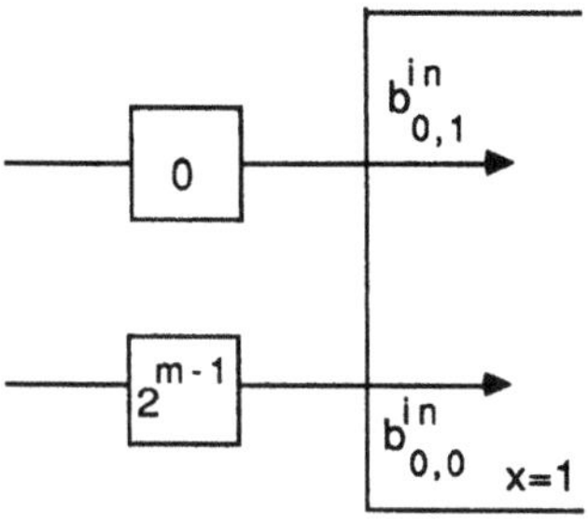

Figure 11.26: Input wiring of cell at $x = 1$.

is matched against $b^{in}_{x-1,1}$ in σ_x. $b_{s-1}(i_{m-1}, \ldots, i_{m-s+1}, k_{m-s}, i_{m-s-1}, \ldots, i_0)$ for $k_{m-s} = 0, 1$ are produced at

$$\langle \sum_{n=0}^{m-s-1} i_n 2^n + k_{m-s} 2^{m-s} + \sum_{n=m-(s-1)+1}^{m-1} i_n 2^{n-1} + 2^m - 2^{m-(s-1)}, s-1 \rangle$$

$$= \langle \sum_{n=0}^{m-s-1} i_n 2^n + k_{m-s} 2^{m-s} + \sum_{n=m-s+2}^{m-1} i_n 2^{n-1} + 2^m - 2^{m-s+1}, s-1 \rangle$$

and they are both used at

$$\langle \sum_{n=0}^{m-s-1} i_n 2^n + \sum_{n=m-s+1}^{m-1} i_n 2^{n-1} + 2^m - 2^{m-s}, s \rangle.$$

The difference is

$$\langle i_{m-s+1} 2^{m-s} - k_{m-s} 2^{m-s} - 2^{m-s} + 2^{m-s+1}, 1 \rangle =$$
$$\langle (i_{m-s+1} - k_{m-s}) 2^{m-s} + 2^{m-s}, 1 \rangle.$$

This yields the following design vectors for the different variables:

$b_{s-1}(i_{m-1}, \ldots, i_{m-s+1}, 0, i_{m-s-1}, \ldots, i_0)$, matched against $b^{in}_{x-1,0}$:

$$\langle 2^{m-s}, 1 \rangle \text{ when } i_{m-s+1} = 0 \qquad \langle 2^{m-s+1}, 1 \rangle \text{ when } i_{m-s+1} = 1$$

$b_{s-1}(i_{m-1}, \ldots, i_{m-s+1}, 1, i_{m-s-1}, \ldots, i_0)$, matched against $b^{in}_{x-1,1}$:

$$\langle 0, 1 \rangle \text{ when } i_{m-s+1} = 0 \qquad \langle 2^{m-s}, 1 \rangle \text{ when } i_{m-s+1} = 1$$

Thus, the input line $b^{in}_{x-1,0}$ is alternately connected to an output of the cell at $x - 1$ through a line with delay $2^{m-s} = 2^{m-x}$ and through a line with delay 2^{m-x+1}. $b^{in}_{x-1,1}$ is alternately connected through a line with zero delay and through a line with delay 2^{m-x}. We can note that the system thus does not have the fixed 1-i/o-property.

Which output line of σ_{x-1} will provide the input to the respective input lines of σ_x at different times? $b_{s-1}(i_{m-1},\ldots,i_{m-s+2},0,k_{m-s},\ldots,i_0)$ is matched against $b^{out}_{x-1,0}$ in σ_{x-1} and $b_{s-1}(i_{m-1},\ldots,i_{m-s+2},1,k_{m-s},\ldots,i_0)$ is matched against $b^{out}_{x-1,1}$. So when $b_{s-1}(i_{m-1},\ldots,i_{m-s+2},0,k_{m-s},\ldots,i_0)$, for $k_{m-s}=0,1$ are recieved at x, then both $b^{in}_{x-1,0}$ and $b^{in}_{x-1,1}$ recieve their inputs from $b^{out}_{x-1,0}$ in σ_{x-1} (but through wires with different delay, so they do not receive the same value), and when $b_{s-1}(i_{m-1},\ldots,i_{m-s+2},1,k_{m-s},\ldots,i_0)$ are recieved, then both $b^{in}_{x-1,0}$ and $b^{in}_{x-1,1}$ recieve their inputs from $b^{out}_{x-1,1}$.

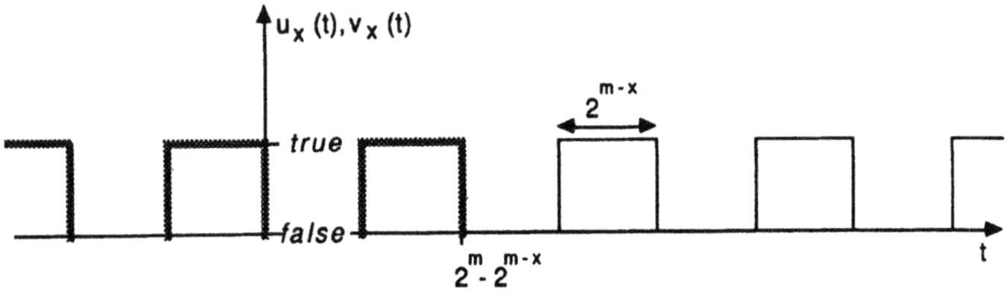

Figure 11.27: $v_x(t)$, $u_x(t)$

At which times are index vectors i with $i_{m-s+1}=0$ and 1, respectively, mapped to $x(i)$? Since $x=s$, the timing function for such index vectors becomes

$$t=\sum_{n=0}^{m-s-1} i_n 2^n + i_{m-s+1}2^{m-s} + \sum_{n=m-s+2}^{m-1} i_n 2^{n-1} + 2^m - 2^{m-s}$$
$$= k_a + i_{m-x+1}2^{m-x} + k_b 2^{m-x+1} + 2^m - 2^{m-x}$$

where $0 \le k_a \le 2^{m-x}-1$ and k_b is a nonnegative integer. From this follows that the following function $v_x\colon T \to \{true, false\}$ defined by:

$$v_x(t)=\begin{cases} false, & t<2^m-2^{m-x} \\ false, & t=k_a+k_b 2^{m-x+1}+2^m-2^{m-x} \\ true, & t=k_a+k_b 2^{m-x+1}+2^m \end{cases}$$

with k_a, k_b as above, is true exactly when an index vector with $i_{m-x+1}=1$ is mapped to x.

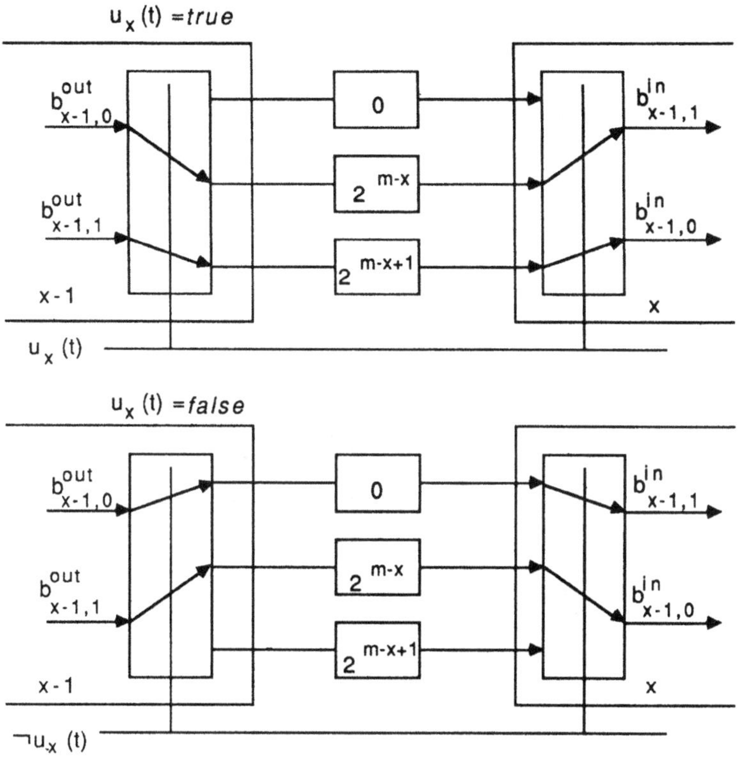

Figure 11.28: Control of communication lines between adjacent cells.

v_x for $t \geq 2^m - 2^{m-x}$ can be periodically extended to a function $u_x: Z \rightarrow \{true, false\}$ with the property that $u_x(t+2^m) = \neg u_x(t)$ for all times t.

Now we have all information necessary to derive how the input lines of σ_x and the output lines of σ_{x-1} should be connected to the communication lines $\langle 0, x-1, x\rangle$, $\langle 2^{m-x}, x-1, x\rangle$, $\langle 2^{m-x+1}, x-1, x\rangle$ at different times. When $i_{m-x+1} = 0$ at x, or when $\neg u_x(t)$, then $b^{in}_{x-1,0}$ is connected to $\langle 2^{m-x}, x-1, x\rangle$. Thus, $b^{out}_{x-1,0}$ must be connected to $\langle 2^{m-x}, x-1, x\rangle$ when $\neg u_x(t+2^{m-x}) = u_x(t)$. $b^{in}_{x-1,1}$ is connected to $\langle 0, x-1, x\rangle$ when $\neg u_x(t)$. Thus, $b^{out}_{x-1,0}$ must be connected to $\langle 0, x-1, x\rangle$ when $\neg u_x(t+0) = \neg u_x(t)$. When $i_{m-x+1} = 1$ at x, or when $u_x(t)$, then $b^{in}_{x-1,0}$ is connected to $\langle 2^{m-x+1}, x-1, x\rangle$. Thus, $b^{out}_{x-1,1}$ must be connected to $\langle 2^{m-x+1}, x-1, x\rangle$ when $u_x(t+2^{m-x+1}) = u_x(t)$. $b^{in}_{x-1,1}$ is connected to $\langle 2^{m-x}, x-1, x\rangle$ when $u_x(t)$. Thus, $b^{out}_{x-1,1}$ must be connected to $\langle 2^{m-x}, x-1, x\rangle$ when $u_x(t+2^{m-x}) = \neg u_x(t)$.

The result of the analysis above is that the input and output line two adjacent cells can be switched between the different lines by the same control signal, as indicated in figure 11.28.

Let us finally mention that a limited pipelining of FFT's is possible on the array. For an infinite time series $\{a_n\}_{n=0}^{\infty}$ it is possible to perform a sliding window Fourier transform, $\sum_{k=0}^{N-1} a(k+\frac{N}{2}l)\omega^{jk}$, for $l = 0, 1, \ldots$, that is: every $N/2$ time the array can start computing a new spectrum.

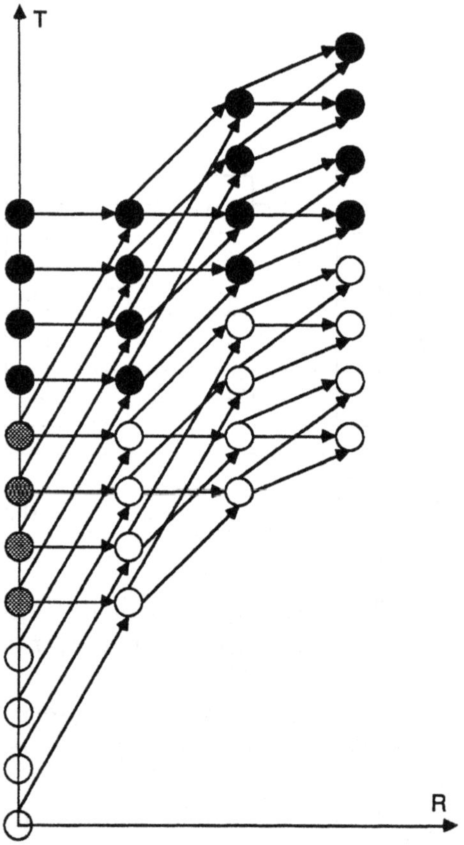

Figure 11.29: Pipelining two sliding window transforms

The correctness of this statement can be intuitively concluded from the space-time diagram in figure 11.29. It can also be deduced by formal means, using the methods developed here.

12. Conclusions

The space-time mapping paradigm seems to be a promising approach to synthesis of fixed parallel hardware, and it is also likely to have applications in compilation techniques for parallel computers. The recent surge of papers on synthesis by space-time mapping methods confirms this statement. Until now, however, most papers in the area have been quite application-oriented. Thus, they have not been concerned with theoretical issues to any further extent than required by the area of application.

Here, we have provided a stringent mathematical formulation of the synthesis process, all the way from a functional specification to the resulting cellular hardware description. What we gain is the following:

- A generalization of previously considered classes of algorithms, such as those defined by URE's or similar systems of recursion equations, that are possible to implement via space-time mappings.
- A classification of the functions that are possible to implement by space-time methods, in a data-independent and thus efficient fashion; these functions are exactly the polynomials over the algebra, where the operations we consider as basic are operators.
- A stringent, but still natural definition of data dependence, that generalizes the notion of data dependence vectors.
- A more general description of communication constraints than has been considered previously. Communication orderings generalize design vectors in very much the same way as sets of cell actions generalize URE's, and they allow the description of irregular and time-varying, but still predetermined communication patterns.
- A generalization of the class of space-time mappings from the simple, linear or affine injective mappings that most authors have considered so far. This generalization is in line with the other generalizations and it makes it possible to describe synthesis of more general, irregular but still fully data-driven, synchronous and thus potentially very efficient parallel systems.
- A formal description, as substitution fields in space-time, of the derived cellular hardware with local programs executing synchronously in each cell. Once a space-time mapping is determined, such a substitution field is established. The mathematical formalization of the transition from algorithm to parallel implementation facilitates the automating of parts of this process and makes it possible to trust the result. It also enhances the understanding and reveals the connections between algorithm, space-time mapping and possible degree of hardwiring in the resulting implementation.

The emphasis of this thesis is on stringent definitions, basic theorems and finding the proper concepts rather than developing algorithms and practical techniques for automated synthesis. When a firm theoretical foundation is laid, however, such algorithms can be formulated in this framework and developed to their full generality. A feasible method, based on the theory in this thesis, for finding time-optimal systolic arrays under fairly general constraints, has been developed since this thesis was written, [Li87]. Furthermore, a general theory makes it possible to find out what is *not* feasible, for instance when excessive combinatorial complexity or even undecidability puts limits to what can be done.

The chapter about applications shows the virtues of a general approach. The treatment of the extensively used model problem of matrix multiplication yields new, interesting solutions exactly because of the freedom in the algorithm description; when the multiplication algorithm is forced into a uniform, less general form, then these perfectly acceptable solutions are lost. The Fast Fourier Transform example is another case, where the previously considered models are insufficient to describe the synthesis process and the resulting implementation. The derived linear array for FFT computations should be of practical interest.

The FFT array design demonstrates the relevance of such a concept as the fixed i/o-property; even though each cell performs the same action at every time, and thus the function unit can be hardwired, the cell receives and sends values on different lines depending on the phase of computation. Thus, the design does not have the fixed i/o-property. The proposed design overcomes this in an efficient fashion, but in general this may be costly. Previous works have neglected this potential problem altogether, simply because their formalisms have been too limited to express cases where it occurs. When a cell *does* have the fixed i/o-property, however, then the input and output lines can be hardwired to the ports of the functional units as well.

We also considered how previously proposed transformation methods for synchronous hardware can be generalized, in a way that fits the rest of the theory developed here. The result is the concept of space-time transformations. These take communication structures into new communication structures. Space-time transformations can be seen as a generalization, where the time is taken into account, of network embeddings. These transformations can also be used to prove equivalence between communication structures, like for instance between the communication structure generated by the perfect shuffle network and the communication ordering for the linear FFT array. This result is interesting in itself, since the perfect shuffle network is a well-known interconnection topology and the linear array provides a VLSI-friendly way to implement all the algorithms that can be carried out on this network.

Space-time mapping methods also seem to be of use when compiling programs for execution on synchronous parallel computers. Such computer systems are common. Pipelined vector units, for instance, are synchronous. SIMD machines work in a lock-step fashion and can thus also be treated as synchronous. The same holds for VLIW architectures. Even more general synchronous architectures, like Concurrent Read - Concurrent Write Parallel Random Access Memory (CRCW PRAM) systems, seem to be realizable; Ranade [Ran87] has proposed an efficient way to emulate a synchronous CRCW PRAM machine on a multiprocessor network with distributed memory. For all these architectures, space-time mapping methods can be applied if the various constraints of the different architectures are taken into account.

Of course, space-time mapping methods applied to compilation are essentially limited to the parts of the algorithms that have a data-independent structure. In an imperative program this corresponds to straight-line code and branching-free loops. Interestingly, these are the parts of such a program that a vectorizing or parallelizing compiler can treat most successfully. It is quite obvious that the theory developed here can be used to describe the transformations performed by such compilers. Hopefully, the theory can be used to find even more powerful compilation techniques.

Another interesting issue is what type of programming language to use for parallel architectures. From a space-time mapping point of view, single-assignment languages are of particular interest. Single-assignment programs without branching and iteration are direct syntactical representations of CCSA-sets. Thus, space-time mapping methods can be applied almost directly to the compilation of such programs.

Existing single-assignment languages, like VAL [Mc82], lack appropriate constructs to express regular computations in a compact, parameterized manner. The "forall" construct in VAL, for instance, does not allow any data dependencies between the assignments that are specified. Therefore, this construct is not sufficient for expressing any of the CCSA-sets considered in chapter 11. It would be interesting to have a more general forall construct that allows parameterized assignments with possible dependencies. Of course, this freedom of expression gives rise to some new difficulties: for instance, one will have to verify that a specified set of assignments really is a CCSA-set (i.e. has the single-assignment property and is causal). This is clearly not doable always.

There is also a need for appropriate constructs to express data-dependent parts of a computation, that is, branching and iteration. Those present in the existing single-assignment languages known to the author seem to be quite limited. This is an interesting field for further research. The reward may be a language that is both suitable for expressing regular algorithms, like for instance those found in numerical linear algebra, and possible to compile efficiently to synchronous parallel architectures.

Acknowledgements

This is the second hardest part of the thesis to write. (The hardest was to find an appropriate title within a limited length.) The reason why I find it so hard to write is that acknowledgements often get so horribly personal; if all persons who have significantly contributed to my development are to be mentioned, then the list would be quite long and I would also run into problems about whom to mention and whom not to mention without stepping on somebody's toe. ('Should that nurse from kindergarten, whose name I have for ever forgotten, be mentioned for giving me access to pedagogical toys by the aid of which I learned to read and write?') Besides, expressing personal affection in public always makes me feel so insecure. In order to keep problems of this type at a minimum, I have decided to keep the acknowledgements strictly professional and only to mention those who have been of importance to me in matters directly related to this thesis. (This does of course not mean that I am not grateful to all other persons who may have expected to be mentioned here. I have not forgotten you, I promise ...)

First I want to thank my advisor Stefan Arnborg for his guidance and support through my years as a graduate student, and for putting me on the track that eventually led to this thesis.

I am also very grateful to Robert Schreiber for making my stay at Stanford University in 1984 possible and for his encouraging support. This visit was very rewarding.

I want to thank my room-mates and graduate student colleagues through the years for providing a creative environment in which it has been a pleasure to work. Especially I want to thank Erik Tidén for many interesting discussions.

I am also grateful for the practical support I have received, without which there certainly would not have been any thesis at all to fill with acknowledgements. I therefore want to thank the personnel at NADA in general, I want to thank the system group of NADA for promptly restoring the source files for this thesis when our computer utilities had a very low possible utilization, and I want to thank Anders Ulfheden and my colleague Lars Langemyr for system assistance during the last days of thesis work when things were getting real tough. When speaking about practicalities, I also want to acknowledge the financial support I have received from STU (the Swedish Board for Technical Development) as a member of the MYPROG research lab.

Finally, and here I without any excuse abandon my intention of keeping these acknowledgements professional, I want to thank all those undoubtedly unlucky ladies who for some reason or another have come to the unpleasant situation of trying to keep a decent table conversation at social occasions with the author of this thesis:

– "So, you are a graduate student. How exciting! What is your subject of study?"

– "Eh, it is a bit hard to explain. First there are some basic definitions ..."

References

[Be72] J. C. Beatty. An axiomatic approach to code optimization for expressions. *J. ACM*, 19(4):613–640, October 1972.

[BL70] G. Birkhoff and J. D. Lipson. Heterogeneous algebras. *J. Combin. Theory*, 8:115–133, January 1970.

[Br74] R. Brent. The parallel evaluation of general arithmetic expressions. *J. ACM*, 21(2):201–206, 1972.

[BG81] R. M. Burstall and J. A. Goguen. Algebras, theories and freeness: an introduction for computer scientists. *Proc. of Marktoberdorf Summer School on Theoretical Foundations of Programming Methodology*, Reidel, August 1981.

[CS83] P. K. Cappello and K. Steiglitz. *Unifying VLSI Array Design with Linear Transformations of Space-Time.* Research Report TRCS83-03, Dept. Comput. Sci., UCSB, 1983.

[C83] M. C. Chen. *Space-Time Algorithms: Semantics and Methodology.* PhD thesis, Dept. Comput. Sci., California Institute of Technology, 1983.

[C85A] M. C. Chen. *A Synthesis Method for Systolic Designs.* Research Report YALEU/DCS/RR-334, Dept. Comput. Sci., Yale University, January 1985.

[C85B] M. C. Chen. *Synthesizing Systolic Designs.* Research Report YALEU/DCS/RR-374, Dept. Comput. Sci., Yale University, March 1985.

[C86A] M. C. Chen. Transformation of parallel programs in Crystal. In H.-J. Kugler editor, *INFORMATION PROCESSING 86*, pages 455–462, Elsevier Publishers B.V. (North-Holland), 1986.

[C86B] M. C. Chen. *A Design Methodology for Synthesizing Parallel Algorithms and Architectures.* Research Report YALEU/DCS/TR-457, Dept. Comput. Sci., Yale University, June 1986.

[C86C] M. C. Chen. *A Synthesis Approach to the Design and Correctness of Systolic Computations.* Research Report YALEU/DCS/TR-477, Dept. Comput. Sci., Yale University, June 1986.

[CM83] M.-Y. Chern and T. Murata. Efficient matrix multiplication on a concurrent data-loading array processor. *Proc. IEEE Int. Conf. on Parallel Processing*, pages 90–94, Bellaire, MI, August 1983.

[CF84] K. Culik II and I. Fris. *Topological Transformations As a Tool in the Design of Systolic Networks.* Report CS-84-11, Dept. Comput. Sci., University of Waterloo, April 1984.

[D84] P. E. Danielsson. Serial/parallel convolvers. *IEEE Trans. Comput.*, C-33:652–667, July 1984.

[DK82] A. L. Davis and R. M. Keller. Data flow program graphs. *Computer*, 15:26–41, February 1982.

[DBA74] G. Dahlquist, Å. Björck and N. Anderson. *Numerical Methods.* Prentice-Hall, Englewood Cliffs, NJ, 1974.

[DI86] J.-M. Delosme and I. Ipsen. Efficient systolic arrays for the solution of Toeplitz systems: an illustration of a methodology for the construction of systolic architectures in VLSI. In W. Moore, A. McCabe and R. Urquhart, editors, *Systolic Arrays*, pages 37–46, Adam Hilger, Bristol and Boston, 1986.

[GJ79] M. R. Garey and D. S. Johnson. *A Guide to the Theory of NP-Completeness.* Freeman, San Francisco, CA, 1979.

[G79] G. Grätzer. *Universal Algebra.* Springer–Verlag, New York, 1979.

[GW70] H. L. Groginsky and G. A. Works. A pipeline fast Fourier transform. *IEEE Trans. Comput.*, C-19(11):1015–1019, November 1970.

[H76] G. Huet. *Résolution d'Équations dans des Languages d'Ordre 1, 2, ..., ω.* PhD thesis, l'Université Paris VII, 1976.

[H80] G. Huet. Confluent reductions: abstract properties and applications to term rewriting systems. *J. ACM*, 27(4):797–821, October 1980.

[HO80] G. Huet and D. C. Oppen. Equations and rewrite rules: a survey. In R. Book, editor, *Formal Languages: Perspectives and Open Problems*, Academic Press, 1980

[HJ81] R. W. Hockney and C. R. Jesshope. *Parallel Computers.* Adam Hilger Ltd., 1981.

[JRK87] H. V. Jagadish, S. K. Rao and T. Kailath. Array architectures for iterative algorithms. *Proc. of the IEEE*, 75(9):1304–1321, September 1987.

[JC82] L. Johnsson and D. Cohen. A formal derivation of array implementation of FFT algorithms. *Proc. USC Workshop on VLSI and Modern Signal Processing*, pages 53–63, November 1982.

[JWCD81] L. Johnsson, U. Weiser, D. Cohen and A. L. Davis. Towards a formal treatment of VLSI arrays. *Proc. Second Caltech Conf. on VLSI*, pages 375–398, January 1981.

[Kah74] G. Kahn. The semantics of a simple language for parallel programming. *Information Processing 74: Proceedings of the IFIP Congress 74*, pages 471–475, 1974.

[Kat82] E. Katona. Cellular algorithms for binary matrix operations. *Proc. CONPAR-81*, pages 203–216, Springer-Verlag, 1981.

[KMW69] R. M. Karp, R. E. Miller and S. Winograd. The organization of computations for uniform recurrence equations. *J. ACM*, 14(3):563–590, July 1969.

[Ke77A] R. M. Keller. *Denotational Models for Parallel Programs with Indeterminate Operators.* Technical Report UUCS-77-103, Dept. Comput. Sci., University of Utah, 1977.

[Ke77B] R. M. Keller. *Semantics of Parallel Program Graphs.* Technical Report UUCS-77-103, Dept. Comput. Sci., University of Utah, 1977.

[Kn69] D. E. Knuth. *The Art of Computer Programming.* Vol. 1, Addison-Wesley, Reading, MA, 1969.

[KM74] D. J. Kuck and Y. Muraoka. Bounds on the parallel evaluation of arithmetic expressions using associativity and commutativity. *Acta Inf.*, 3:203–216, 1974.

[KMC72] D. J. Kuck, Y. Muraoka and S. C. Chen. On the number of operations simultaneously executable in Fortran-like programs and their resulting speedup. *IEEE Trans. Comput.*, C-21:1293–1310, December 1972.

[Ku86] S.-Y. Kung. VLSI array processors. In W. Moore, A. McCabe and R. Urquhart, editors, *Systolic Arrays*, pages 7–24, Adam Hilger, Bristol and Boston, 1986.

[KLe80] H. T. Kung and C. E. Leiserson. Algorithms for VLSI processor arrays. Chapter 8.3 in C. Mead and L. Conway, *Introduction to VLSI systems*, Addison-Wesley, Reading, MA, 1980.

[KLi83] H. T. Kung and W. T. Lin. *An Algebra for VLSI Algorithm Design.* Technical Report, Dept. Comput. Sci., Carnegie-Mellon University, PA, 1983.

[La74] L. Lamport. The parallel execution of DO loops. *Comm. ACM* 17:83–93, February 1974.

[LS81] C. E. Leiserson and J. B. Saxe. Optimizing synchronous systems. *Proc. of 22nd Ann. FOCA Symposium*, pages 23–36, 1981.

[Le83] H. Lev-Ari. *Modular Computing Networks: a New Methodology for Analysis and Design of Parallel Algorithms/Architectures.* ISI Report 29, Integrated Systems Inc., December 1983.

[Le84] H. Lev-Ari. *Canonical Realizations of Completely Regular Modular Computing Networks.* ISI Report 41, Integrated Systems Inc., May 1984.

[LW85] G.-J. Li and B. W. Wah. The design of optimal systolic arrays. *IEEE Trans. Comput.*, C-34:66–77, January 1985.

[Li83] B. Lisper. *Description and Synthesis of Systolic Arrays.* Technical Report TRITA-NA-8318, NADA, KTH, 1983.

[Li85] B. Lisper. *Hardware Synthesis from Specification with Polynomials.* Technical Report TRITA-NA-8506, NADA, KTH, 1985.

[Li87] B. Lisper. *Time-optimal Synthesis of Systolic Arrays with Pipelined Cells.* Research Report YALEU/DCS/TR-560, Dept. Comput. Sci., Yale University, September 1987.

[Li88] B. Lisper. Synthesis and equivalence of concurrent systems. *Theoretical Computer Science*, 58:183–199, 1988.

[Mc82] J. R. McGraw. The VAL language: description and analysis. *ACM Trans. Program. Lang. Syst.*, 4(1):44–82, January 1982.

[Ma74] Z. Manna. *Mathematical Theory of Computation.* McGraw-Hill, New York, 1974.

[MaWa84] Z. Manna and R. Waldinger. *The Logical Basis for Computer Programming, Volume II: Deductive Systems.* Preliminary version, to be published by Addison-Wesley.

[MaWa85] Z. Manna and R. Waldinger. *The Logical Basis for Computer Programming, Volume I: Deductive Reasoning.* Addison-Wesley, Reading, MA, 1985.

[MR82] R. G. Melhem and W. C. Rheinboldt. *A Mathematical Model for the Verification of Systolic Networks.* Technical Report ICMA-82-47, Institute for Computational Mathematics & Appl., University of Pittsburgh, 1982.

[Me83] R. G. Melhem. *Formal Verification of a Systolic System for Finite Element Stiffness Matrices.* Technical Report ICMA-83-56, Institute for Computational Mathematics & Appl., Univ. of Pittsburgh, 1983.

[MiWi84] W. L. Miranker and A. Winkler. Spacetime representations of computational structures. *Computing*, 32:93–114, 1984.

[Mo82] D. I. Moldovan. On the analysis and synthesis of VLSI algorithms. *IEEE Trans. Comput.*, C-31:1121–1126, October 1982.

[MF86] D. I. Moldovan and J. A. B. Fortes. Partitioning and mapping algorithms in fixed size systolic arrays. *IEEE Trans. Comput.*, C-35:1–12, January 1986.

[Mu71] Y. Muraoka. *Parallelism Exposure and Exploitation in Programs.* Ph. D. thesis, Dept. Comput. Sci., University of Illinois at Urbana-Champaign, 1971.

[NPW81] M. Nielsen, G. Plotkin and G. Winskel. Petri nets, event structures and domains, part I. *Theoretical Computer Science*, 13:85–108, 1981.

[PA85] K. Pingali and Arvind. Efficient demand-driven evaluation, part 1. *ACM Trans. Program. Lang. Syst.*, 7:311–333, 1985.

[PV80] F. P. Preparata and J. E. Vuillemin. Area-time optimal VLSI networks for multiplying matrices. *Inform. Process. Lett.*, 11(2):77–80, 1980.

[Q83] P. Quinton. *The Systematic Design of Systolic Arrays.* Research Report RR 216, INRIA, Rennes, July 1983.

[QG85] P. Quinton and P. Gachet. *Automatic Design of Systolic Chips.* Research Report RR 450, INRIA, Rennes, October 1985.

[Raj86] S. V. Rajopadye. *Synthesis, Verification and Optimization of Systolic Arrays.* PhD thesis, Dept. Comput. Sci., University of Utah, December 1986.

[RFS82A] I. V. Ramakrishnan, D. S. Fussell and A. Silberschatz. *Towards a Characterization of Programs for a Model of VLSI Array-Processors.* Technical Report TR-202, Dept. Comput. Sci., University of Texas at Austin, July 1982.

[RFS82B] I. V. Ramakrishnan, D. S. Fussell and A. Silberschatz. *A Linear Array Matrix Multiplication Algorithm.* Technical Report TR-210, Dept. Comput. Sci., University of Texas at Austin, September 1982.

[RFS86] I. V. Ramakrishnan, D. S. Fussell and A. Silberschatz. Mapping homogeneous graphs on linear arrays. *IEEE Trans. Comput.*, C-35(3)189–209, March 1986.

[Ran87] A. Ranade. How to emulate shared memory. In *Proc. IEEE Symposium on Foundations of Computer Science*, pages 185–194, October 1987.

[RK88] S. K. Rao and T. Kailath. Regular iterative algorithms and their implementation on processor arrays. *Proc. of the IEEE*, 76(3):259–269, March 1988.

[Sco82] D. S. Scott. Domains for denotational semantics. *Proc. ICALP*, pages 577–611, 1982.

[Sch83] R. Schreiber. Private communication.

[St71] H. S. Stone. Parallel processing with the perfect shuffle. *IEEE Trans. Comput.*, C-20:153–161, February 1971.

[Sw86] E. Swartzlander. Systolic FFT processors. In W. Moore, A. McCabe and R. Urquhart, editors, *Systolic Arrays*, pages 133–140, Adam Hilger, Bristol and Boston, 1986.

[U75] J. D. Ullman. NP-complete scheduling problems. *J. Comput. System Sci.*, 10:384–393, 1975.

[U84] J. D. Ullman. *Computational Aspects of VLSI.* Computer Science Press, Rockville, MD, 1984.

[VRF83] P. J. Varman, I. V. Ramakrishnan and D. S. Fussell. *Robust Matrix-Multiplication Algorithms for VLSI.* Technical Report

TR-221, Dept. Comput. Sci., University of Texas at Austin, March 1983.

[WD81] U. Weiser and A. L. Davis. A wavefront tool for VLSI design. In H. T. Kung, B. Sproull and G. Steele, editors, *VLSI Systems and Computations*, pages 226–234, Springer-Verlag, Berlin, 1981.

[W81] U. Weiser. *Mathematical and Graphical Tools for the Creation of Computational Arrays.* PhD thesis, Dept. Comput. Sci., University of Utah, 1981.

Appendix A. Closures

For completeness, we in this appendix give general definitions and prove some properties of closures of sets. In the following we assume a universe U. All sets considered will be subsets of U.

Definition A1: Consider the predicate p on sets. For any set A, $C_p(A)$, *the closure of A with respect to p*, is defined by:
c1. $p(C_p(A))$
c2. $A \subseteq C_p(A)$
c3. $p(B) \wedge A \subseteq B \implies C_p(A) \subseteq B$

Proposition A1: *If $C_p(A)$ exists, then it is unique.*

Proof. Assume $C_p(A)$, $C'_p(A)$ satisfies c1-c3. Because of c1 and c2, from c3 $C_p(A) \subseteq C'_p(A)$ and $C'_p(A) \subseteq C_p(A)$, thus $C_p(A) = C'_p(A)$. ∎

Proposition A2: $p(A) \iff A = C_p(A)$

Proof.
$\Longrightarrow$: if $p(A)$, then A trivially satisfies c1-c3.
$\Longleftarrow$: directly from c1. ∎

In the following we consider only such predicates p and sets A for which $C_p(A)$ always exists.

Proposition A3: $C_p(A) \cap C_p(B) \supseteq C_p(A \cap B)$

Proof. $A \cap B \subseteq A \subseteq C_p(A)$ and $A \cap B \subseteq B \subseteq C_p(B)$. By c1 and c3 follows that $C_p(A \cap B) \subseteq C_p(A)$ and $C_p(A \cap B) \subseteq C_p(B)$. Thus, $C_p(A \cap B) \subseteq C_p(A) \cap C_p(B)$. ∎

Proposition A4: $C_p(A) \cup C_p(B) \subseteq C_p(A \cup B)$

Proof. $A \subseteq A \cup B \subseteq C_p(A \cup B)$, so by c3 $p(C_p(A \cup B)) \wedge A \subseteq C_p(A \cup B) \implies C_p(A) \subseteq C_p(A \cup B)$. By the same reason $C_p(B) \subseteq C_p(A \cup B)$. Thus, $C_p(A) \cup C_p(B) \subseteq C_p(A \cup B)$. ∎

Lemma A1: *If $p(C_p(A) \cup C_p(B))$, then $C_p(A) \cup C_p(B) = C_p(A \cup B)$.*

Proof. It remains to prove $C_p(A \cup B) \subseteq C_p(A) \cup C_p(B)$. If $A \subseteq C_p(A)$ and $B \subseteq C_p(B)$, then $A \cup B \subseteq C_p(A) \cup C_p(B)$. If $p(C_p(A) \cup C_p(B))$, then c3 gives $C_p(A \cup B) \subseteq C_p(A) \cup C_p(B)$. ∎

Now we will consider predicates $q(a, A)$ where $a \in U$ and $A \subseteq U$. We call such a predicate *monotone* iff for all a and A holds that $q(a, A) \wedge A \subseteq B \implies q(a, B)$. *A pointwise predicate* p on sets is defined by

$p(A) \iff \forall a[a \in A \implies q(a, A)]$. In the following p is assumed to be pointwise. We call p monotone if q is monotone.

Lemma A2: *If p is monotone, then $p(A) \wedge p(B) \implies p(A \cup B)$.*

Proof.

$$p(A) \wedge p(B) \iff \forall a[a \in A \implies q(a, A)] \wedge \forall a[a \in B \implies q(a, B)].$$

Since p and thus q is monotone, this implies

$$\begin{aligned} & \forall a[a \in A \implies q(a, A \cup B)] \wedge \forall a[a \in B \implies q(a, A \cup B)] \\ \iff & \forall a[(a \in A \implies q(a, A \cup B)) \wedge (a \in B \implies q(a, A \cup B))] \\ \iff & \forall a[(a \in A \vee a \in B) \implies q(a, A \cup B)] \\ \iff & p(A \cup B). \end{aligned}$$

■

Proposition A5: *For monotone predicates p, $C_p(A) \cup C_p(B) = C_p(A \cup B)$.*

Proof. Since $p(C_p(A))$ and $p(C_p(B))$, lemma A2 implies $p(C_p(A) \cup C_p(B))$. Lemma A1 then gives the result. ■

Now we will restrict the scope further. For every binary relation R on the universe U, we consider the predicate $p_R(A) \iff \forall a \in A[sp(a, R) \subseteq A]$. Note that every such predicate is monotone. The closure $C_{p_R}(A)$ will for notational convenience be written $C(R, A)$ in the following.

Proposition A6: $C(R, A) = \{\, a' \mid \exists a \in A[a' R^* a] \,\}$.

Proof.

$\supseteq$: Denote $\{\, a' \mid \exists a \in A[a' R^n a] \,\}$ by $R^n(A)$ and $\{\, a' \mid \exists a \in A[a' R^* a] \,\}$ by $R^*(A)$. Let us prove that $R^n(A) \subseteq C(R, A)$ for all $n \in N$. This is done by induction on n:

$n = 0$: $R^0(A) = A \subseteq C(R, A)$.

$n > 0$: Assume $R^n(A) \subseteq C(R, A)$. For each $a' \in R^n(A)$ holds, since then $a' \in C(R, A)$, that $sp(a', R) = \{\, a'' \mid a'' R a' \,\} \subseteq C(R, A)$. Thus,

$$\bigcup(\{\, a'' \mid a'' R a' \,\} \mid a' \in R^n(A)) \subseteq C(R, A)$$

and since

$$\bigcup(\{\, a'' \mid a'' R a' \,\} \mid a' \in R^n(A)) = R^{n+1}(A)$$

the inductive step is proved.

Since $R^*(A) = \bigcup(\, R^n(A) \mid n \in N\,)$, it follows that $R^*(A) \subseteq C(R, A)$.

$\subseteq$: Let us prove that $p_R(R^*(A))$, that is: $sp(a, R) \subseteq R^*(A)$ for all $a \in R^*(A)$. The inclusion will then follow, since $A = R^0 \subseteq R^*(A)$. Now,

$$a \in R^*(A) \iff \exists a' \in A \exists n \in N[aR^n a'].$$

Thus, for every a'' such that $a''Ra$ holds that $a''R^{n+1}a'$ for some $n \in N$. Therefore, $a'' \in R^{n+1}(A) \subseteq R^*(A)$. This implies $p_R(R^*(A))$. ■

Appendix B. Basic properties of binary relations

Lemma B1: *Let A be any set and let R be a binary relation over A. Then, for all $a, b \in A$, $aR^+b \iff aRb$ or there exists an $a' \in A$ such that aR^+a' and $a'Rb$.*

Proof. We show the two directions separately. First, $aR^+b \iff \exists n \in Z^+[aR^nb]$.

$\Longrightarrow$: Assume aR^+b. Then $\exists n \in Z^+[aR^nb]$. We differ between the cases $n = 1$ and $n > 1$.

$n = 1$: In this case aRb.

$n > 1$: In this case it must exist an $a' \in A$ such that $aR^{n-1}a'$ and $a'Rb$. $aR^{n-1}a'$ and $n > 1 \Longrightarrow aR^+a'$.

$\Longleftarrow$: $aRb \Longrightarrow aR^+b$ directly. If there exists $a' \in A$ such that aR^+a' and $a'Rb$, then aR^na' for some $n > 0$. This implies $aR^{n+1}b$ which in turn implies aR^+b. ∎

Lemma B2: *Let A be any set and let R be a binary relation over A. Then, for all $a, b \in A$, $aR^+b \iff aRb$ or there exists $a' \in A$ such that aRa' and $a'R^+b$.*

Proof. The proof is exactly analogous to that of lemma B1 and is therefore omitted. ∎

Lemma B3: *Let $\prec$ be a well-founded relation over A. Then $\prec^+$ is finite-downward iff $sp(a, \prec)$ is finite for all $a \in A$.*

Proof.

$\Longrightarrow$: We prove that if there exists an $a \in A$ such that $sp(a, \prec)$ is infinite then $\prec^+$ cannot be finite-downward. Assume that there is an $a \in A$ such that $sp(a, \prec)$ is infinite. Then, since $\prec$ is a subrelation to $\prec^+$, there is an a such that $sp(a, \prec^+)$ is infinite, and $\prec^+$ cannot be finite-downward.

$\Longleftarrow$: This implication is the contrapositive to König's lemma. König's lemma is usually stated for trees ([Kn69], theorem K, p. 381), but it also holds for partial orders. ∎

Lemma B4: *Let A, B, C be sets, let f be a function $A \rightarrow B$ that is onto and let g be a function $A \rightarrow C$. Let R_B be a binary relation on B and let R_C be a binary relation on C. If for all $a, a' \in A$ holds that $f(a)R_Bf(a') \Longrightarrow g(a)R_Cg(a')$, then for all $a, a' \in A$, $f(a)R_B^+f(a') \Longrightarrow g(a)R_C^+g(a')$.*

Proof. Assume that for all $a, a' \in A$ holds that $f(a)R_Bf(a') \Longrightarrow g(a)R_Cg(a')$. $bR_B^+b' \iff \exists n \in Z^+[bR_B^nb']$, and similarly for R_C^+. Thus, if we under the assumption above can show that for all $n \in Z^+$ and for all $a, a' \in A$ holds that $f(a)R_B^n f(a') \implies g(a)R_C^n g(a')$, then the desired result will follow. We now show this by induction on n over the positive integers. We take the inductive sentence to be "for all $a, a' \in A$, $f(a)R_B^n f(a') \implies g(a)R_C^n g(a')$".

$= 1$: The inductive sentence follows directly from the assumptions in the lemma.

> 1: Assume that the inductive sentence holds for n. Let us show that it then also must hold for $n+1$. Consider any $a, a' \in A$ such that $f(a)R_B^{n+1} f(a')$. Then

$$\begin{aligned}
f(a)R_B^{n+1} f(a') &\implies \exists b'' \in B[f(a)R_B^n b'' \wedge b'' R_B f(a')] \\
&\implies (f \text{ onto, thus } \exists a'' \in A[f(a'') = b'']) \\
&\implies \exists a'' \in A[f(a)R_B^n f(a'') \wedge f(a'')R_B f(a')] \\
&\implies (\text{by assumption \& induction hypothesis}) \\
&\implies \exists a'' \in A[g(a)R_C^n g(a'') \wedge g(a'')R_C g(a')] \\
&\implies g(a)R_C^{n+1} g(a').
\end{aligned}$$

■

Lemma B5: *Let A, B be sets, and let f be a function $f\colon A \to B$. Let R_A be a binary relation on A, and R_B be a binary relation on B. If, for all $a, a' \in A$, $aR_A a' \implies f(a)R_B f(a')$, then $aR_A^+ a' \implies f(a)R_B^+ f(a')$ for all $a, a' \in A$.*

Proof. The lemma follows directly from lemma B4 above. (Select f as the identity function on B and g as f in this lemma.) ■

The following corollary follows immediately from lemma B5, since $a = a' \implies f(a) = f(a')$:

Corollary B1: *Let A, B be sets, and let f be a function $A \to B$. Let R_A be a binary relation on A and let R_B be a binary relation on B. If for all $a, a' \in A$, $aR_A a' \implies f(a)R_B f(a')$, then for all $a, a' \in A$, $aR_A^* a' \implies f(a)R_B^* f(a')$.*

The following lemma considers the restriction of relations over a universe U to a subset A of U:

Lemma B6: *For any relation R over U and for all $A \subseteq U$, $R|_A^+ \subseteq R^+|_A$*

Proof. The lemma follows if we can prove for all $n \geq 1$ that $R|_A^n \subseteq R^n|_A$
This is done by induction over n:

$n = 1$: $R|_A^1 = R|_A = R^1|_A \subseteq R^1|_A$.

$n > 1$: Assume as induction hypothesis that $R|_A^n \subseteq R^n|_A$. Let us show that then $R|_A^{n+1} \subseteq R^{n+1}|_A$, or that for all $a, a' \in A$ holds that $a\ R|_A^{n+1}\ a' \implies a\ R^{n+1}|_A\ a'$. So assume that $a\ R|_A^{n+1}\ a'$ for $a, a' \in A$. Now

$$
\begin{aligned}
a\ R|_A^{n+1}\ a' &\iff \exists a'' \in A[a\ R|_A^n\ a'' \wedge a''\ R|_A\ a'] \\
&\implies \text{(by the induction hypothesis)} \\
&\implies \exists a'' \in A[a\ R^n|_A\ a'' \wedge a''\ R|_A\ a'] \\
&\implies \exists a''[aR^n a'' \wedge a'' R a'] \\
&\implies aR^{n+1}a' \\
&\implies a\ R^{n+1}|_A\ a'
\end{aligned}
$$

where the last implication follows since $a, a' \in A$.

Let us now reconsider the predicates p_R on subsets of U, defined by $p_R(A) \iff \forall a \in A[sp(a, R) \subseteq A]$ for all $A \subseteq U$ and $R \subseteq U \times U$. For the restriction of relations R to a subset A such that $p_R(A)$, lemma B6 can be sharpened:

Lemma B7: *For any relation R over U and for all $A \subseteq U$ such that $p_R(A)$, $R|_A^+ = R^+|_A$.*

Proof. Lemma B6 yields $R|_A^+ \subseteq R^+|_A$. Thus, it suffices to show $R^+|_A \subseteq R|_A^+$. This follows if we can prove for all $n \geq 1$ that $R^n|_A \subseteq R|_A^n$, which is done by induction over n:

$n = 1$: $R^1|_A = R|_A = R|_A^1 \subseteq R|_A^1$.

$n > 1$: Assume as induction hypothesis that $R^n|_A \subseteq R|_A^n$. Let us show that then $R^{n+1}|_A \subseteq R|_A^{n+1}$, or that for all $a, a' \in A$ holds that $a\ R^{n+1}|_A\ a' \implies a\ R|_A^{n+1}\ a'$. So assume that $a\ R^{n+1}|_A\ a'$ for all $a, a' \in A$. Then

$$
\begin{aligned}
a\ R^{n+1}|_A\ a' &\iff \exists a''[aR^n a'' \wedge a'' R a'] \\
&\implies (a' \in A \wedge a'' R a' \implies a'' \in sp(a', R) \subseteq A) \\
&\implies \exists a'' \in A[aR^n a'' \wedge a'' R a'] \\
&\implies \exists a'' \in A[a\ R^n|_A\ a'' \wedge a''\ R|_A\ a'] \\
&\implies \text{(by the induction hypothesis)}
\end{aligned}
$$

$$\implies \exists a'' \in A[a\, R|_A^n\, a'' \wedge a''\, R|_A\, a']$$
$$\implies a\, R|_A^{n+1}\, a'.$$

■

Lemma B8: *Let A be a set and let R, R' be binary relations over A. If R^+, R'^+ are finite-downward, then $(R \cap R')^+$ is finite-downward.*

Proof. From proposition A3 follows that $(R \cap R')^+ \subseteq R^+ \cap R'^+$. So since R^+, R'^+ are irreflexive, $(R \cap R')^+$ must be irreflexive, and since $(R \cap R')^+$ by definition is transitive, it is a strict partial order. $sp(a, R^+)$ and $sp(a, R'^+)$ are finite for any $a \in A$, so $sp(a, R^+ \cap R'^+)$ must be finite. This in turn implies that $sp(a, (R \cap R')^+)$ is finite. Thus, $(R \cap R')^+$ is finite-downward. ■

Lemma B9: *Let A be a set and let R, R' be binary relations over A such that $R \subseteq R'$. If R' is finite-downward, then R^+ is finite-downward.*

Proof. $R \subseteq R'$ and R' is transitive, therefore

$$R^+ \subseteq R'. \tag{1}$$

R' is finite-downward, thus a strict partial order and irreflexive. By (1), R^+ must then be irreflexive, and it is by definition transitive. Therefore, R^+ is a strict partial order. Moreover, $sp(a, R')$ is finite for all $a \in A$. By (1), $sp(a, R^+) \subseteq sp(a, R')$ and it is therefore also finite for all $b \in B$. ■

Lemma B10: *Let $\prec_1$ be a well-founded relation on a set A and let $\prec_2$ be a relation on $A' \subseteq A$, such that $\prec_2^+$ is a total strict partial order on A' with least element a_0. Then*

$$\prec_1 \cup \prec_2 \textit{ well-founded on } A \iff \forall a, a' \in A[a \prec_1^+ a' \implies a' \not\prec_2^+ a].$$

Proof. Assume that $\prec_1$, $\prec_2$ are as in the formulation above. For brevity we introduce the following notation: we denote $\prec_1 \cup \prec_2$ by $\prec_\cup$. If there is an infinite decreasing sequence with respect to the relation $\prec$, ending in a, we write $-\infty \prec^+ a$. If not, we write $-\infty \not\prec^+ a$. We prove the two directions separately:

$\implies$: We prove the equivalent implication

$$\neg\forall a, a' \in A[a \prec_1^+ a' \implies a' \not\prec_2^+ a] \implies \prec_\cup \text{ not well-founded on } A:$$

$$\begin{aligned}
\neg\forall a, a' \in A[a \prec_1^+ a' \implies a' \not\prec_2^+ a] &\implies \\
\exists a, a' \in A[a \prec_1^+ a' \wedge a' \prec_2^+ a] &\implies \\
\exists a \in A[a\ (\prec_1^+ \cup \prec_2^+)\ a] &\implies \\
\text{(since proposition A4 implies } \prec_1^+ \cup \prec_2^+ \subseteq \prec_\cup^+) &\implies \\
\exists a \in A[a \prec_\cup^+ a] &\implies \\
\exists a \in A[-\infty \prec_\cup^+ a]&.
\end{aligned}$$

$\Longleftarrow$: Assume that for all $a, a' \in A$

$$a \prec_1^+ a' \implies a' \not\prec_2^+ a. \tag{1}$$

The proof is by induction over $\prec_2^+$ (that is well-founded, since it is a total strict partial order on A' with a least element a_0). The inductive sentence to be proved for all $a \in A'$ is "$-\infty \not\prec_\cup^+ a$".

Base case: Assume $-\infty \prec_\cup^+ a_0$. Then it cannot be the case that $-\infty \prec_1^+ a_0$, since that would imply that $\prec_1$ is not well-founded. Thus, there must be an $a' \in A'$ such that $a' \neq a_0$, and either

$$-\infty \prec_\cup^+ a' \prec_2^+ a_0 \tag{2}$$

or there exists a "greatest" $a'' \in A'$ (with respect to an infinite sequence ending in a_0), such that $a'' \neq a_0$ and

$$-\infty \prec_\cup^+ a' \prec_2^+ a'' \prec_1^+ a_0. \tag{3}$$

(2) can be ruled out immediately since it contradicts that a_0 is the least element in A' with respect to $\prec_2^+$. Since a_0 is the least element and since $\prec_2^+$ is total, (3) implies $a_0 \prec_2^+ a''$ and $a'' \prec_1^+ a_0$. But this contradicts the assumption (1). Thus, it must be the case that $-\infty \not\prec_\cup^+ a_0$.

Induction step: Consider any $a \neq a_0$ in A'. Assume as induction hypothesis that for all a' such that $a' \prec_2^+ a$

$$-\infty \not\prec_\cup a'. \tag{4}$$

Assume that

$$-\infty \prec_\cup^+ a. \tag{5}$$

As before, it cannot be the case that $-\infty \prec_1^+ a$, since then $\prec_1$ would not be well-founded. Thus, there must be an $a' \in A'$ such that $a' \neq a$ and either

$$-\infty \prec_\cup^+ a' \prec_2^+ a \tag{6}$$

or there exists, as in (3), an $a'' \in A'$ such that $a'' \neq a$ and

$$-\infty \prec_{\cup}^{+} a' \prec_{2}^{+} a'' \prec_{1}^{+} a. \tag{7}$$

(6) directly violates the induction hypothesis (4). If (7) is to hold, then there are two possibilities, since $a, a' \in A$ and since $\prec_2^+$ is total. The first is that $a' \prec_2^+ a$. But this contradicts (4). The second is that $a \prec_2^+ a'$. But then $a \prec_2^+ a'$ and $a' \prec_1^+ a$ which also contradicts (1). Thus, (5) is impossible and we must have $-\infty \not\prec_{\cup}^{+} a$. ■

Appendix C. Associativity of function composition

In this appendix we will prove, that the well-known associativity property $(f \circ g) \circ h = f \circ (g \circ h)$ for composition of functions of one variable also holds for composition of functions of several variables.

Proposition C1: *Let f be a function $\prod(A_i \mid i \in I) \to A_f$, let g_i be a function $\prod(A_k \mid k \in I_i) \to A_i$ for all $i \in I$ and let h_k be a function $\prod(A_l \mid l \in I_k) \to A_k$ for all $k \in I_i$. Then*

$$f \circ \langle (g_i \circ \langle h_k \mid k \in I_i \rangle) \mid i \in I \rangle =$$
$$(f \circ \langle g_i \mid i \in I \rangle) \circ \langle h_k \mid k \in \bigcup(I_i \mid i \in I) \rangle.$$

Proof. First, define $J_i = \bigcup(I_k \mid k \in I_i)$ for all $i \in I$, define $J = \bigcup(J_i \mid k \in I)$ and $K = \bigcup(I_i \mid i \in I)$. Both functions for which the theorem is stated above are functions $\prod(A_j \mid j \in J) \to A_f$. Furthermore, for all $i \in I$ holds that $g_i \circ \langle h_k \mid k \in I_i \rangle$ is a function $\prod(A_j \mid j \in J_i) \to A_i$ and that $f \circ \langle g_i \mid i \in I \rangle$ is a function $\prod(A_k \mid k \in K) \to A_f$. Now, for all $a \in \prod(A_j \mid j \in J)$,

$$\begin{aligned} f \circ \langle (g_i \circ \langle h_k \mid k \in I_i \rangle) \mid i \in I \rangle(a) &= f((g_i \circ \langle h_k \mid k \in I_i \rangle)(a|_{J_i}) \mid i \in I) \\ &= f(g_i(h_k(a|_{J_i}|_{I_k}) \mid k \in I_i) \mid i \in I) \\ &= f(g_i(h_k(a|_{I_k}) \mid k \in I_i) \mid i \in I). \end{aligned}$$

On the other hand

$$(f \circ \langle g_i \mid i \in I \rangle) \circ \langle h_k \mid k \in K \rangle(a) =$$
$$(f \circ \langle g_i \mid i \in I \rangle)(h_k(a|_{I_k}) \mid k \in K) =$$
$$f(g_i(\langle h_k(a|_{I_k}) \mid k \in K \rangle|_{I_i}) \mid i \in I) =$$
$$f(g_i(h_k(a|_{I_k}) \mid k \in I_i) \mid i \in I).$$

■

Appendix D. Basic properties of expressions and substitutions

Lemma D1: *(Substitutions preserve sort) For all substitutions* $\sigma \in \mathcal{S}^{(X)}(\mathcal{A})$ *and all* $\mathbf{p} \in \mathcal{E}^{(X)}(\mathcal{A})$, $S(\sigma(\mathbf{p})) = S(\mathbf{p})$.

Proof.
1a. $\mathbf{p} \in X$:
$\mathbf{p} \notin dom(\sigma)$: $\sigma(\mathbf{p}) = \mathbf{p} \implies S(\sigma(\mathbf{p})) = S(\mathbf{p})$.
$\mathbf{p} \in dom(\sigma)$: Directly from definition 2.12, $S(\sigma(\mathbf{p})) = S(\mathbf{p})$.
1b. $\mathbf{p} = f \bullet \langle\rangle$: $\sigma(\mathbf{p}) = \sigma(f \bullet \langle\rangle) = f \bullet \langle\rangle = \mathbf{p} \implies S(\sigma(\mathbf{p})) = S(\mathbf{p})$.
2. $\mathbf{p} = f \bullet \langle \mathbf{p}_i \mid i \in I_f \rangle$: $\sigma(\mathbf{p}) = f \bullet \langle \sigma(\mathbf{p}_i) \mid i \in I_f \rangle$. Thus, $S(\sigma(\mathbf{p})) = S_f = S(\mathbf{p})$. ■

Proposition D1: *For every algebra* $\mathcal{A}$ *and all sets of variables* X', X,

$$X' \subseteq X \iff \mathcal{E}^{(X')}(\mathcal{A}) \subseteq \mathcal{E}^{(X)}(\mathcal{A}).$$

Proof. The two directions are shown separately.

$\implies$: Assume $X' \subseteq X$. We prove by induction over $\mathcal{E}^{(X')}(\mathcal{A})$ that $\mathbf{p} \in \mathcal{E}^{(X)}(\mathcal{A})$ for all $\mathbf{p} \in \mathcal{E}^{(X')}(\mathcal{A})$:
1a. $\mathbf{p} \in X'$: Then, since $X' \subseteq X$, it follows that $\mathbf{p} \in X \subseteq \mathcal{E}^{(X)}(\mathcal{A})$.
1b. $\mathbf{p} = f \bullet \langle\rangle$: Then $\mathbf{p} \in \mathcal{E}^{(X)}(\mathcal{A})$ for all sets of variables X.
2. $\mathbf{p} = f \bullet \langle \mathbf{p}_i \mid i \in I_f \rangle$: Assume that $\mathbf{p}_i \in \mathcal{E}^{(X)}(\mathcal{A})$ for all $i \in I_f$. $\mathbf{p}$ fulfils 2 in definition 2.8 of $\mathcal{E}^{(X')}(\mathcal{A})$. Thus, it is also constructed according to 2 for $\mathcal{E}^{(X)}(\mathcal{A})$, and $\mathbf{p} \in \mathcal{E}^{(X)}(\mathcal{A})$.

$\impliedby$: Let us prove $\neg(X' \subseteq X) \implies \neg(\mathcal{E}^{(X')}(\mathcal{A}) \subseteq \mathcal{E}^{(X)}(\mathcal{A}))$. Assume $\neg(X' \subseteq X)$. Then there exists an x such that $x \in X'$ and $x \notin X$. From 1 in definition 2.8, $x \in \mathcal{E}^{(X')}(\mathcal{A})$. From 3, $x \notin \mathcal{E}^{(X)}(\mathcal{A})$. ■

Lemma D2: *For any* $\mathbf{p} \in \mathcal{E}^{(X)}(\mathcal{A})$, $ec(\mathbf{p}) = sp(\mathbf{p}, \prec_e^*)$.

Proof. The evaluable closure is defined by a monotone predicate. Thus, proposition A6 can be applied to $ec(\{\mathbf{p}\}) = ec(\mathbf{p})$ and the result follows immediately. ■

The following proposition gives an alternative characterization of *varset*, cf. [Be72]:

Proposition D2: *For every* $E \subseteq \mathcal{E}^{(X)}(\mathcal{A})$, $varset(E) = ec(E) \cap X$.

Proof. The proposition is first shown inductively for expressions, taking the inductive sentence to be $varset(\mathbf{p}) = ec(\{\mathbf{p}\}) \cap X$. We will use the alternative definition of the evaluable closure given in theorem 4.1.

vs1. $\mathbf{p} \in \mathcal{E}^{(X)}(\mathcal{A})$:

1a. $\mathbf{p} \in X$: $varset(\mathbf{p}) = \{\mathbf{p}\} = \{\mathbf{p}\} \cap X = ec(\{\mathbf{p}\}) \cap X$.

1b. $\mathbf{p} = f \bullet \langle\rangle$: $varset(\mathbf{p}) = \bigcup(\, varset(\mathbf{p}_i) \mid i \in \emptyset\,) = \emptyset = \{\mathbf{p}\} \cap X = ec(\{\mathbf{p}\}) \cap X$.

2. $\mathbf{p} = f \bullet \langle\, \mathbf{p}_i \mid i \in I_f \,\rangle$: Assume as induction hypothesis that $varset(\mathbf{p}_i) = ec(\{\mathbf{p}_i\}) \cap X$ for all $i \in I_f$. Then

$$\begin{aligned} varset(\mathbf{p}) &= \bigcup(\, varset(\mathbf{p}_i) \mid i \in I_f\,) \\ &= \bigcup(\, ec(\{\mathbf{p}_i\}) \cap X \mid i \in I_f\,) \\ &= \bigcup(\, ec(\{\mathbf{p}_i\}) \mid i \in I_f\,) \cap X \\ &= (\bigcup(\, ec(\{\mathbf{p}_i\}) \mid i \in I_f\,) \cup \{\mathbf{p}\}) \cap X \\ &= ec(\{\mathbf{p}\}) \cap X. \end{aligned}$$

vs2. $E \subseteq \mathcal{E}^{(X)}(\mathcal{A})$:

$$\begin{aligned} varset(E) &= \bigcup(\, varset(\mathbf{p}) \mid \mathbf{p} \in E\,) \\ &= \bigcup(\, ec(\{\mathbf{p}\}) \cap X \mid \mathbf{p} \in E\,) \\ &= \bigcup(\, ec(\{\mathbf{p}\}) \mid \mathbf{p} \in E\,) \cap X \\ &= ec(E) \cap X. \end{aligned}$$

■

Lemma D3: *For any* $\mathbf{p} \in \mathcal{E}^{(X)}(\mathcal{A})$ *and any* $X' \subseteq X$,

$$varset(\mathbf{p}) \subseteq X' \iff \mathbf{p} \in \mathcal{E}^{(X')}(\mathcal{A}).$$

Proof. The proof is by induction over $\mathcal{E}^{(X)}(\mathcal{A})$:

1a. $\mathbf{p} \in X$:

$\Longrightarrow$: $varset(\mathbf{p}) \subseteq X' \implies \mathbf{p} \in X' \implies \mathbf{p} \in \mathcal{E}^{(X')}(\mathcal{A})$.

$\Longleftarrow$: Since $\mathbf{p} \in X$, $\mathbf{p}$ is not a compound expression. $\mathbf{p} \in \mathcal{E}^{(X')}(\mathcal{A})$, $\mathbf{p}$ not compound and 3 in definition 2.8 $\implies \mathbf{p} \in X'$.

1b. $\mathbf{p} = f \bullet \langle\rangle$: $varset(\mathbf{p}) = \emptyset \subseteq X'$. $\mathbf{p} = f \bullet \langle\rangle \in \mathcal{E}^{(X')}(\mathcal{A})$ for all X'.

2. $\mathbf{p} = f \bullet \langle\, \mathbf{p}_i \mid i \in I_f \,\rangle$: Assume as induction hypothesis that

$$\forall i \in I_f [varset(\mathbf{p}_i) \subseteq X' \iff \mathbf{p}_i \in \mathcal{E}^{(X')}(\mathcal{A})].$$

This implies

$$\forall i \in I_f[varset(\mathbf{p}_i) \subseteq X'] \iff \forall i \in I_f[\mathbf{p}_i \in \mathcal{E}^{(X')}(\mathcal{A})].$$

Since $varset(\mathbf{p}) = \bigcup(\, varset(\mathbf{p}_i) \mid i \in I_f\,)$ it follows that

$$\begin{aligned} varset(\mathbf{p}) \subseteq X' &\iff \forall i \in I_f[varset(\mathbf{p}_i) \subseteq X'] \\ &\iff \forall i \in I_f[\mathbf{p}_i \in \mathcal{E}^{(X')}(\mathcal{A})] \\ &\iff \mathbf{p} \in \mathcal{E}^{(X')}(\mathcal{A}) \end{aligned}$$

where in the last equivalence, since $\mathbf{p}$ is a compound expression in $\mathcal{E}^{(X)}(\mathcal{A})$, $\Longrightarrow$ follows from 2 in definition 2.8 and $\Longleftarrow$ from 3. ■

Lemma D4: $\mathbf{p}' \prec_e^* \mathbf{p} \Longrightarrow varset(\mathbf{p}') \subseteq varset(\mathbf{p})$.

Proof. Consider any $\mathbf{p}''$ such that $\mathbf{p}'' \prec_e \mathbf{p}$. Then $\mathbf{p} = f \bullet \langle \mathbf{p}_i \mid i \in I_f \rangle$ and $\mathbf{p}'' = \mathbf{p}_i$ for some $i \in I_f$. Since $varset(\mathbf{p}) = \bigcup(\, varset(\mathbf{p}_i) \mid i \in I_f\,)$, it follows that $varset(\mathbf{p}'') \subseteq varset(\mathbf{p})$. So $\mathbf{p}'' \prec_e \mathbf{p} \Longrightarrow varset(\mathbf{p}'') \subseteq varset(\mathbf{p})$. Then corollary B1 can be applied and since $\subseteq$ is transitive and reflexive we obtain, for any $\mathbf{p}'$, $\mathbf{p}' \prec_e^* \mathbf{p} \Longrightarrow varset(\mathbf{p}') \subseteq varset(\mathbf{p})$. ■

Lemma D5: *For any* $\mathbf{p} \in \mathcal{E}^{(X)}(\mathcal{A})$ *and any* $\sigma \in \mathcal{S}^{(X)}(\mathcal{A})$

$$varset(\mathbf{p}) \cap dom(\sigma) = \emptyset \iff \sigma(\mathbf{p}) = \mathbf{p}.$$

This lemma is well-known from similar developments, see for instance [MaWa84] or [H76]. Therefore we do not prove it here.

Lemma D6: *For any* $\mathbf{p} \in \mathcal{E}^{(X)}(\mathcal{A})$ *and any* $\sigma \in \mathcal{S}^{(X)}(\mathcal{A})$

$$varset(\sigma(\mathbf{p})) = \bigcup(\, varset(\sigma(x)) \mid x \in varset(\mathbf{p})\,).$$

Proof. The lemma is shown by induction over $\mathcal{E}^{(X)}(\mathcal{A})$:

1a. $\mathbf{p} \in X$: $varset(\mathbf{p}) = \{\mathbf{p}\}$, so $varset(\sigma(\mathbf{p})) = \bigcup(\, varset(\sigma(x)) \mid x \in varset(\mathbf{p})\,)$.

1b. $\mathbf{p} = f \bullet \langle\rangle$: Then $varset(\mathbf{p}) = varset(\sigma(\mathbf{p})) = \emptyset = \bigcup(\, varset(\sigma(x)) \mid x \in \emptyset\,) = \bigcup(\, varset(\sigma(x)) \mid x \in varset(\mathbf{p})\,)$.

2. $\mathbf{p} = f \bullet \langle \mathbf{p}_i \mid i \in I_f \rangle$: assume as induction hypothesis, that for all i in I_f

$$varset(\sigma(\mathbf{p}_i)) = \bigcup(\, varset(\sigma(x)) \mid x \in varset(\mathbf{p}_i)\,).$$

Then

$$\begin{aligned} varset(\sigma(\mathbf{p})) &= \bigcup(\, varset(\sigma(\mathbf{p}_i)) \mid i \in I_f \,) \\ &= \text{ (by the induction hypothesis)} \\ &= \bigcup(\bigcup(\, varset(\sigma(x)) \mid x \in varset(\mathbf{p}_i)\,) \mid i \in I_f \,) \\ &= \bigcup(\, varset(\sigma(x)) \mid x \in \bigcup(\, varset(\mathbf{p}_i) \mid i \in I_f \,)) \\ &= \bigcup(\, varset(\sigma(x)) \mid x \in varset(\mathbf{p})\,). \end{aligned}$$

■

Lemma D7: *For any* $\mathbf{p} \in \mathcal{E}^{(X)}(\mathcal{A})$ *and any* $\sigma \in \mathcal{S}^{(X)}(\mathcal{A})$

$$varset(\sigma(\mathbf{p})) = (varset(\mathbf{p}) \setminus dom(\sigma)) \cup range(\sigma|_{varset(\mathbf{p})}).$$

Proof. The proof is by induction over $\mathcal{E}^{(X)}(\mathcal{A})$:

1a. $\mathbf{p} \in X$: There are two cases:

$\mathbf{p} \notin dom(\sigma)$: Then $\sigma(\mathbf{p}) = \mathbf{p}$, so $varset(\sigma(\mathbf{p})) = varset(\mathbf{p}) = \{\mathbf{p}\}$. Also $varset(\mathbf{p}) \setminus dom(\sigma) = \{\mathbf{p}\} \setminus dom(\sigma) = \{\mathbf{p}\}$. Furthermore

$$range(\sigma|_{varset(\mathbf{p})}) = range(\sigma|_{\{\mathbf{p}\}}) = range(\emptyset) = \emptyset,$$

so in this case $varset(\sigma(\mathbf{p})) = \{\mathbf{p}\} \cup \emptyset = (varset(\mathbf{p}) \setminus dom(\sigma)) \cup range(\sigma|_{varset(\mathbf{p})})$.

$\mathbf{p} \in dom(\sigma)$: Then $varset(\mathbf{p}) \setminus dom(\sigma) = \emptyset$, and

$$\begin{aligned} range(\sigma|_{varset(\mathbf{p})}) &= \bigcup(\, varset(\sigma(x')) \mid x' \in varset(\mathbf{p})\,) \\ &= \bigcup(\, varset(\sigma(x')) \mid x' = \mathbf{p}\,) \\ &= varset(\sigma(\mathbf{p})). \end{aligned}$$

Thus , $varset(\sigma(\mathbf{p})) = \emptyset \cup varset(\sigma(\mathbf{p})) = (varset(\mathbf{p}) \setminus dom(\sigma)) \cup range(\sigma|_{varset(\mathbf{p})})$.

1b. $\mathbf{p} = f \bullet \langle\rangle$: Then $varset(\sigma(\mathbf{p})) = varset(\mathbf{p}) = \emptyset$, $varset(\mathbf{p}) \setminus dom(\sigma) = \emptyset \setminus dom(\sigma) = \emptyset$ and $range(\sigma|_{varset(\mathbf{p})}) = range(\sigma|_{\emptyset}) = range(\emptyset) = \emptyset$. Thus, the set equality holds also in this case.

2. $\mathbf{p} = f \bullet \langle\, \mathbf{p}_i \mid i \in I_f \,\rangle$: assume as induction hypothesis, that for all $i \in I_f$

$$varset(\sigma(\mathbf{p}_i)) = (varset(\sigma(\mathbf{p}_i)) \setminus dom(\sigma)) \cup range(\sigma|_{varset(\mathbf{p}_i)}).$$

Then

$$\begin{aligned}
varset(\sigma(\mathbf{p})e) &= \\
\bigcup(\, varset(\sigma(\mathbf{p}_i)) \mid i \in I_f\,) &= \\
\text{(by the induction hypothesis)} &= \\
\bigcup(\,(varset(\mathbf{p}_i) \setminus dom(\sigma)) \cup range(\sigma|_{varset(\mathbf{p}_i)}) \mid i \in I_f\,) &= \\
\bigcup(\, varset(\mathbf{p}_i) \setminus dom(\sigma) \mid i \in I_f\,) \cup \bigcup(\, range(\sigma|_{varset(\mathbf{p}_i)}) \mid i \in I_f\,) &= \\
(\bigcup(\, varset(\mathbf{p}_i) \mid i \in I_f\,) \setminus dom(\sigma)) \cup range(\sigma|_{\bigcup(\,varset(\mathbf{p}_i) \mid i \in I_f\,)}) &= \\
(varset(\mathbf{p}) \setminus dom(\sigma)) \cup range(\sigma|_{varset(\mathbf{p})})&.
\end{aligned}$$

■

The following lemma, found in for instance [MaWa84], follows immediately as a corollary to lemma D7.

Lemma D8: *For any* $\mathbf{p} \in \mathcal{E}^{(X)}(\mathcal{A})$ *and any* $\sigma \in \mathcal{S}^{(X)}(\mathcal{A})$

$$varset(\sigma(\mathbf{p})) \subseteq (varset(\mathbf{p}) \setminus dom(\sigma)) \cup range(\sigma).$$

This lemma also follows immediately from lemma D7.

Lemma D9: *For any* $\mathbf{p} \in \mathcal{E}^{(X)}(\mathcal{A})$ *and any* $\sigma \in \mathcal{S}^{(X)}(\mathcal{A})$

$$varset(\mathbf{p}) \setminus dom(\sigma) \subseteq varset(\sigma(\mathbf{p})).$$

The following lemma follows from lemma 5.1, [H76].

Lemma D10: *(Substitution restriction) For any* $\mathbf{p} \in \mathcal{E}^{(X)}(\mathcal{A})$ *and any* $\sigma \in \mathcal{S}^{(X)}(\mathcal{A})$

$$\sigma(\mathbf{p}) = \sigma|_{varset(\mathbf{p})}(\mathbf{p}).$$

Lemma D11: *For any* $\mathbf{p} \in \mathcal{E}^{(X)}(\mathcal{A})$, *any* $X' \subseteq X$ *and any* $\sigma \in \mathcal{S}^{(X)}(\mathcal{A})$

$$varset(\mathbf{p}) \subseteq X' \cup dom(\sigma) \wedge range(\sigma|_{varset(\mathbf{p})}) \subseteq X' \implies \sigma(\mathbf{p}) \in \mathcal{E}^{(X')}(\mathcal{A}).$$

Proof. From lemma D3 follows that $\sigma(\mathbf{p}) \in \mathcal{E}^{(X')}(\mathcal{A})$ iff $varset(\sigma(\mathbf{p})) \subseteq X'$. So we will prove the equivalent implication

$$varset(\mathbf{p}) \subseteq X' \cup dom(\sigma) \wedge range(\sigma|_{varset(\mathbf{p})}) \subseteq X' \implies varset(\sigma(\mathbf{p})) \subseteq X'$$

instead. By lemma D7 holds that

$$varset(\sigma(\mathbf{p})) = (varset(\mathbf{p}) \setminus dom(\sigma)) \cup range(\sigma|_{varset(\mathbf{p})}). \qquad (1)$$

If $varset(\mathbf{p}) \subseteq X' \cup dom(\sigma)$, then

$$varset(\mathbf{p}) \setminus dom(\sigma) \subseteq X'. \tag{2}$$

If also

$$range(\sigma|_{varset(\mathbf{p})}) \subseteq X'$$

then (1) and (2) yields

$$varset(\sigma(\mathbf{p})) \subseteq X' \cup X' = X'.$$

■

Lemma D12: *For any* $\mathbf{p} \in \mathcal{E}^{(X)}(\mathcal{A})$, *for any* $x \in X$ *and for any* $\sigma \in \mathcal{S}^{(X)}(\mathcal{A})$

$$x \notin varset(\mathbf{p}) \wedge x \in varset(\sigma(\mathbf{p})) \implies$$
$$\exists x' \in dom(\sigma)[x' \in varset(\mathbf{p}) \wedge x \in varset(\sigma(x'))].$$

Proof. The proof is by induction over $\mathcal{E}^{(X)}(\mathcal{A})$, taking the inductive sentence to be the above.

1a. $\mathbf{p} \in X$: There are two cases.

$\mathbf{p} \notin dom(\sigma)$: $\sigma(\mathbf{p}) = \mathbf{p}$, so if $x \notin varset(\mathbf{p})$ then $x \notin varset(\sigma(\mathbf{p}))$ as well. Thus, the implication always holds in this case.

$\mathbf{p} \in dom(\sigma)$: Assume, for any $x \in X$, that $x \notin varset(\mathbf{p}) \wedge x \in varset(\sigma(\mathbf{p}))$. Choose $x' = \mathbf{p}$. Then the following holds:

$$x' = \mathbf{p} \in dom(\sigma)$$
$$x' \in \{x'\} = \{\mathbf{p}\} = varset(\mathbf{p})$$
$$x \in varset(\sigma(\mathbf{p})) = varset(\sigma(x')).$$

So there exists an x' that fulfils the requirements and the implication holds.

1b. $\mathbf{p} = f \bullet \langle\rangle$: Then $varset(\mathbf{p}) = varset(\sigma(\mathbf{p})) = \emptyset$, there is no x in $varset(\sigma(\mathbf{p}))$ and the implication holds.

2. $\mathbf{p} = f \bullet \langle \mathbf{p}_i \mid i \in I_f \rangle$: Assume as induction hypothesis, that

$$\forall i \in I_f[x \notin varset(\mathbf{p}_i) \wedge x \in varset(\sigma(\mathbf{p}_i)) \implies$$
$$\exists x' \in dom(\sigma)[x' \in varset(\mathbf{p}_i) \wedge x \in varset(\sigma(x'))]]$$

which implies

$$\exists i \in I_f[x \notin varset(\mathbf{p}_i) \wedge x \in varset(\sigma(\mathbf{p}_i)) \implies$$
$$\exists x' \in dom(\sigma)[x' \in varset(\mathbf{p}_i) \wedge x \in varset(\sigma(x'))]]$$

which, in turn, implies

$$\exists i \in I_f[x \notin varset(\mathbf{p}_i) \wedge x \in varset(\sigma(\mathbf{p}_i))] \implies$$
$$\exists i \in I_f[\exists x' \in dom(\sigma)[x' \in varset(\mathbf{p}_i) \wedge x \in varset(\sigma(x'))]].$$

Let us show that the inductive sentence holds for $\mathbf{p}$ as well. So assume, for some $x \in X$, that $x \notin varset(\mathbf{p}) \wedge x \in varset(\sigma(\mathbf{p}))$. Then, since $\mathbf{p} = f \bullet \langle \mathbf{p}_i \mid i \in I_f \rangle$,

$$\begin{aligned}
x \notin \bigcup(\, varset(\mathbf{p}_i) \mid i \in I_f \,) \wedge x \in \bigcup(\, varset(\sigma(\mathbf{p}_i)) \mid i \in I_f \,) &\iff \\
\forall i \in I_f[x \notin varset(\mathbf{p}_i)] \wedge \exists i \in I_f[x \in varset(\sigma(\mathbf{p}_i))] &\implies \\
\exists i \in I_f[x \notin varset(\mathbf{p}_i) \wedge x \in varset(\sigma(\mathbf{p}_i))] &\implies \\
\text{(by the induction hypothesis)} &\implies \\
\exists i \in I_f[\exists x' \in dom(\sigma)[x' \in varset(\mathbf{p}_i) \wedge x \in varset(\sigma(x'))]] &\iff \\
\text{(since } varset(\mathbf{p}) = \bigcup(\, varset(\mathbf{p}_i) \mid i \in I_f \,)) &\iff \\
\exists x' \in dom(\sigma)[x' \in varset(\mathbf{p}) \wedge x \in varset(\sigma(x'))].&
\end{aligned}$$

■

Theorem D1: *(Substitutions vs. function composition): For all* $\mathbf{p} \in \mathcal{E}^{(X)}(\mathcal{A})$, $X' \subseteq X$ *and substitutions* $\sigma \in \mathcal{S}^{(X)}(\mathcal{A})$,

$$\phi(\sigma|_{X'}(\mathbf{p})) = \phi(\mathbf{p}) \circ \langle\, \phi(\sigma(x)) \mid x \in X' \,\rangle.$$

Proof. The proof is by induction over $\mathcal{E}^{(X)}(\mathcal{A})$:
1a. $\mathbf{p} \in X$: Then

$$\begin{aligned}
\phi(\mathbf{p}) \circ \langle\, \phi(\sigma(x)) \mid x \in X' \,\rangle &= e^X_{\mathbf{p}} \circ \langle\, \phi(\sigma(x)) \mid x \in X' \,\rangle \\
&= e^X_{\mathbf{p}} \circ \langle\, \phi(\sigma(x))^X \mid x \in X \rangle.
\end{aligned}$$

For all $a \in \prod(\, A_x \mid x \in X \,)$

$$\begin{aligned}
e^X_{\mathbf{p}} \circ \langle\, \phi(\sigma(x))^X \mid x \in X \,\rangle(a) &= e^X_{\mathbf{p}}(\, \phi(\sigma(x))^X(a) \mid x \in X \,) \\
&= \begin{cases} e^X_{\mathbf{p}}(a), & \mathbf{p} \notin X' \\ \phi(\sigma(\mathbf{p}))(a), & \mathbf{p} \in X'. \end{cases}
\end{aligned}$$

$\mathbf{p} \notin X'$: $\phi(\sigma|_{X'}(\mathbf{p})) = \phi(\mathbf{p}) = e^X_{\mathbf{p}}$. For all $a \in \prod(\, A_x \mid x \in X\,)$

$$\begin{aligned}\phi(\mathbf{p}) \circ \langle\, \phi(\sigma(x)) \mid x \in X' \,\rangle(a) &= e^X_{\mathbf{p}} \circ \langle\, \phi(\sigma(x)) \mid x \in X' \,\rangle(a)\\ &= e^X_{\mathbf{p}}(a)\\ &= \phi(\sigma|_{X'}(\mathbf{p}))(a).\end{aligned}$$

$\mathbf{p} \in X'$: $\phi(\sigma|_{X'}(\mathbf{p})) = \phi(\sigma(\mathbf{p}))$. For all $a \in \prod(\, A_x \mid x \in X\,)$

$$\begin{aligned}\phi(\mathbf{p}) \circ \langle\, \phi(\sigma(x)) \mid x \in X' \,\rangle(a) &= e^X_{\mathbf{p}} \circ \langle\, \phi(\sigma(x)) \mid x \in X' \,\rangle(a)\\ &= \phi((\sigma(\mathbf{p}))(a))\\ &= \phi(\sigma|_{X'}(\mathbf{p}))(a).\end{aligned}$$

1b. $\mathbf{p} = f \bullet \langle\rangle$: $\phi(\sigma|_{X'}(\mathbf{p})) = \phi(\mathbf{p}) = f \circ \langle\rangle$. By proposition C1,

$$\begin{aligned}\phi(\mathbf{p}) \circ \langle\, \phi(\sigma(x)) \mid x \in X' \,\rangle &= (f \circ \langle\rangle) \circ \langle\, \phi(\sigma(x)) \mid x \in X' \,\rangle\\ &= \text{(for any } g_i\text{)}\\ &= f \circ \langle\, g_i \circ \langle\, \phi(\sigma(x)) \mid x \in X' \,\rangle \mid i \in \emptyset \,\rangle\\ &= f \circ \langle\rangle.\end{aligned}$$

2. $\mathbf{p} = f \bullet \langle\, \mathbf{p}_i \mid i \in I_f \,\rangle$: Assume as induction hypothesis that

$$\forall i \in I_f[\phi(\sigma|_{X'}(\mathbf{p}_i)) = \phi(\mathbf{p}_i) \circ \langle\, \phi(\sigma(x)) \mid x \in X' \,\rangle].$$

Then

$$\begin{aligned}\phi(\sigma|_{X'}(\mathbf{p})) &= \phi(\sigma|_{X'}(f \bullet \langle\, \mathbf{p}_i \mid i \in I_f \,\rangle))\\ &= \phi(f \bullet \langle\, \sigma|_{X'}(\mathbf{p}_i) \mid i \in I_f \,\rangle)\\ &= f \circ \langle\, \phi(\sigma|_{X'}(\mathbf{p}_i)) \mid i \in I_f \,\rangle\\ &= \text{(by the induction hypothesis)}\\ &= f \circ \langle\, (\phi(\mathbf{p}_i) \circ \langle\, \phi(\sigma(x)) \mid x \in X' \,\rangle) \mid i \in I_f \,\rangle\\ &= \text{(by proposition C1)}\\ &= (f \circ \langle\, \phi(\mathbf{p}_i) \mid i \in I_f \,\rangle) \circ \langle\, \phi(\sigma(x)) \mid x \in X' \,\rangle\\ &= \phi(\mathbf{p}) \circ \langle\, \phi(\sigma(x)) \mid x \in X' \,\rangle.\end{aligned}$$

■

Theorem 2.1: *For all substitutions* σ_1, σ_2 *and expressions* $\mathbf{p}$, $\sigma_1\sigma_2(\mathbf{p}) = \sigma_1(\sigma_2(\mathbf{p}))$.

Proof. Consider

$$\sigma_1\sigma_2 = \{\, x \leftarrow \sigma_1(\sigma_2(x)) \mid x \in dom(\sigma_2) \,\} \cup \sigma_1|_{dom(\sigma_1)\setminus dom(\sigma_2)}.$$

The proof is by induction over $\mathcal{E}^{(X)}(\mathcal{A})$. We will show that for all possible expressions $\mathbf{p} \in \mathcal{E}^{(X)}(\mathcal{A})$, $\sigma_1\sigma_2$ applied to $\mathbf{p}$ will produce the same result as $\sigma_1(\sigma_2(\mathbf{p}))$.

1a. $\mathbf{p} \in X$: There are three cases:

$\mathbf{p} \notin dom(\sigma_1) \cup dom(\sigma_2)$: Then $\sigma_1(\sigma_2(\mathbf{p})) = \sigma_2(\mathbf{p}) = \mathbf{p}$, and $\sigma_1\sigma_2(\mathbf{p}) = \mathbf{p}$.

$\mathbf{p} \in dom(\sigma_1) \setminus dom(\sigma_2)$: Then $\sigma_1(\sigma_2(\mathbf{p})) = \sigma_1(\mathbf{p})$, and $\sigma_1\sigma_2(\mathbf{p}) = \sigma_2(\mathbf{p})$.

$\mathbf{p} \in dom(\sigma_2)$: Then, immediately, $\sigma_1(\sigma_2(\mathbf{p})) = \sigma_1\sigma_2(\mathbf{p})$.

1b. $\mathbf{p} = f \bullet \langle\rangle$: Then $\sigma'(\mathbf{p}) = \mathbf{p}$ for any σ', so $\sigma_1(\sigma_2(\mathbf{p})) = \mathbf{p} = \sigma_1\sigma_2(\mathbf{p})$.

2. $\mathbf{p} = f \bullet \langle\, \mathbf{p}_i \mid i \in I_f \,\rangle$: Assume as induction hypothesis, that for all $i \in I_f$

$$\sigma_1(\sigma_2(\mathbf{p}_i)) = \sigma_1\sigma_2(\mathbf{p}_i).$$

Then

$$\begin{aligned}
\sigma_1(\sigma_2(\mathbf{p})) &= \sigma_1(\sigma_2(f \bullet \langle\, \mathbf{p}_i \mid i \in I_f \,\rangle)) \\
&= \sigma_1(f \bullet \langle\, \sigma_2(\mathbf{p}_i) \mid i \in I_f \,\rangle) \\
&= f \bullet \langle\, \sigma_1(\sigma_2(\mathbf{p}_i)) \mid i \in I_f \,\rangle \\
&= \text{ (by the induction hypothesis)} \\
&= f \bullet \langle\, \sigma_1\sigma_2(\mathbf{p}_i) \mid i \in I_f \,\rangle \\
&= \sigma_1\sigma_2(f \bullet \langle\, \mathbf{p}_i \mid i \in I_f \,\rangle) \\
&= \sigma_1\sigma_2(\mathbf{p}).
\end{aligned}$$

■

Appendix E. Proofs of theorems of chapter 3

Theorem 3.1: *(if-conversion) Let $f: A \to A_f$ be realizable in $\mathcal{A} = \langle \mathcal{S}; F \rangle$. Assume that there exists a representation function $\eta_f: A \to \mathcal{P}^{(X)}(\mathcal{A})$ such that $\eta_f(A) = \{p_1, \ldots, p_n\}$ is finite. Assume further that each condition $\eta_f(a) = p_i$ is expressible as a polynomial c_i in $\mathcal{P}^{(X)}(\langle \{B\} \cup \mathcal{S}; \{if_{A_f}\} \cup F \rangle)$. Then f is structurally data-independent in $\langle \{B\} \cup \mathcal{S}; \{if_{A_f}\} \cup F \rangle$.*

Proof. We will construct a polynomial in $\mathcal{P}^{(X)}(\langle \{B\} \cup \mathcal{S}; \{if_{A_f}\} \cup F \rangle)$ that is equal to f. This will prove the structural data-independence of f. Consider the following sequence of polynomials:

$$q_1 = p_1$$

$$i = 2, \ldots, n: \qquad q_i = if_{A_f}(c_i, p_i, q_{i-1})$$

We now claim that $q_n = f$. This will be proved by induction over i. The inductive sentence is:

$$\bigvee_{j=1}^{i} c_j(a) \implies f(a) = q_i(a).$$

$i = 1$: $q_1(a) = p_1(a)$, and according to the definition of η_f, $c_1(a) \implies f(a) = \eta_f(a)(a) = p_1(a)$.

$i > 1$: Assume that the inductive sentence holds for i. Assume further, that $\bigvee_{j=1}^{i+1} c_j(a)$ is true.

$$q_{i+1}(a) = if_{A_f}(c_{i+1}, p_{i+1}, q_i)(a) = if_{A_f}(c_{i+1}(a), p_{i+1}(a), q_i(a)).$$

If $c_{i+1}(a)$, then $f(a) = \eta_f(a)(a) = p_{i+1}(a) = q_{i+1}(a)$. If not $c_{i+1}(a)$, then $\bigvee_{j=1}^{i} c_j(a)$ and, according to the induction hypothesis, $f(a) = q_i(a) = q_{i+1}(a)$.

Especially, the inductive sentence holds for n. Since $\bigvee_{j=1}^{n} c_j(a)$ always is true, it follows that $f(a) = q_n(a)$ for all $a \in A$, that is, $f = q_n$. ∎

Appendix F. Proofs of theorems of chapter 4

Theorem 4.1: *The evaluable closure can be alternatively defined by:*

1. *For every expression* $\mathbf{p} \in \mathcal{E}^{(X)}(\mathcal{A})$:
   1a. *If* $\mathbf{p} \in X$, *then* $ec(\{\mathbf{p}\}) = \{\mathbf{p}\}$.
   1b. *If* $\mathbf{p} = f \bullet \langle\, \mathbf{p}_i \mid i \in I_f \,\rangle$, *then* $ec(\{\mathbf{p}\}) = \bigcup(\, ec(\{\mathbf{p}_i\}) \mid i \in I_f \,) \cup \{\mathbf{p}\}$.
2. *For any* $E \subseteq \mathcal{E}^{(X)}(\mathcal{A})$, $ec(E) = \bigcup(\, ec(\{\mathbf{p}\}) \mid \mathbf{p} \in E \,)$.

Proof. We prove that $ec(E)$, as defined above, really has the closure properties in each case, with respect to the defining predicate in definition 4.1. Note that this predicate is pointwise and monotone.

1. In the case of a singleton $\{\mathbf{p}\}$, the proof is inductive. The inductive sentence is "$\mathbf{p} \in ec(\{\mathbf{p}\})$ and $ec(\{\mathbf{p}\})$ is evaluable". If this holds for all $\mathbf{p}$, then, by proposition A2, $ec(\{\mathbf{p}\})$ is equal to its evaluable closure for all $\mathbf{p}$.
   1a. $\mathbf{p} \in X$: $ec(\{\mathbf{p}\})$ is evaluable, since it contains no compound expressions. $\mathbf{p} \in \{\mathbf{p}\} = ec(\{\mathbf{p}\})$.
   1b. $\mathbf{p} = f \bullet \langle\rangle$: $ec(\{\mathbf{p}\}) = \bigcup(\, ec(\{\mathbf{p}_i\}) \mid i \in \emptyset \,) \cup \{\mathbf{p}\} = \{\mathbf{p}\}$ is evaluable, since there are no $\mathbf{p}_i$ in $\langle\, \mathbf{p}_i \mid i \in \emptyset \,\rangle = \langle\rangle$. $\mathbf{p} \in \{\mathbf{p}\} = ec(\{\mathbf{p}\})$.
   2. $\mathbf{p} = f \bullet \langle\, \mathbf{p}_i \mid i \in I_f \,\rangle$: Assume that the inductive sentence is true for $\mathbf{p}_i$ and for all $i \in I_f$. Then, by lemma A2, $\bigcup(\, ec(\{\mathbf{p}_i\}) \mid i \in I_f \,)$ is evaluable. Furthermore, all $\mathbf{p}_i \in \bigcup(\, ec(\{\mathbf{p}_i\}) \mid i \in I_f \,)$. Thus, $\bigcup(\, ec(\{\mathbf{p}_i\}) \mid i \in I_f \,) \cup \{\mathbf{p}\}$ is evaluable. Also, $\mathbf{p} \in \bigcup(\, ec(\{\mathbf{p}_i\}) \mid i \in I_f \,) \cup \{\mathbf{p}\}$.
2. $E \subseteq \mathcal{E}^{(X)}(\mathcal{A})$: The claim is trivially true for $E = \emptyset$. When E is a singleton it is already proved. For other E the claim follows from proposition A5. ∎

Before proving theorem 4.2 we prove the following lemma, which is a partial result:

Lemma F1: $\prec_e$ *is a well-founded relation.*

Proof. $\prec_e$ is a well-founded relation if the following induction principle holds (it is equivalent to the no infinite decreasing sequence condition, see for instance [MaWa85]):

$$\forall \mathbf{p} \in \mathcal{E}^{(X)}(\mathcal{A})[\forall \mathbf{p}' \in \mathcal{E}^{(X)}(\mathcal{A})[\mathbf{p}' \prec_e \mathbf{p} \implies q(\mathbf{p}')] \implies q(\mathbf{p})] \\ \implies \forall \mathbf{p} \in \mathcal{E}^{(X)}(\mathcal{A})[q(\mathbf{p})] \qquad (1)$$

(1) follows from the induction principle for expressions in definition 2.10 if the premise in (1):

$$\forall \mathbf{p} \in \mathcal{E}^{(X)}(\mathcal{A})[\forall \mathbf{p}' \in \mathcal{E}^{(X)}(\mathcal{A})[\mathbf{p}' \prec_e \mathbf{p} \implies q(\mathbf{p}')] \implies q(\mathbf{p})] \qquad (2)$$

implies the premises 1a, 1b, 2 in definition 2.10 when $\mathbf{p}$ is a variable, constant, or compound expression respectively. Let us prove that this is the case. For brevity, define for all $\mathbf{p} \in \mathcal{E}^{(X)}(\mathcal{A})$

$$r(\mathbf{p}) \iff (\forall \mathbf{p}' \in \mathcal{E}^{(X)}(\mathcal{A})[\mathbf{p}' \prec_e \mathbf{p} \implies q(\mathbf{p}')] \implies q(\mathbf{p})).$$

$\mathbf{p} \in X$: In this case there is no expression $\mathbf{p}'$ such that $\mathbf{p}' \prec_e \mathbf{p}$. $r(\mathbf{p})$ reduces to $q(\mathbf{p})$, or 1a in definition 2.10.
$\mathbf{p} = f \bullet \langle\rangle$: Also in this case $r(\mathbf{p}) \iff q(\mathbf{p})$, by the same reason as above. This is exactly 1b.
$\mathbf{p} = f \bullet \langle \mathbf{p}_i \mid i \in I_f \rangle$: $\mathbf{p}' \prec_e \mathbf{p}$ iff $\mathbf{p}' = \mathbf{p}_i$ for some $i \in I_f$. Thus,

$$\begin{aligned} r(\mathbf{p}) &\iff \forall \mathbf{p}' \in \mathcal{E}^{(X)}(\mathcal{A})[\exists i \in I_f[\mathbf{p}_i = \mathbf{p}'] \implies q(\mathbf{p}')] \implies q(\mathbf{p}) \\ &\iff \forall \mathbf{p}' \in \mathcal{E}^{(X)}(\mathcal{A})[\forall i \in I_f[\mathbf{p}_i = \mathbf{p}' \implies q(\mathbf{p}_i)]] \implies q(\mathbf{p}) \\ &\iff \forall i \in I_f[\exists \mathbf{p}' \in \mathcal{E}^{(X)}(\mathcal{A})[\mathbf{p}_i = \mathbf{p}'] \implies q(\mathbf{p}_i)] \implies q(\mathbf{p}) \\ &\iff \forall i \in I_f[q(\mathbf{p}_i)] \implies q(\mathbf{p}) \end{aligned}$$

which is exactly 2 in definition 2.10. ∎

Theorem 4.2: *$\prec_e^+$ is finite-downward.*

Proof. By lemma F1, $\prec_e$ is well-founded. According to lemma B3, it then suffices to show that $sp(\mathbf{p}, \prec_e)$ is finite for all $\mathbf{p} \in \mathcal{E}^{(X)}(\mathcal{A})$. There are three cases to consider:

$\mathbf{p} \in X$. Then $sp(\mathbf{p}, \prec_e) = \emptyset$.
$\mathbf{p} = f \bullet \langle\rangle$: Also in this case, $sp(\mathbf{p}, \prec_e) = \emptyset$.
$\mathbf{p} = f \bullet \langle \mathbf{p}_i \mid i \in I_f \rangle$. Then $sp(\mathbf{p}, \prec_e) = \{ \mathbf{p}_i \mid i \in I_f \}$. This set is finite, since I_f is finite. ∎

Lemma 4.1: *For any evaluable $E \subseteq \mathcal{E}^{(X)}(\mathcal{A})$, $\prec_e|_E^+ = \prec_e^+|_E$.*

Proof. The predicate "being evaluable" on SoE's is of the form required in lemma B7. Therefore, the result follows directly from that lemma. ∎

Appendix G. Proofs of theorems of chapter 5

First we show the following lemma:

Lemma G1: *Let $\psi: D \rightarrow E$ be onto, and let $\prec \in \mathcal{R}(D, \psi, E)$. Then the following holds:*

R1. $\forall \mathrm{p}, \mathrm{p}' \in D[\mathrm{p}' \prec \mathrm{p} \implies \psi(\mathrm{p}') \prec_e \psi(\mathrm{p})]$.

R2. $\forall \mathbf{p}, \mathbf{p}' \in E[\mathbf{p}' \prec_e \mathbf{p} \implies \exists \mathrm{p}, \mathrm{p}' \in D[\psi(\mathrm{p}') = \mathbf{p}' \wedge \psi(\mathrm{p}) = \mathbf{p} \wedge \mathrm{p}' \prec \mathrm{p}]]$.

Proof. Let $\prec$ be given by the family of selector functions $(\, s_{\mathbf{p}} \mid \mathbf{p} \in comp(D, \psi)\,)$.

R1: Consider two arbitrary $\mathrm{p}, \mathrm{p}' \in D$ such that $\mathrm{p}' \prec \mathrm{p}$. By definition 5.4, it holds for some compound expression $\mathbf{p} \in E$ that $\psi(\mathrm{p}) = \mathbf{p} = f \bullet \langle\, \mathbf{p}_i \mid i \in I_f \,\rangle$ and $\exists i \in I_f[s_{\mathbf{p}}(i) = \mathrm{p}']$. Thus, by the definition of selector functions, $\psi(s_{\mathbf{p}}(i)) = \psi(\mathrm{p}') = \mathbf{p}_i$ and from the definition of $\prec_e$ we have $\mathbf{p}_i \prec_e \mathbf{p}$. Since $\psi(\mathrm{p}') = \mathbf{p}_i$ and $\psi(\mathrm{p}) = \mathbf{p}$ the implication follows.

R2: Consider two arbitrary expressions $\mathbf{p}, \mathbf{p}'$ such that $\mathbf{p}' \prec_e \mathbf{p}$. Then $\mathbf{p} = f \bullet \langle\, \mathbf{p}_i \mid i \in I_f \,\rangle$ and $\exists i \in I_f[\mathbf{p}' = \mathbf{p}_i]$. Since $\psi: D \rightarrow E$ is onto, there exists a $\mathrm{p} \in D$ such that $\psi(\mathrm{p}) = \mathbf{p}$. There also exists a $\mathrm{p}' \in D$ such that $\psi(\mathrm{p}') = \mathbf{p}_i$, namely $s_{\mathbf{p}}(i)$. By definition 5.4 it holds that $\mathrm{p}' \prec \mathrm{p}$. ■

Proposition 5.1: *Assume that $\psi: D \rightarrow E$ is onto and $1-1$. Then, $\mathcal{R}(D, \psi, E)$ has exactly one element $\prec$ given by*

$$\mathrm{p} \prec \mathrm{p}' \iff \psi(\mathrm{p}) \prec_e \psi(\mathrm{p}')$$

for every two $\mathrm{p}, \mathrm{p}' \in D$.

Proof. We first show the equivalence. The two directions are shown separately.

$\Longrightarrow$: Directly from R1 in lemma G1.

$\Longleftarrow$: Since ψ is $1-1$ it has an inverse ψ^{-1}, and since it is onto ψ, ψ^{-1} are bijections. R2 reduces to:

$$\forall \mathbf{p}, \mathbf{p}' \in E[\mathbf{p} \prec_e \mathbf{p}' \implies \psi^{-1}(\mathbf{p}) \prec \psi^{-1}(\mathbf{p}')].$$

Especially, we can for all $\mathrm{p}, \mathrm{p}' \in D$ choose $\mathbf{p} = \psi(\mathrm{p})$ and $\mathbf{p}' = \psi(\mathrm{p}')$, which gives the desired result.

That $\mathcal{R}(D, \psi, E)$ has exactly one element now follows from the equivalence and the fact that ψ is a bijection. ■

Theorem 5.1: *For every* $\prec \in \mathcal{R}(D, \psi, E)$, $\prec^+$ *is finite-downward.*

Proof. By R1, $\prec$ is monotonically mapped into $\prec_e$ by ψ. By lemma F1, $\prec_e$ is well-founded. It follows that $\prec$ is well-founded (see for instance [MaWa85]). According to lemma B3 it then suffices to show that $sp(\mathbf{p}, \prec_e)$ is finite for all $\mathbf{p} \in \mathcal{E}^{(X)}(\mathcal{A})$, in order to show the theorem. There are three cases:

$\psi(\mathrm{p}) \in X$. Then $sp(\mathrm{p}, \prec) = \emptyset$.
$\psi(\mathrm{p}) = f \bullet \langle\,\rangle$: Also in this case, $sp(\mathrm{p}, \prec) = \emptyset$.
$\psi(\mathrm{p}) = f \bullet \langle\, \mathbf{p}_i \mid i \in I_f \,\rangle$. Then $sp(\mathrm{p}, \prec) = \{\, s_{\mathrm{p}}(i) \mid i \in I_f \,\}$, for some selector function $s_{\mathrm{p}} : I_f \to D$. This set is finite since I_f is finite. ■

Proposition 5.3: *Let* $CN = \langle D, R, f \rangle$ *be a causal and well-formed computation network. For every* $\mathrm{p} \in D$, $o(\mathrm{p}) = \phi \circ \psi(\mathrm{p})$.

Proof. Since CN is causal, R is finite-downward and thus well-founded. The result is proved by well-founded induction over R. There are two cases: $f(\mathrm{p}) \in X$ and $f(\mathrm{p}) \in F$. Since CN is well-formed, it holds that $f(\mathrm{p}) \in X \implies sp(\mathrm{p}, R) = \emptyset$. Thus, no induction hypothesis is needed in this case:

1. $f(\mathrm{p}) \in X$: $o(\mathrm{p}) = e^X_{f(\mathrm{p})} = \phi(f(\mathrm{p})) = \phi \circ \psi(\mathrm{p})$.
2. $f(\mathrm{p}) \in F$: assume as induction hypothesis, that $o(\mathrm{p}') = \phi \circ \psi(\mathrm{p}')$ for all $\mathrm{p}' \in sp(\mathrm{p}, R)$. Since CN is well-formed, $sp(\mathrm{p}, R) = \{\, s_{\mathrm{p}}(i) \mid i \in I_{f(\mathrm{p})} \,\}$. Now,

$$\begin{aligned} o(\mathrm{p}) &= f(\mathrm{p}) \circ \langle\, o(s_{\mathrm{p}}(i)) \mid i \in I_{f(\mathrm{p})} \,\rangle = \text{ (by the induction hypothesis)} \\ &= f(\mathrm{p}) \circ \langle\, \phi \circ \psi(s_{\mathrm{p}}(i)) \mid i \in I_{f(\mathrm{p})} \,\rangle \\ &= \phi(f(\mathrm{p}) \bullet \langle\, \psi(s_{\mathrm{p}}(i)) \mid i \in I_{f(\mathrm{p})} \,\rangle) \\ &= \phi(\psi(\mathrm{p})) = \phi \circ \psi(\mathrm{p}). \end{aligned}$$

Note that the proof in 2 works also when $n(f(\mathrm{p})) = 0$. ■

Theorem 5.2: *Let $\mathcal{A} = \langle \mathcal{S}; F \rangle$.*

1. *For every causal and well-formed computation network $\langle D, R, f \rangle$, D is a set of computational events of $\psi(D)$ and $R \in \mathcal{R}(D, \psi, \psi(D))$.*
2. *Let $E \subseteq \mathcal{E}^{(X)}(\mathcal{A})$ be an evaluable set of expressions. Let $\psi: D \to E$ be onto. For every $\prec \in \mathcal{R}(D, \psi, E)$, $\langle D, \prec, f \rangle$ is a causal and well-formed computation network in X over $\mathcal{A}$, where f is given by:*

$$\psi(\mathrm{p}) = \begin{cases} f(\mathrm{p}), & \psi(\mathrm{p}) \in X \\ f', & \psi(\mathrm{p}) = f' \bullet \langle \mathbf{p}_i \mid i \in I_{f'} \rangle. \end{cases}$$

Proof.

1. By proposition 5.2, $\psi(D)$ is evaluable. Apparently $\psi: D \to \psi(D)$ is onto, thus D is a SoCE of $\psi(D)$. What about R? From wf3, if $\psi(\mathrm{p}) = f' \bullet \langle \mathbf{p}_i \mid i \in I_{f'} \rangle$ and $n(f') > 0$, then $f' = f(\mathrm{p})$ and there is a function $s_\mathrm{p}: I_{f(\mathrm{p})} \to sp(\mathrm{p}, R)$ such that $\langle \mathbf{p}_i \mid i \in I_{f'} \rangle = \langle \psi(s_\mathrm{p}(i)) \mid i \in I_{f(\mathrm{p})} \rangle$. Thus, $\psi(s_\mathrm{p}(i)) = \mathbf{p}_i$ for all $i \in I_{f'}$, so s_p is a selector function according to definition 5.4. Wf3 yields $s_\mathrm{p}(i) R \mathrm{p}$ for all $i \in I_{f(\mathrm{p})}$, so R fulfils 1 in definition 5.4. Since s_p is onto, there is no other p' such that $\mathrm{p}' R \mathrm{p}$. By wf1, wf2 in definition 5.8, there is no p' such that $\mathrm{p}' R \mathrm{p}$ when $\psi(\mathrm{p})$ is a variable or a constant. Thus, R fulfils 2 in definition 5.4.
2. Trivially $\langle D, \prec, f \rangle$ is a computation network in X over $\mathcal{A}$.
   Causality: By theorem 5.1 holds that if $\prec \in \mathcal{R}(D, \psi, E)$, then $\prec$ is finite-downward. Thus, $\langle D, \prec, f \rangle$ is causal.
   Well-formedness: Since $X \cap F = \emptyset$, it holds that $f(\mathrm{p}) \in X \implies \psi(\mathrm{p}) \in X$ and $f(\mathrm{p}) \in F \implies \psi(\mathrm{p}) = f(\mathrm{p}) \bullet \langle \mathbf{p}_i \mid i \in I_{f(\mathrm{p})} \rangle$.
   wf1. $f(\mathrm{p}) \in X \implies \psi(\mathrm{p}) \in X \implies sp(\mathrm{p}, \prec) = \emptyset$, according to definition 5.4.
   wf2. $f(\mathrm{p}) \in F \wedge n(f(\mathrm{p})) = 0 \implies \psi(\mathrm{p}) = f(\mathrm{p}) \bullet \langle \rangle \implies sp(\mathrm{p}, \prec) = \emptyset$.
   wf3. $f(\mathrm{p}) \in F \wedge n(f(\mathrm{p})) > 0 \implies \psi(\mathrm{p}) = f(\mathrm{p}) \bullet \langle \mathbf{p}_i \mid i \in I_{f(\mathrm{p})} \rangle$. From 1 in definition 5.4, there exists an $s_\mathrm{p}: I_{f(\mathrm{p})} \to D$ such that $\psi(s_\mathrm{p}(i)) = \mathbf{p}_i$ for all $i \in I_{f(\mathrm{p})}$. 1 and 2 in definition 5.4 gives $sp(\mathrm{p}, \prec) = s_\mathrm{p}(I_{f(\mathrm{p})})$, so s_p is onto when considered to be a function to $sp(\mathrm{p}, \prec)$. Finally, since $f(\mathrm{p}) \bullet \langle \mathbf{p}_i \mid i \in I_{f(\mathrm{p})} \rangle \in \mathcal{E}^{(X)}(\mathcal{A})$, $S_{f(s_\mathrm{p}(i))} = S(\psi(s_\mathrm{p}(i))) = S(\mathbf{p}_i) = S_{if(\mathrm{p})}$ for all $i \in I_{f(\mathrm{p})}$. (Here the first equality follows from the definition of the sort function S on formal expressions, definition 2.8. The third equality follows from the inductive definition of $\mathcal{E}^{(X)}(\mathcal{A})$, where a formal expression is built up according to the sorts of the arguments of the operators involved.)

■

Appendix H. Proofs of theorems of chapter 6

Lemma 6.1: *(fi-transitivity): fi is transitive with respect to its first two arguments, that is:*

$$fi(\langle A, R_A\rangle, \langle B, R_B\rangle, f) \wedge fi(\langle B, R_B\rangle, \langle C, R_C\rangle, g) \implies fi(\langle A, R_A\rangle, \langle C, R_C\rangle, g \circ f).$$

Proof. Assume that $fi(\langle A, R_A\rangle, \langle B, R_B\rangle, f)$ and $fi(\langle B, R_B\rangle, \langle C, R_C\rangle, g)$ hold. From these premises we want to show fi1$(g \circ f)$, fi2$(\langle A, R_A\rangle, \langle C, R_C\rangle, f)$ and fi3(R_A, R_C). These properties are independent of each other. Therefore we show them separately.

fi1$(g \circ f)$, $[a]\varepsilon_{g\circ f}$ is finite for all $a \in A$: The premises are fi1(f): $[a]\varepsilon_f$ is finite for all $a \in A$ and fi1(g): $[b]\varepsilon_g$ is finite for all $b \in B$. $A/\varepsilon_{g\circ f}$ is a "coarsening" of A/ε_f. For every $[a]\varepsilon_{g\circ f} \in A/\varepsilon_{g\circ f}$ holds that

$$[a]\varepsilon_{g\circ f} = \bigcup([a']\varepsilon_f \mid f(a') \in [f(a)]\varepsilon_g).$$

This is a union of finitely many finite sets and thus it is finite.

fi2$(\langle A, R_A\rangle, \langle C, R_C\rangle, g \circ f)$: for all $c, c' \in C$,

$$cR_Cc' \iff (c \neq c' \wedge \exists a, a' \in A[aR_Aa' \wedge g \circ f(a) = c \wedge g \circ f(a') = c']).$$

The premises are fi2$(\langle A, R_A\rangle, \langle B, R_B\rangle, f)$: for all $b, b' \in B$

$$bR_Bb' \iff (b \neq b' \wedge \exists a, a' \in A[aR_Aa' \wedge f(a) = b \wedge f(a') = b']) \qquad (1)$$

and fi2$(\langle B, R_B\rangle, \langle C, R_C\rangle, g)$: for all $c, c' \in C$

$$cR_Cc' \iff (c \neq c' \wedge \exists b, b' \in B[bR_Bb' \wedge g(b) = c \wedge g(b') = c']). \qquad (2)$$

We differ between two subcases:

a. $c \notin g(B)$ or $c' \notin g(B)$: Then, by (2), $\neg(cR_Cc')$. Also, there are no $a, a' \in A$ such that $g \circ f(a) = c$ and $g \circ f(a') = c'$. Thus, both sides of the equivalence in fi2$(\langle A, R_A\rangle, \langle C, R_C\rangle, g \circ f)$ are false and the equivalence holds.

b. $c, c' \in g(B)$: Again there are two subcases.

   ba. For all $b \in [c]\varepsilon_g$, there is no $a \in A$ such that $f(a) = b$, or similarly for all $b' \in [c']\varepsilon_g$: then (1) yields $\neg(bR_Bb')$. (2) then implies $\neg(cR_Cc')$, so the left hand side of the equivalence in fi2$(\langle A, R_A\rangle, \langle C, R_C\rangle, g \circ f)$ is false. On the other hand there are no $a, a' \in A$ such that $g \circ f(a) = c$ and $g \circ f(a') = c'$. Thus, the right side of the equivalence is also false and the equivalence holds.

bb. There are $a, a' \in A$ such that $g \circ f(a) = c$ and $g \circ f(a') = c'$: For such c, c' we can set $b = f(a)$ and $b' = f(a')$ in (2) and replace the existential quantification over b, b' with quantification over a, a':

$$\begin{aligned} cR_C c' &\iff \\ c \neq c' &\wedge \\ \exists a, a' \in A[f(a) R_B f(a') \wedge g(f(a)) &= c \wedge g(f(a')) = c']. \end{aligned} \tag{3}$$

In the same manner we can instantiate $b = f(a'')$ and $b' = f(a''')$ and re-quantify over a'', a''' in fi2$(\langle A, R_A\rangle, \langle B, R_B\rangle, f)$: for all $a'', a''' \in A$,

$$\begin{aligned} f(a'') R_B f(a''') &\iff \\ f(a'') \neq f(a''') &\wedge \\ \exists a, a' \in A[a R_A a' \wedge f(a) = f(a'') &\wedge f(a') = f(a''')]. \end{aligned} \tag{4}$$

If we rename a, a' in (4) to a'', a''' and instantiate the universally quantified a'', a''' in (4) to a, a' in (3), then we can substitute the right-hand side in the resulting equivalence for $f(a) R_B f(a')$ in (3), obtaining

$$\begin{aligned} cR_C c' \iff c \neq c' \wedge \\ \exists a, a' \in A[f(a) \neq f(a') \wedge \\ \exists a'', a''' \in A[a'' R_A a''' \wedge \\ f(a'') = f(a) \wedge \\ f(a''') = f(a')] \wedge \\ g(f(a)) = c \wedge g(f(a')) = c']. \end{aligned} \tag{5}$$

$(c \neq c' \wedge g(f(a)) = c \wedge g(f(a')) = c') \implies f(a) \neq f(a')$, so $f(a) \neq f(a')$ can be removed in (5). Since $f(a'') = f(a)$ and $f(a''') = f(a')$, we can substitute $f(a'')$ for $f(a)$ and $f(a''')$ for $f(a')$ in the two last conjuncts and remove the equalities, thereby totally eliminating a, a'. The result is

$$\begin{aligned} cR_C c' &\iff \\ c \neq c' &\wedge \\ \exists a'', a''' \in A[a'' R_A a''' \wedge g(f(a'')) &= c \wedge g(f(a''')) = c']. \end{aligned}$$

This is exactly fi2$(\langle A, R_A\rangle, \langle C, R_C\rangle, g \circ f)$.

fi3(R_A, R_C), R_A^+ finite-downward $\Longrightarrow$ R_C^+ finite-downward: This follows immediately from the transitivity of $\Longrightarrow$. ■

Lemma 6.4: *If fi($\langle A, R_A\rangle, \langle B, R_B\rangle, f$), $R'_A \subseteq R_A$ and if R_B^+ is finite-downward, then there exists an R'_B such that fi($\langle A, R'_A\rangle, \langle B, R'_B\rangle, f$) and $R'_B \subseteq R_B$.*

Proof. Assume that *fi*($\langle A, R_A\rangle, \langle B, R_B\rangle, f$) holds, that $R'_A \subseteq R_A$ and that R_B^+ is finite-downward. Define the relation R'_B on B by fi2($\langle A, R_A\rangle, \langle B, R'_B\rangle, f$): for all $b, b' \in B$,

$$bR'_Bb' \iff (b \neq b' \wedge \exists a, a' \in A[aR'_Aa' \wedge f(a) = b \wedge f(a') = b']). \quad (1)$$

Let us show that $R'_B \subseteq R_B$. $R'_A \subseteq R_A$, thus

$$\forall a, a' \in A[aR'_Aa' \Longrightarrow aR_Aa']. \quad (2)$$

fi2($\langle A, R_A\rangle, \langle B, R_B\rangle, f$) implies, for all $b, b' \in B$, that

$$bR_Bb' \iff (b \neq b' \wedge \exists a, a' \in A[aR_Aa' \wedge f(a) = b \wedge f(a') = b']). \quad (3)$$

(2) in (1) yields, for all $b, b' \in B$, that

$$bR'_Bb' \Longrightarrow (b \neq b' \wedge \exists a, a' \in A[aR_Aa' \wedge f(a) = b \wedge f(a') = b']). \quad (4)$$

(4) and (3) now gives

$$\forall b, b' \in B[bR'_Bb' \Longrightarrow bR_Bb'],$$

or $R'_B \subseteq R_B$. Let us now show fi1(f) and fi3(R'_A, R'_B):

fi1(f): Follows directly from *fi*($\langle A, R_A\rangle, \langle B, R_B\rangle, f$).

fi3(R'_A, R'_B): Let us show that $R_B'^+$ is finite-downward. Then the implication will follow trivially. We have already shown $R'_B \subseteq R_B$ which implies $R'_B \subseteq R_B^+$. R_B^+ is finite-downward. Lemma B9 then gives the desired result. ■

Lemma 6.5: *If R_A is irreflexive and if f is a bijection $A \to B$, then there exists a relation R_B on B such that fi($\langle A, R_A\rangle, \langle B, R_B\rangle, f$) and f is a graph isomorphism $\langle A, R_A\rangle \to \langle B, R_B\rangle$.*

Proof. Assume that R_A is irreflexive and that f is a bijection $A \to B$. Then, since f^{-1} is 1-1,

$$f^{-1}(b)R_Af^{-1}(b') \Longrightarrow f^{-1}(b) \neq f^{-1}(b') \Longrightarrow b \neq b' \quad (1)$$

for all $b, b' \in B$. Define R_B on B by fi2$(\langle A, R_A\rangle, \langle B, R_B\rangle, f)$. Then, for all $b, b' \in B$,

$$\begin{aligned} bR_Bb' &\iff b \neq b' \wedge \exists a, a' \in A[f(a) = b \wedge f(a') = b' \wedge aR_Aa'] \\ &\iff \text{(choose } a = f^{-1}(b),\ a' = f^{-1}(b')) \\ &\iff b \neq b' \wedge f^{-1}(b)R_Af^{-1}(b') \\ &\iff \text{(by (1))} \\ &\iff f^{-1}(b)R_Af^{-1}(b'). \end{aligned}$$

Thus, f^{-1} is a graph isomorphism $\langle B, R_B\rangle \to \langle A, R_A\rangle$ which implies that f is a graph isomorphism $\langle A, R_A\rangle \to \langle B, R_B\rangle$. It remains to check that fi1, fi3 are fulfilled:

fi1(f): Follows directly, since f is 1-1.

fi3(R_A, R_B): Follows immediately, since f is a graph isomorphism from $\langle A, R_A\rangle$ to $\langle B, R_B\rangle$. ■

Lemma H1: *For any recursion scheme* $\mathbf{F}$, $sp(\langle x, \mathbf{p}\rangle, \prec_{\mathbf{F}})$ *is finite for all* $\langle x, \mathbf{p}\rangle \in \mathbf{F}$.

Proof. $\langle x', \mathbf{p}'\rangle \prec_{\mathbf{F}} \langle x, \mathbf{p}\rangle$ iff $x' \in varset(\mathbf{p})$. There are only finitely many such x'. $\mathbf{F}$ is a partial function, so if both $\langle x', \mathbf{p}'\rangle$ and $\langle x', \mathbf{p}''\rangle$ belongs to $\mathbf{F}$, then $\mathbf{p}' = \mathbf{p}''$ which implies $\langle x', \mathbf{p}'\rangle = \langle x', \mathbf{p}''\rangle$. Thus, there are only finitely many $\langle x', \mathbf{p}'\rangle$ in $sp(\langle x, \mathbf{p}\rangle, \prec_{\mathbf{F}})$. ■

Lemma 6.6: $\langle x, \mathbf{F}(x)\rangle$ *is minimal with respect to* $\prec_{\mathbf{F}}$ *iff* $varset(\mathbf{F}(x)) \subseteq X_f(\mathbf{F})$.

Proof.

$$\begin{aligned} \langle x, \mathbf{F}(x)\rangle \text{ minimal w.r.t. } \prec_{\mathbf{F}} &\iff \\ \text{(from definition 6.5)} &\iff \\ \neg\exists\langle x', \mathbf{F}(x')\rangle \in \mathbf{F}[x' \in varset(\mathbf{F}(x))] &\iff \\ \neg\exists x' \in X_a(\mathbf{F})[x' \in varset(\mathbf{F}(x))] &\iff \\ \neg\exists x'[x' \in X_a(\mathbf{F}) \wedge x' \in varset(\mathbf{F}(x))] &\iff \\ \forall x'[x' \notin X_a(\mathbf{F}) \vee x' \notin varset(\mathbf{F}(x))] &\iff \\ \forall x'[x' \in X_f(\mathbf{F}) \vee x' \notin varset(\mathbf{F}(x))] &\iff \\ \forall x'[x' \in varset(\mathbf{F}(x)) \implies x' \in X_f(\mathbf{F})] &\iff \\ varset(\mathbf{F}(x)) \subseteq X_f(\mathbf{F}). & \end{aligned}$$

■

Theorem 6.1: *Let* $\mathbf{F}$ *be a causal recursion scheme in* X. *For any* $x \in X$, $o_{\mathbf{F}}(x) = \phi \circ \psi_{\mathbf{F}}(x)$.

Proof. There are two cases:
$x \in X_f(\mathbf{F})$: $o_{\mathbf{F}}(x) = e_x^X = \phi(x) = \phi(\psi_{\mathbf{F}}(x)) = \phi \circ \psi_{\mathbf{F}}(x)$.
$x \in X_a(\mathbf{F})$: Every $x \in X_a(\mathbf{F})$ corresponds to exactly one $\langle x, \mathbf{p} \rangle \in \mathbf{F}$ and vice versa. Thus, we can reason about $\langle x, \mathbf{p} \rangle$ instead of x. Especially, since $\mathbf{F}$ is causal, we can perform well-founded induction over $\prec_{\mathbf{F}}$:

1. $\langle x, \mathbf{p} \rangle$ minimal: Then, by lemma 6.6, $\mathit{varset}(\mathbf{p}) \subseteq X_f(\mathbf{F})$. Thus, $\psi_{\mathbf{F}}(x') = x'$ for all $x' \in \mathit{varset}(\mathbf{p})$, and

$$\begin{aligned} o_{\mathbf{F}}(x) &= \phi(\mathbf{p}) \circ \langle\, o_{\mathbf{F}}(x') \mid x' \in \mathit{varset}(\mathbf{p}) \,\rangle \\ &= \phi(\mathbf{p}) \circ \langle\, e_{x'}^X \mid x' \in \mathit{varset}(\mathbf{p}) \,\rangle \\ &= \phi(\mathbf{p}) \circ \langle\, \phi(x') \mid x' \in \mathit{varset}(\mathbf{p}) \,\rangle \\ &= \phi(\mathbf{p}) \circ \langle\, \phi(\psi_{\mathbf{F}}(x')) \mid x' \in \mathit{varset}(\mathbf{p}) \,\rangle \\ &= \text{ (by theorem D1)} \\ &= \phi(\psi_{\mathbf{F}}(\mathbf{p})|_{\mathit{varset}(\mathbf{p})}) \\ &= \text{ (by lemma D10)} \\ &= \phi(\psi_{\mathbf{F}}(\mathbf{p})) \\ &= \phi(\psi_{\mathbf{F}}(x)) \\ &= \phi \circ \psi_{\mathbf{F}}(x). \end{aligned}$$

2. Assume as induction hypothesis, that for all $\langle x', \mathbf{p}' \rangle \prec_{\mathbf{F}} \langle x, \mathbf{p} \rangle$, that is: for all $x' \in \mathit{varset}(\mathbf{p}) \cup X_a(\mathbf{F})$, $o_{\mathbf{F}}(x') = \phi \circ \psi_{\mathbf{F}}(x')$. Then

$$\begin{aligned} o_{\mathbf{F}}(x) &= \phi(\mathbf{p}) \circ \langle\, o_{\mathbf{F}}(x') \mid x' \in \mathit{varset}(\mathbf{p}) \,\rangle \\ &= \text{ (by the induction hypothesis)} \\ &= \phi(\mathbf{p}) \circ \langle\, \phi \circ \psi_{\mathbf{F}}(x') \mid x' \in \mathit{varset}(\mathbf{p}) \,\rangle \\ &= \text{ (by theorem D1)} \\ &= \phi(\psi_{\mathbf{F}}|_{\mathit{varset}(\mathbf{p})}(\mathbf{p})) \\ &= \text{ (by lemma D10)} \\ &= \phi(\psi_{\mathbf{F}}(\mathbf{p})) \\ &= \phi(\psi_{\mathbf{F}}(x)) \\ &= \phi \circ \psi_{\mathbf{F}}(x). \end{aligned}$$

■

Lemma H2: *For any causal recursion scheme* $\mathbf{F}$ *in* X, $range(\psi_{\mathbf{F}}) \subseteq X_f(\mathbf{F})$.

Proof. The result follows if, for all $x \in X$, $varset(\psi_{\mathbf{F}}(x)) \subseteq X_f(\mathbf{F})$. Let us prove this. There are two cases:

$x \in X_f(\mathbf{F})$: $varset(\psi_{\mathbf{F}}(x)) = varset(x) = \{x\} \subseteq X_f(\mathbf{F})$.

$x \in X_a(\mathbf{F})$: The proof is by induction over $\prec_{\mathbf{F}}$.

1. $\langle x, \mathbf{F}(x)\rangle$ minimal: Then $varset(\mathbf{F}(x)) \subseteq X_f(\mathbf{F})$, and

$$\begin{aligned}\psi_{\mathbf{F}}(x) &= \psi_{\mathbf{F}}(\mathbf{F}(x)) \\ &= \text{(by lemma D10)} \\ &= \psi_{\mathbf{F}}|_{varset(\mathbf{F}(x))}(\mathbf{F}(x)) \\ &= \{\, x' \leftarrow x' \mid x' \in varset(\mathbf{F}(x)) \,\}(\mathbf{F}(x)) \\ &= \mathbf{F}(x),\end{aligned}$$

so $varset(\psi_{\mathbf{F}}(x)) = varset(\mathbf{F}(x)) \subseteq X_f(\mathbf{F})$.

2. Assume as induction hypothesis that for all $x' \in varset(\mathbf{F}(x)) \cap X_a(\mathbf{F})$ holds that $varset(\psi_{\mathbf{F}}(x')) \subseteq X_f(\mathbf{F})$. Then

$$\begin{aligned}&range(\psi_{\mathbf{F}}|_{varset(\mathbf{F}(x))}) = \\ &\bigcup(\, varset(\psi_{\mathbf{F}}(x')) \mid x' \in varset(\mathbf{F}(x)) \,) = \\ &\bigcup(\, varset(\psi_{\mathbf{F}}(x')) \mid x' \in varset(\mathbf{F}(x)) \cap X_f(\mathbf{F}) \,) \cup \\ &\bigcup(\, varset(\psi_{\mathbf{F}}(x')) \mid x' \in varset(\mathbf{F}(x)) \cap X_a(\mathbf{F}) \,) \subseteq \\ &X_f(\mathbf{F}) \cup X_f(\mathbf{F}) = \\ &X_f(\mathbf{F}).\end{aligned}$$

Now $\psi_{\mathbf{F}}(x) = \psi_{\mathbf{F}}(\mathbf{F}(x))$. Thus, by lemma D7,

$$\begin{aligned}&varset(\psi_{\mathbf{F}}(x)) = \\ &varset(\psi_{\mathbf{F}}(\mathbf{F}(x))) = \\ &(varset(\mathbf{F}(x)) \setminus dom(\psi_{\mathbf{F}})) \cup range(\psi_{\mathbf{F}}|_{varset(\mathbf{F}(x))}) = \\ &(\psi_{\mathbf{F}} \text{ total} \implies dom(\psi_{\mathbf{F}}) \supseteq varset(\mathbf{F}(x))) = \\ &\emptyset \cup range(\psi_{\mathbf{F}}|_{varset(\mathbf{F}(x))}) \subseteq \\ &X_f(\mathbf{F}).\end{aligned}$$

■

Theorem 6.2: *Let* $\mathbf{F}$ *be a causal recursion scheme in* X. *For any* $x \in X$,

$$\psi_{\mathbf{F}}(x) \in \mathcal{E}^{(X_f(\mathbf{F}))}(\mathcal{A}).$$

Proof. By lemma H2, $varset(\psi_{\mathbf{F}}(x)) \subseteq X_f(\mathbf{F})$ for all $x \in X$. The theorem then follows directly from lemma D3. ∎

Theorem 6.3: *Let* $\langle M, \prec_M \rangle$ *be a cs-assignment structure. For all* m, m' *in* M,

$$m \prec_M m' \iff m \neq m' \wedge dom(m) \cap range(m') \neq \emptyset.$$

Proof. By definition,

$$\begin{aligned} & m \prec_M m' \iff \\ & m \neq m' \wedge \exists y, y' \in \mathbf{F}[\varphi_M(y) = m \wedge \varphi_M(y') = m' \wedge y \prec_{\mathbf{F}} y'] \iff \\ & m \neq m' \wedge \exists y, y' \in \mathbf{F}[y \in m \wedge y' \in m' \wedge y \prec_{\mathbf{F}} y'] \iff \\ & m \neq m' \wedge \exists x, x' \in X_b(\mathbf{F})[\langle x, \mathbf{F}(x)\rangle \in m \wedge \langle x', \mathbf{F}(x')\rangle \in m' \wedge \\ & \qquad \langle x, \mathbf{F}(x)\rangle \prec_{\mathbf{F}} \langle x', \mathbf{F}(x')\rangle]. \end{aligned}$$

Now, $\langle x, \mathbf{F}(x)\rangle \in m \iff x \in dom(m)$ and $\langle x, \mathbf{F}(x)\rangle \prec_{\mathbf{F}} \langle x', \mathbf{F}(x')\rangle \iff x \in varset(\mathbf{F}(x'))$. Thus,

$$\begin{gathered} m \prec_M m' \iff \\ m \neq m' \wedge \\ \exists x, x' \in X_b(\mathbf{F})[x \in dom(m) \wedge \langle x', \mathbf{F}(x')\rangle \in m' \wedge x \in varset(\mathbf{F}(x'))]. \end{gathered}$$

Furthermore, $range(m') = \bigcup(varset(\mathbf{F}(x')) \mid \langle x', \mathbf{F}(x')\rangle \in m')$. Thus,

$$x \in range(m') \iff \exists x' \in X_b(\mathbf{F})[\langle x', \mathbf{F}(x')\rangle \in m' \wedge x \in varset(\mathbf{F}(x'))],$$

and

$$\begin{aligned} m \prec_M m' &\iff m \neq m' \wedge \exists x \in X_b(\mathbf{F})[x \in dom(m) \wedge x \in range(m'))] \\ &\iff m \neq m' \wedge dom(m) \cap range(m') \neq \emptyset. \end{aligned}$$

∎

Theorem 6.4: *For any CCSA structure $\langle M, \prec_M\rangle$ derived from* $\mathbf{F}$, *$\langle P(M), \prec_{P(M)}\rangle$ is an isomorphic pure CCSA structure derived from $p(\mathbf{F})$. p is an isomorphism $M \to P(M)$, and $o_{p(\mathbf{F})}(p(m)) = o_{\mathbf{F}}(m)$ for all $m \in M$.*

Proof. For any $\langle x, \mathbf{p}\rangle \in \mathbf{F}$, denote $\varphi_M(\langle x, \mathbf{p}\rangle)$ by m_x. Hence $p(\langle x, \mathbf{p}\rangle) = \langle x, p(m_x)(\mathbf{p})\rangle$. If we define $\prec_p$ on $p(\mathbf{F})$ by:

$$y \prec_{\mathbf{F}} y' \iff p(y) \prec_p p(y')$$

for all $y, y' \in \mathbf{F}$, then p is a graph isomorphism $\langle \mathbf{F}, \prec_{\mathbf{F}}\rangle \to \langle p(\mathbf{F}), \prec_p\rangle$.

Let us now consider $p(\mathbf{F})$. We will show the following: first, that $p(\mathbf{F})$ is causal, that is: that $\prec^+_{p(\mathbf{F})}$ given by definition 6.5 is finite-downward. Second, that $P(M) = \{p(m) \mid m \in M\}$ (where p is extended to sets of assignments) is a CCSA-set derived from $p(\mathbf{F})$. Third, that p extended to sets is a graph isomorphism $\langle M, \prec_M\rangle \to \langle P(M), \prec_{P(M)}\rangle$, where $\prec_{P(M)}$ is given by definition 6.8. Fourth, that $p(\mathbf{F})$ is pure, and finally fifth, that $o_{p(\mathbf{F})}(p(m)) = o_{\mathbf{F}}(m)$ for all $m \in \mathbf{F}$.

1. $\prec^+_{p(\mathbf{F})}$ *finite-downward*: Consider $\prec^+_p$. It is finite-downward since $\prec^+_{\mathbf{F}}$ on $\mathbf{F}$ is finite-downward. Let us show that $\prec_{p(\mathbf{F})} \subseteq \prec^+_p$, which by lemma B9 will yield the desired result. This is shown by well-founded induction over $\prec_{\mathbf{F}}$, taking the inductive sentence to be

$$\forall y' \in \mathbf{F}[p(y') \prec_{p(\mathbf{F})} p(y) \implies p(y') \prec^+_p p(y)].$$

(All assignments in $p(\mathbf{F})$ can be written as $p(y)$, where $y \in \mathbf{F}$.)

- $y = \langle x, \mathbf{p}\rangle$ minimal w.r.t. $\prec_{\mathbf{F}}$: Then, by lemma 6.6, $\mathit{varset}(\mathbf{p}) \subseteq X_f(\mathbf{F})$. Since $X_a(p(\mathbf{F})) = X_a(\mathbf{F})$ (which follows from the definition of p), $X_f(p(\mathbf{F})) = X_f(\mathbf{F})$. Thus,

$$\begin{aligned} \mathit{varset}(\mathbf{p}) \cap \mathit{dom}(m_x) = \emptyset &\implies \\ p(\langle x, \mathbf{p}\rangle) = \langle x, m_x(\mathbf{p})\rangle = \langle x, \mathbf{p}\rangle &\implies \\ \forall \langle x', \mathbf{p}'\rangle \in p(\mathbf{F})[x' \notin \mathit{varset}(\mathbf{p})] &\iff \\ \neg\exists \langle x', \mathbf{p}'\rangle \in p(\mathbf{F})[\langle x', \mathbf{p}'\rangle \prec_{p(\mathbf{F})} \langle x, \mathbf{p}\rangle] &\iff \\ \neg\exists y' \in \mathbf{F}[p(y') \prec_{p(\mathbf{F})} p(y)], \end{aligned}$$

  where $p(y') = \langle x', \mathbf{p}'\rangle$ and $p(y) = \langle x, \mathbf{p}\rangle$. So in this case there is no y' such that $p(y') \prec_{p(\mathbf{F})} p(y)$ and the implication holds trivially.

- Inductive step: assume as induction hypothesis, that for all y' such that $y' \prec_{\mathbf{F}} y$

$$\forall y'' \in p(\mathbf{F})[p(y'') \prec_{p(\mathbf{F})} p(y') \implies p(y'') \prec_p^+ p(y')].$$

Let us prove that then, for all $y'' \in \mathbf{F}$, $p(y'') \prec_{p(\mathbf{F})} p(y) \implies p(y'') \prec_p^+ p(y)$. So assume, for an arbitrary y'', that $p(y'') \prec_{p(\mathbf{F})} p(y)$. Then, if we set $y'' = \langle x'', \mathbf{p}'' \rangle$ and $y = \langle x, \mathbf{p} \rangle$,

$$p(\langle x'', \mathbf{p}'' \rangle) \prec_{p(\mathbf{F})} p(\langle x, \mathbf{p} \rangle) \implies x'' \in \mathit{varset}(p(m_x)(\mathbf{p})).$$

There are two cases:
$x'' \in \mathit{varset}(\mathbf{p})$: Then

$$\begin{aligned} \langle x'', \mathbf{p}'' \rangle \prec_{\mathbf{F}} \langle x, \mathbf{p} \rangle &\implies p(\langle x'', \mathbf{p}'' \rangle) \prec_p p(\langle x, \mathbf{p} \rangle) \\ &\implies p(\langle x'', \mathbf{p}'' \rangle) \prec_p^+ p(\langle x, \mathbf{p} \rangle). \end{aligned}$$

$x'' \notin \mathit{varset}(\mathbf{p})$: Then

$$x'' \notin \mathit{varset}(\mathbf{p}) \wedge x'' \in \mathit{varset}(p(m_x)(\mathbf{p}))$$

which by lemma D12 implies that there is a

$$p(\langle x', \mathbf{p}' \rangle) = \langle x', p(m_x)(\mathbf{p}') \rangle$$

in $p(m_x)$, such that

$$x' \in \mathit{varset}(\mathbf{p}) \tag{1}$$

and

$$x'' \in \mathit{varset}(p(m_x)(\mathbf{p}')). \tag{2}$$

From (2),

$$p(\langle x'', \mathbf{p}'' \rangle) \prec_{p(\mathbf{F})} p(\langle x', \mathbf{p}' \rangle). \tag{3}$$

From (1)

$$\langle x', \mathbf{p}' \rangle \prec_{\mathbf{F}} \langle x, \mathbf{p} \rangle \tag{4}$$

so the induction hypothesis can be applied to (3), which yields

$$p(\langle x'', \mathbf{p}'' \rangle) \prec_p^+ p(\langle x', \mathbf{p}' \rangle). \tag{5}$$

Furthermore, (4) implies

$$p(\langle x', \mathbf{p}' \rangle) \prec_p p(\langle x, \mathbf{p} \rangle)$$

which together with (5) gives

$$p(\langle x'', \mathbf{p}'' \rangle) \prec_p^+ p(\langle x, \mathbf{p} \rangle),$$

which is what we wanted to show.

2. *P(M) SoMA derived from p(***F***)*: First, note that $P(M) = \{ p(m) \mid m \in M \}$ is a partition of $p(\mathbf{F})$, since $p(m) = \bigcup(p(y) \mid y \in m)$ for every $m \in M$ and M is a partition of $\mathbf{F}$. Let us show that there is a $\prec_{P(M)}$ on $P(M)$ such that $fi(\langle p(\mathbf{F}), \prec_{p(\mathbf{F})} \rangle, \langle P(M), \prec_{P(M)} \rangle, \varphi_{P(M)})$:
fi1($\varphi_{P(M)}$): M is a CCSA-set derived from $\mathbf{F}$, thus $fi(\langle \mathbf{F}, \prec_{\mathbf{F}} \rangle, \langle M, \prec_M \rangle, \varphi_M)$ holds. This implies that every $m \in M$ is finite. Since $f: \mathbf{F} \rightarrow p(\mathbf{F})$ is 1-1, every $p(m)$ in $P(M)$ is finite.
fi2($\langle p(\mathbf{F}), \prec_{p(\mathbf{F})} \rangle, \langle P(M), \prec_{P(M)} \rangle, \varphi_{P(M)}$): Define $\prec_{P(M)}$ so this holds, i.e.: for all $m, m' \in P(M)$,

$$\begin{gathered} m \prec_{P(M)} m' \iff \\ m \neq m' \wedge \\ \exists y, y' \in p(\mathbf{F})[y \prec_{p(\mathbf{F})} y' \wedge \varphi_{P(M)}(y) = m \wedge \varphi_{P(M)}(y') = m']. \end{gathered}$$

fi3($\prec_{p(\mathbf{F})}, \prec_{P(M)}$): According to 1, $\prec_{p(\mathbf{F})}^+$ is finite-downward. Thus, we must prove that $\prec_{P(M)}^+$ as defined above is finite-downward. Define $\prec_p$ on $P(M)$ by

$$p(m) \prec_p p(m') \iff m \prec_M m'$$

for all $m, m' \in M$. $\prec_p$ is isomorphic to $\prec_M$ on $\mathbf{F}$. By lemma B5, $\prec_p^+$ is then isomorphic to $\prec_M^+$. Thus, $\prec_p^+$ is finite-downward. So if we can prove that $\prec_{P(M)} \subseteq \prec_p^+$, then we have by lemma B9 proved that $\prec_{P(M)}^+$ is finite-downward as well. Below we will actually prove the somewhat stronger result $\prec_{P(M)} \subseteq \prec_p$. From theorem 6.3 follows that

$$p(m) \prec_p p(m') \iff m \neq m' \wedge dom(m) \cap range(m') \neq \emptyset.$$

On the other hand,

$$\begin{gathered} p(m) \prec_{P(M)} p(m') \iff \\ p(m) \neq p(m') \wedge dom(p(m)) \cap range(p(m')) \neq \emptyset. \end{gathered}$$

Since p is 1-1, it holds that $m \neq m' \iff p(m) \neq p(m')$, so

$$dom(p(m)) \cap range(p(m')) \subseteq dom(m) \cap range(m') \implies \prec_{P(M)} \subseteq \prec_p.$$

Furthermore, $dom(p(m)) = dom(m)$ (since p preserves the first component of its argument), so it suffices to show

$$range(p(m')) \subseteq range(m')$$

or equivalently

$$\forall \langle x, \mathbf{p}\rangle \in m'[varset(p(m')(\mathbf{p})) \subseteq range(m')].$$

This is proved for all $\langle x, \mathbf{p}\rangle$ in $\mathbf{F}$ by induction over $\prec_{\mathbf{F}}|_{\varphi_M(\langle x,\mathbf{p}\rangle)}$, the restriction of $\prec_{\mathbf{F}}$ to $\varphi_M(\langle x, \mathbf{p}\rangle) = m_x$. The inductive sentence is

$$varset(p(m_x)(\mathbf{p})) \subseteq range(m_x).$$

- $\langle x, \mathbf{p}\rangle$ minimal in m_x: then

$$\begin{aligned} \neg\exists\langle x', \mathbf{p}'\rangle \in m_x[x' \in varset(\mathbf{p})] &\iff \\ varset(\mathbf{p}) \cap dom(m_x) = \emptyset &\iff \\ varset(\mathbf{p}) \cap dom(p(m_x)) = \emptyset &\iff \\ \text{(by lemma D5)} &\iff \\ p(m_x)(\mathbf{p}) = \mathbf{p} &\implies \\ varset(p(m_x)(\mathbf{p})) = varset(\mathbf{p}) \subseteq range(m_x). \end{aligned}$$

- Inductive step. Assume as induction hypothesis, that for all $\langle x', \mathbf{p}'\rangle$ such that $\langle x', \mathbf{p}'\rangle \prec_{\mathbf{F}}|_{m_x} \langle x, \mathbf{p}\rangle$,

$$varset(p(m_x)(\mathbf{p}')) \subseteq range(m_x).$$

  Let us show that $varset(p(m_x)(\mathbf{p})) \subseteq range(m_x)$ as well. Consider an arbitrary $x' \in varset(\mathbf{p})$. There are two cases:

  $x' \notin dom(m_x)$: Since $dom(p(m_x)) = dom(m_x)$ it holds that $p(m_x)(x') = x'$, and since x' is in $varset(\mathbf{p})$ it is in $range(m_x)$ as well. Thus, $varset(x') = \{x'\} \subseteq range(m_x)$.

  $x' \in dom(m_x)$: Then there is a $\mathbf{p}'$ such that $\langle x', \mathbf{p}'\rangle \in m_x$. $\langle x', p(m_x)(\mathbf{p}')\rangle = p(\langle x', \mathbf{p}'\rangle) \in p(m_x)$, so $p(m_x)(x') = p(m_x)(\mathbf{p}')$ and $varset(p(m_x)(x')) = varset(p(m_x)(\mathbf{p}'))$. Furthermore, since $\langle x', \mathbf{p}'\rangle \prec_{\mathbf{F}}|_{m_x} \langle x, \mathbf{p}\rangle$, we can apply the induction hypothesis to obtain $varset(p(m_x)(x')) \subseteq range(m_x)$.

So in every case holds that $varset(p(m_x)(x')) \subseteq range(m_x)$. Thus, by lemma D6,

$$varset(p(m_x)(\mathbf{p})) = \bigcup(\, varset(p(m_x)(x)) \mid x \in varset(\mathbf{p})\,)$$
$$\subseteq range(m_x).$$

3. *p graph isomorphism* $\langle M, \prec_M \rangle \rightarrow \langle P(M), \prec_{P(M)} \rangle$: This is shown by proving that $\prec_{P(M)} = \prec_p$. We have already proved $\prec_{P(M)} \subseteq \prec_p$, so it remains to prove $\prec_p \subseteq \prec_{P(M)}$. There are two cases:

   $m = m'$: Then $p(m) \not\prec_p p(m')$ and $p(m) \not\prec_{P(M)} p(m')$.

   $m \neq m'$: Then $p(m) \prec_p p(m') \iff dom(m) \cap range(m') \neq \emptyset$, and $p(m) \prec_{P(M)} p(m') \iff dom(p(m)) \cap range(p(m')) \neq \emptyset$. So if we can prove that for all m, m' such that $m \neq m'$

   $$dom(m) \cap range(m') \subseteq dom(p(m)) \cap range(p(m')),$$

   or equivalently

   $$dom(m) \cap range(m') \subseteq dom(m) \cap range(p(m')),$$

   then the desired result follows. This set inclusion follows if, for all $\langle x, \mathbf{p} \rangle \in m'$,

   $$dom(m) \cap varset(\mathbf{p}) \subseteq dom(m) \cap varset(p(m')(\mathbf{p})).$$

   Since $m \neq m'$, it holds that $dom(m) \cap dom(m') = \emptyset$, which implies $dom(m) \cap dom(p(m')) = \emptyset$. Hence $dom(m) \setminus dom(p(m')) = dom(m)$. Lemma D9 yields

   $$varset(\mathbf{p}) \setminus dom(p(m')) \subseteq varset(p(m')(\mathbf{p})),$$

   so

   $$\begin{aligned} dom(m) \cap varset(\mathbf{p}) &= (dom(m) \cap varset(\mathbf{p})) \setminus dom(p(m')) \\ &= dom(m) \cap (varset(\mathbf{p}) \setminus dom(p(m'))) \\ &\subseteq dom(m) \cap varset(p(m')(\mathbf{p})). \end{aligned}$$

4. $p(\mathbf{F})$ *pure*: This is the case if $range(p(m)) \cap dom(p(m)) = \emptyset$ or, equivalently, $range(p(m)) \cap dom(m) = \emptyset$, for all $m \in M$. This follows if for all $\langle x, \mathbf{p} \rangle$ in m

   $$varset(p(m)(\mathbf{p})) \cap dom(m) = \emptyset.$$

   We now prove this by induction over $\prec_{\mathbf{F}}|_m$:

- $y = \langle x, \mathbf{p} \rangle$ minimal w.r.t. $\prec_{\mathbf{F}}|_m$: Then $varset(\mathbf{p}) \cap dom(m) = \emptyset$, so $p(m)(\mathbf{p}) = \mathbf{p}$ which implies $varset(p(m)(\mathbf{p})) = varset(\mathbf{p})$. Thus, $varset(p(m)(\mathbf{p})) \cap dom(m) = \emptyset$ in this case.
- Inductive step: assume as induction hypothesis that it holds, for all $\langle x', \mathbf{p}' \rangle$ such that $\langle x', \mathbf{p}' \rangle \prec_{\mathbf{F}}|_m \langle x, \mathbf{p} \rangle$, that

$$varset(p(m)(\mathbf{p}')) \cap dom(m) = \emptyset.$$

By lemma D7,

$$\begin{aligned} varset(p(m)(\mathbf{p})) &= \\ (varset(\mathbf{p}) \setminus dom(p(m))) \cup range(p(m)|_{varset(\mathbf{p})}) &= \\ (varset(\mathbf{p}) \setminus dom(m)) \cup range(p(m)|_{varset(\mathbf{p})}). \end{aligned}$$

Furthermore, since $p(\langle x, \mathbf{p} \rangle) = \langle x, p(m)(\mathbf{p}) \rangle$ for all $\langle x, \mathbf{p} \rangle$ in m,

$$\begin{aligned} range(p(m)|_{varset(\mathbf{p})}) &= \\ \bigcup(varset(p(m)(\mathbf{p}')) \mid \langle x', \mathbf{p}' \rangle \in m \wedge x' \in varset(\mathbf{p})) &= \\ \bigcup(varset(p(m)(\mathbf{p}')) \mid \langle x', \mathbf{p}' \rangle \prec_{\mathbf{F}}|_m \langle x, \mathbf{p} \rangle). \end{aligned}$$

Thus,

$$\begin{aligned} varset(p(m)(\mathbf{p})) \cap dom(m) &= \\ ((varset(\mathbf{p}) \setminus dom(m)) \cup \\ \bigcup(varset(p(m)(\mathbf{p}')) \mid \langle x', \mathbf{p}' \rangle \prec_{\mathbf{F}}|_m \langle x, \mathbf{p} \rangle)) \\ \cap dom(m) &= \\ ((varset(\mathbf{p}) \setminus dom(m)) \cap dom(m)) \cup \\ (\bigcup(varset(p(m)(\mathbf{p}')) \mid \langle x', \mathbf{p}' \rangle \prec_{\mathbf{F}}|_m \langle x, \mathbf{p} \rangle) \cap dom(m)) &= \\ \emptyset \cup \bigcup(varset(p(m)(\mathbf{p}')) \cap dom(m) \mid \langle x', \mathbf{p}' \rangle \prec_{\mathbf{F}}|_m \langle x, \mathbf{p} \rangle) &= \\ \text{(by the induction hypothesis)} &= \\ \bigcup(\emptyset \mid \langle x', \mathbf{p}' \rangle \prec_{\mathbf{F}}|_m \langle x, \mathbf{p} \rangle) &= \\ \emptyset. \end{aligned}$$

5. $o_{p(\mathbf{F})}(p(m)) = o_{\mathbf{F}}(m)$ *for all* $m \in M$: We know that $dom(p(m)) = dom(m)$ for all $m \in M$. So the result follows if, for all $m \in M$ and for all $x \in dom(m)$, $o_{\mathbf{F}}(x) = o_{p(\mathbf{F})}(x)$. By theorem 6.1, this will always be the case if $\psi_{\mathbf{F}}(x) = \psi_{p(\mathbf{F})}(x)$. We will prove this by induction: first, for all $m \in M$, over $\prec_{\mathbf{F}}|_m$ and under the assumption that $\psi_{\mathbf{F}}(x) = \psi_{p(\mathbf{F})}(x)$ for all $x \in range(m) \setminus dom(m)$, and then over

$\prec_M$ using this partial result. So let us prove the following, for any $m \in M$:

$$\forall x' \in range(m) \setminus dom(m)[\psi_{\mathbf{F}}(x') = \psi_{p(\mathbf{F})}(x')] \implies \forall \langle x, \mathbf{p} \rangle \in m[\psi_{\mathbf{F}}(x) = \psi_{p(\mathbf{F})}(x)].$$

Assume that for all $x' \in range(m) \setminus dom(m)$ holds that

$$\psi_{\mathbf{F}}(x') = \psi_{p(\mathbf{F})}(x') \tag{1}$$

and let us prove by induction that the desired result follows.

- $\langle x, \mathbf{p} \rangle$ minimal w.r.t. $\prec_{\mathbf{F}}|_m$: Then

  $$varset(\mathbf{p}) \cap dom(m) = varset(\mathbf{p}) \cap dom(p(m)) = \emptyset,$$

  so $p(m)(\mathbf{p}) = \mathbf{p}$ and

  $$\psi_{p(\mathbf{F})}(x) = \psi_{p(\mathbf{F})}(p(m)(\mathbf{p})) = \psi_{p(\mathbf{F})}(\mathbf{p}).$$

  By definition it holds that $varset(\mathbf{p}) \subseteq range(m)$. Since $varset(\mathbf{p}) \cap dom(m) = \emptyset$, it follows that $varset(\mathbf{p}) \subseteq range(m) \setminus dom(m)$. So by the assumption (1), $\psi_{\mathbf{F}}(x') = \psi_{p(\mathbf{F})}(x')$ for all x' in $varset(\mathbf{p})$. Thus,

  $$\psi_{p(\mathbf{F})}(x) = \psi_{p(\mathbf{F})}(\mathbf{p}) = \psi_{p(\mathbf{F})}|_{varset(\mathbf{p})}(\mathbf{p}) = \psi_{\mathbf{F}}|_{varset(\mathbf{p})}(\mathbf{p}) = \psi_{\mathbf{F}}(\mathbf{p}) = \psi_{\mathbf{F}}(x),$$

  Where lemma D10 has been used twice.
- Inductive step: assume as induction hypothesis, that it holds, for all $\langle x', \mathbf{p}' \rangle$ such that $\langle x', \mathbf{p}' \rangle \prec_{\mathbf{F}}|_m \langle x, \mathbf{p} \rangle$, that

  $$\psi_{\mathbf{F}}(x') = \psi_{p(\mathbf{F})}(x').$$

  Now

  $$\begin{aligned}
  \psi_{p(\mathbf{F})}(x) &= \psi_{p(\mathbf{F})}(p(m)(\mathbf{p})) \\
  &= \text{(by definition, } dom(\psi_{p(\mathbf{F})}) = X_a(p(\mathbf{F})) = X_a(\mathbf{F})) \\
  &= (\{ x' \leftarrow \psi_{p(\mathbf{F})} p(m)(x') \mid x' \in dom(p(m)) \} \\
  &\quad \cup \psi_{p(\mathbf{F})}|_{X_a(\mathbf{F}) \setminus dom(p(m))})(\mathbf{p}) \\
  &= \text{(by lemma D10)} \\
  &= (\{ x' \leftarrow \psi_{p(\mathbf{F})} p(m)(x') \mid x' \in dom(p(m)) \} \\
  &\quad \cup \psi_{p(\mathbf{F})}|_{X_a(\mathbf{F}) \setminus dom(p(m))})|_{varset(\mathbf{p})}(\mathbf{p}) \\
  &= (\{ x' \leftarrow \psi_{p(\mathbf{F})} p(m)(x') \mid x' \in dom(p(m)) \}|_{varset(\mathbf{p})} \\
  &\quad \cup \psi_{p(\mathbf{F})}|_{X_a(\mathbf{F}) \setminus dom(p(m))}|_{varset(\mathbf{p})})(\mathbf{p}).
  \end{aligned} \tag{2}$$

Furthermore,

$$\begin{aligned}
\{\, x' \leftarrow \psi_{p(\mathbf{F})} p(m)(x') \mid x' \in dom(p(m)) \,\}|_{varset(\mathbf{p})} &= \\
\{\, x' \leftarrow \psi_{p(\mathbf{F})} p(m)(x') \mid x' \in dom(p(m)) \cap varset(\mathbf{p}) \,\} &= \\
\{\, x' \leftarrow \psi_{p(\mathbf{F})} p(m)(\mathbf{p}') \mid \langle x', \mathbf{p}' \rangle \prec_{\mathbf{F}}|_m \langle x, \mathbf{p} \rangle \,\} &= \\
\{\, x' \leftarrow \psi_{p(\mathbf{F})}(x') \mid \langle x', \mathbf{p}' \rangle \prec_{\mathbf{F}}|_m \langle x, \mathbf{p} \rangle \,\} &= \\
\text{(by the induction hypothesis)} &= \\
\{\, x' \leftarrow \psi_{\mathbf{F}}(x') \mid \langle x', \mathbf{p}' \rangle \prec_{\mathbf{F}}|_m \langle x, \mathbf{p} \rangle \,\} &= \\
\psi_{\mathbf{F}}|_{varset(\mathbf{p}) \cap dom(m)}. & \qquad (3)
\end{aligned}$$

Also

$$\begin{aligned}
\psi_{p(\mathbf{F})}|_{X_a(\mathbf{F}) \setminus dom(p(m))}|_{varset(\mathbf{p})} &= \psi_{p(\mathbf{F})}|_{(X_a(\mathbf{F}) \setminus dom(p(m))) \cap varset(\mathbf{p})} \\
&= \psi_{p(\mathbf{F})}|_{varset(\mathbf{p}) \setminus dom(p(m))} \\
&= \text{(since } varset(\mathbf{p}) \setminus dom(p(m)) \subseteq \\
&\quad\ range(m) \setminus dom(m)\text{, and (1))} \\
&= \psi_{\mathbf{F}}|_{varset(\mathbf{p}) \setminus dom(p(m))}. \qquad (4)
\end{aligned}$$

(2), (3) and (4) now gives

$$\begin{aligned}
\psi_{p(\mathbf{F})}(x) &= (\psi_{\mathbf{F}}|_{varset(\mathbf{p}) \cap dom(m)} \cup \psi_{\mathbf{F}}|_{varset(\mathbf{p}) \setminus dom(p(m))})(\mathbf{p}) \\
&= (\psi_{\mathbf{F}}|_{(varset(\mathbf{p}) \cap dom(m)) \cup (varset(\mathbf{p}) \setminus dom(p(m)))})(\mathbf{p}) \\
&= \psi_{\mathbf{F}}|_{varset(\mathbf{p})}(\mathbf{p}) \\
&= \text{(by lemma D10)} \\
&= \psi_{\mathbf{F}}(\mathbf{p})
\end{aligned}$$

which is the desired result.

Let us now use the result proved for all $m \in M$ above;

$$\begin{aligned}
\forall x' \in range(m) \setminus dom(m)[\psi_{\mathbf{F}}(x') = \psi_{p(\mathbf{F})}(x')] \implies \\
\forall \langle x, \mathbf{p} \rangle \in m[\psi_{\mathbf{F}}(x) = \psi_{p(\mathbf{F})}(x)] \qquad (1)
\end{aligned}$$

to prove, by induction over $\prec_M$, that $o_{p(\mathbf{F})}(p(m)) = o_{\mathbf{F}}(m)$ for all $m \in M$. We will prove this by proving that for all $\langle x, \mathbf{p} \rangle$ in any m holds that $\psi_{\mathbf{F}}(x) = \psi_{p(\mathbf{F})}(x)$. By theorem 6.1, this implies $o_{p(\mathbf{F})}(p(m)) = o_{\mathbf{F}}(m)$.

- m minimal w.r.t. $\prec_M$: Then $range(m) \setminus dom(m) \subseteq X_f(\mathbf{F})$, so for all $x' \in range(m) \setminus dom(m)$ holds that $\psi_{\mathbf{F}}(x') = x' = \psi_{p(\mathbf{F})}(x')$. From (1) follows that $\psi_{\mathbf{F}}(x) = \psi_{p(\mathbf{F})}(x)$ for all $\langle x, \mathbf{p} \rangle \in m$.

- Inductive step: assume as induction hypothesis, that for all m', such that $m' \prec_M m$, holds that $\psi_{\mathbf{F}}(x') = \psi_{\mathbf{p}(\mathbf{F})}(x')$, for all $\langle x', m' \rangle$ in m'. Theorem 6.3 implies that

$$range(m) \setminus dom(m) \subseteq \bigcup(\, dom(m') \mid m' \prec_M m\,) \cup X_f(\mathbf{F}) =$$
$$\bigcup(\, x' \mid \langle x', \mathbf{p}' \rangle \in m' \wedge m' \prec_M m\,) \cup X_f(\mathbf{F}).$$

For all $x' \in X_f(\mathbf{F})$, it holds that $\psi_{\mathbf{F}}(x') = x' = \psi_{\mathbf{p}(\mathbf{F})}(x')$. For all $x' \in \bigcup(\, x'' \mid \langle x'', \mathbf{p}'' \rangle \in m' \wedge m' \prec_M m\,)$, we can apply the induction hypothesis and obtain $\psi_{\mathbf{F}}(x') = \psi_{\mathbf{p}(\mathbf{F})}(x')$. So $\psi_{\mathbf{F}}(x') = \psi_{\mathbf{p}(\mathbf{F})}(x')$ for all $x' \in range(m) \setminus dom(m)$. Therefore, from (1), follows that $\psi_{\mathbf{F}}(x) = \psi_{\mathbf{p}(\mathbf{F})}(x)$ for all $\langle x, \mathbf{p} \rangle \in m$. ∎

Theorem 6.5: *Let M be a subset of $\mathcal{S}^{(X)}(\mathcal{A})$, such that all $m \in M$ are finite and $m \neq m' \implies dom(m) \cap dom(m') = \emptyset$ for all $m, m' \in M$. Let the relation $\prec_M$ on M be given by*

$$m \prec_M m' \iff dom(m) \cap range(m') \neq \emptyset$$

for all $m, m' \in M$. If $\prec_M^+$ is finite-downward, then $\mathbf{F}_M = \bigcup(\, m \mid m \in M\,)$ is a causal recursion scheme in X over $\mathcal{A}$ and $\langle M, \prec_M \rangle$ is a pure CCSA structure derived from $\langle \mathbf{F}_M, \prec_{\mathbf{F}_M} \rangle$.

Proof. $\mathbf{F}_M$ *recursion scheme*: Since $M \subseteq \mathcal{S}^{(X)}(\mathcal{A})$, it holds that every $m \in M$ is a partial function $X \to \mathcal{E}^{(X)}(\mathcal{A})$ and that $S(x) = S(m(x))$ for all x where $m(x)$ is defined. Since the domains of every two distinct $m, m' \in M$ are disjoint, the union $\mathbf{F}_M = \bigcup(\, m \mid m \in M\,)$ is also a partial function $X \to \mathcal{E}^{(X)}(\mathcal{A})$. It fulfils the sort constraint for substitutions, since all $m \in M$ do. Thus, $\mathbf{F}_M$ is a recursion scheme in X over $\mathcal{A}$.
$\mathbf{F}_M$ *causal*: Consider the relation $\prec_{\mathbf{F}_M}$ on $\mathbf{F}_M$ given by

$$\langle x, \mathbf{p} \rangle \prec_{\mathbf{F}_M} \langle x', \mathbf{p}' \rangle \iff x \in varset(\mathbf{p}') \tag{1}$$

for all $\langle x, \mathbf{p} \rangle, \langle x', \mathbf{p}' \rangle \in \mathbf{F}_M$. M is a partition of $\mathbf{F}_M$. Consider the natural mapping $\varphi_M : \mathbf{F}_M \to M$. Denote $\varphi_M(\langle x, \mathbf{p} \rangle)$ by m_x, for all $\langle x, \mathbf{p} \rangle \in \mathbf{F}_M$. For all $\langle x, \mathbf{p} \rangle, \langle x', \mathbf{p}' \rangle \in \mathbf{F}_M$

$$m_x \prec_M m_{x'} \iff \exists x''[x \in dom(m_x) \wedge x'' \in range(m_{x'})]. \tag{2}$$

From (1) and (2) we can conclude, that for all $y, y' \in \mathbf{F}_M$

$$y \prec_{\mathbf{F}_M} y' \implies \varphi_M(y) \prec_M \varphi_M(y').$$

$\prec_M^+$ is finite-downward and thus $\prec_M$ is well-founded. This implies (see, for instance, [MaWa85]) that $\prec_{\mathbf{F}_M}$ is well-founded. $sp(\langle x, \mathbf{p}\rangle, \prec_{\mathbf{F}_M})$ is finite for all $\langle x, \mathbf{p}\rangle \in \mathbf{F}_M$, according to lemma H1. By lemma B3 follows that $\prec_{\mathbf{F}_M}$ is finite-downward.

$\langle M, \prec_M\rangle$ *derived from* $\langle \mathbf{F}_M, \prec_{\mathbf{F}_M}\rangle$: We must show, that $fi(\langle \mathbf{F}_M, \prec_{\mathbf{F}_M}\rangle, \langle M, \prec_M\rangle, \varphi_M)$ holds.

fi1(φ_M): all $m \in M$ are assumed to be finite in the formulation of the theorem.

fi2($\langle \mathbf{F}_M, \prec_{\mathbf{F}_M}\rangle, \langle M, \prec_M\rangle, \varphi_M$): define $\prec$ on M by fi2($\langle \mathbf{F}_M, \prec_{\mathbf{F}_M}\rangle, \langle M, \prec\rangle, \varphi_M$). Then

$$\begin{aligned}
m_x \prec m_{x'} &\iff \exists\langle x'', \mathbf{p}''\rangle, \langle x''', \mathbf{p}'''\rangle \in \mathbf{F}_M[\langle x'', \mathbf{p}''\rangle \in m_x \wedge \\
&\qquad\qquad \langle x''', \mathbf{p}'''\rangle \in m_{x'} \wedge \\
&\qquad\qquad \langle x'', \mathbf{p}''\rangle \prec_{\mathbf{F}_M} \langle x''', \mathbf{p}'''\rangle] \\
&\iff \exists\langle x'', \mathbf{p}''\rangle, \langle x''', \mathbf{p}'''\rangle \in \mathbf{F}_M[\langle x'', \mathbf{p}''\rangle \in m_x \wedge \\
&\qquad\qquad \langle x''', \mathbf{p}'''\rangle \in m_{x'} \wedge \\
&\qquad\qquad x'' \in varset(\mathbf{p}''')] \\
&\iff \exists x''[x'' \in dom(m_x) \wedge x'' \in range(m_{x'})] \\
&\iff m_x \prec_M m_{x'}.
\end{aligned}$$

fi3($\prec_{\mathbf{F}_M}, \prec_M$): Follows immediately, since $\prec_M^+$ is finite-downward.

Let us finally note that since $m \prec_M m' \iff dom(m) \cap range(m') \neq \emptyset$ for all $m, m' \in M$, proposition 6.1 implies that M is pure. ∎

Theorem 6.6: *For every causal recursion scheme* $\mathbf{F}$ *over* $\mathcal{A}$, $g(\mathbf{F})$ *is a causal guarded recursion scheme over* $i(\mathcal{A})$ *such that* $o_{g(\mathbf{F})}(x) = o_{\mathbf{F}}(x)$ *for all* $x \in X_a(\mathbf{F})$. *Furthermore,* g *is a graph isomorphism* $\langle \mathbf{F}, \prec_{\mathbf{F}}\rangle \to \langle g(\mathbf{F}), \prec_{g(\mathbf{F})}\rangle$.

Proof. Consider the causal recursion scheme $\mathbf{F}$ in X over $\mathcal{A}$. $g: \mathbf{F} \to g(\mathbf{F})$ is a bijection. Also, since $varset(i_{S(\mathbf{p})} \bullet \mathbf{p}) = varset(\mathbf{p})$ for any $\mathbf{p}$, $varset(g(\mathbf{F})(x)) = varset(\mathbf{F}(x))$ for all $x \in X_a(\mathbf{F})$. Thus, for any $y, y' \in \mathbf{F}$, $y \prec_{\mathbf{F}} y' \iff g(y) \prec_{g(\mathbf{F})} g(y')$. Therefore, g is a graph isomorphism from $\langle \mathbf{F}, \prec_{\mathbf{F}}\rangle$ to $\langle g(\mathbf{F}), \prec_{g(\mathbf{F})}\rangle$. We now prove that for all $x \in X_a(\mathbf{F})$ holds that $o_{\mathbf{F}}(x) = o_{g(\mathbf{F})}(x)$, by induction over the well-founded relation $\prec_{\mathbf{F}}$:

1. $\langle x, \mathbf{F}(x)\rangle$ minimal w.r.t. $\prec_{\mathbf{F}}$: Then $varset(\mathbf{F}(x)) \subseteq X_f(\mathbf{F})$. Thus, $\mathbf{F}(x) \notin X_a(\mathbf{F})$ which implies $g(\mathbf{F})(x) = \mathbf{F}(x)$. Furthermore, it holds for all $x' \in X_f(\mathbf{F})$ that $o_{\mathbf{F}}(x') = e^X_{x'} = o_{g(\mathbf{F})}(x')$, so

$$\begin{aligned} o_{g(\mathbf{F})}(x) &= \phi(g(\mathbf{F})(x)) \circ \langle\, o_{g(\mathbf{F})}(x') \mid x' \in varset(g(\mathbf{F})(x))\,\rangle \\ &= \phi(\mathbf{F}(x)) \circ \langle\, o_{\mathbf{F}}(x') \mid x' \in varset(\mathbf{F}(x))\,\rangle \\ &= o_{\mathbf{F}}(x). \end{aligned}$$

2. Inductive step: assume as induction hypothesis that for all $\langle x', \mathbf{p}'\rangle$ such that $\langle x', \mathbf{p}'\rangle \prec_{\mathbf{F}} \langle x, \mathbf{p}\rangle$ holds that $o_{g(\mathbf{F})}(x') = o_{\mathbf{F}}(x')$. Then

$$\begin{aligned} o_{g(\mathbf{F})}(x) &= \phi(g(\mathbf{F})(x)) \circ \langle\, o_{g(\mathbf{F})}(x') \mid x' \in varset(g(\mathbf{F})(x))\,\rangle \\ &= \ \text{(by the induction hypothesis)} \\ &= \phi(g(\mathbf{F})(x)) \circ \langle\, o_{\mathbf{F}}(x') \mid x' \in varset(g(\mathbf{F})(x))\,\rangle \\ &= \ (varset(g(\mathbf{F})(x)) = varset(\mathbf{F}(x))) \\ &= \phi(g(\mathbf{F})(x)) \circ \langle\, o_{\mathbf{F}}(x') \mid x' \in varset(\mathbf{F}(x))\,\rangle. \end{aligned} \tag{1}$$

   There are two cases:

   $\mathbf{F}(x) \notin X_a(\mathbf{F})$: Then $g(\mathbf{F})(x) = \mathbf{F}(x)$ and from (1)

$$\begin{aligned} o_{g(\mathbf{F})}(x) &= \phi(g(\mathbf{F})(x)) \circ \langle\, o_{\mathbf{F}}(x') \mid x' \in varset(\mathbf{F}(x))\,\rangle \\ &= \phi(\mathbf{F}(x)) \circ \langle\, o_{\mathbf{F}}(x') \mid x' \in varset(\mathbf{F}(x))\,\rangle \\ &= o_{\mathbf{F}}(x). \end{aligned}$$

   $\mathbf{F}(x) \in X_a(\mathbf{F})$: Then $g(\mathbf{F})(x) = i_{S(x)} \bullet \mathbf{F}(x)$ and from (1)

$$\begin{aligned} o_{g(\mathbf{F})}(x) &= \phi(i_{S(x)} \bullet \mathbf{F}(x)) \circ \langle\, o_{\mathbf{F}}(x') \mid x' \in varset(\mathbf{F}(x))\,\rangle \\ &= (i_{S(x)} \circ \phi(\mathbf{F}(x))) \circ \langle\, o_{\mathbf{F}}(x') \mid x' \in varset(\mathbf{F}(x))\,\rangle \\ &= i_{S(x)} \circ (\phi(\mathbf{F}(x)) \circ \langle\, o_{\mathbf{F}}(x') \mid x' \in varset(\mathbf{F}(x))\,\rangle) \\ &= \phi(\mathbf{F}(x)) \circ \langle\, o_{\mathbf{F}}(x') \mid x' \in varset(\mathbf{F}(x))\,\rangle \\ &= o_{\mathbf{F}}(x). \end{aligned}$$

■

The following lemma is needed when proving the results concerning the concepts defined in definition 6.14. In the following, $\langle M, \prec_M\rangle$ is a guarded CCSA structure derived from a recursion scheme $\mathbf{F}$ in X over $\mathcal{A}$.

Lemma H3: *For any $\langle m, \mathbf{p}\rangle \in D(M)$, $\psi_M(\langle m, \mathbf{p}\rangle)$ is compound iff $\mathbf{p}$ is compound.*

Proof.

$\Longrightarrow$: Assume that $\psi_M(\langle m, \mathbf{p}\rangle)$ is compound. By c1 follows that $\mathbf{p} \notin X_a(\mathbf{F})$, since $\mathbf{F}$ is guarded. Furthermore, it cannot be the case that $\mathbf{p} \in X_f(\mathbf{F})$, since then

$$\psi_M(\langle m, \mathbf{p}\rangle) = \psi_{\mathbf{F}}(\mathbf{p}) = \mathbf{p}$$

that would imply $\psi_M(\langle m, \mathbf{p}\rangle) \in X_f(\mathbf{F})$, which contradicts the assumption. Thus, $\mathbf{p}$ must be compound.

$\Longleftarrow$: Assume that $\mathbf{p}$ is compound, that is: $\mathbf{p} = f \bullet \langle\, \mathbf{p}_i \mid i \in I_f\,\rangle$. Then

$$\psi_M(\langle m, \mathbf{p}\rangle) = \psi_{\mathbf{F}}(\mathbf{p}) = \psi_{\mathbf{F}}(f \bullet \langle\, \mathbf{p}_i \mid i \in I_f\,\rangle) = f \bullet \langle\, \psi_{\mathbf{F}}(\mathbf{p}_i) \mid i \in I_f\,\rangle$$

which is compound. ∎

Lemma 6.7: *$E(M)$ is an evaluable subset of $\mathcal{E}^{(X_f(\mathbf{F}))}(\mathcal{A})$. $D(M)$ is a set of computational events of $E(M)$. $SF(M)$ is a family of selector functions with respect to $D(M)$ and ψ_M.*

Proof. $E(M)$ evaluable subset of $\mathcal{E}^{(X_f(\mathbf{F}))}(\mathcal{A})$:

$E \subseteq \mathcal{E}^{(X_f(\mathbf{F}))}(\mathcal{A})$: From c1, c2 and lemma D2 follows, that for all $\langle m, \mathbf{p}'\rangle \in D(M)$ there is a $\mathbf{p} \in \mathcal{E}^{(X)}(\mathcal{A})$ such that $\mathbf{p}' \prec_e^* \mathbf{p}$. Lemma D3 implies $\mathit{varset}(\mathbf{p}) \subseteq X$, so from lemma D4 follows that $\mathit{varset}(\mathbf{p}') \subseteq X$. Thus, by lemma D7,

$$\begin{aligned}
\mathit{varset}(\psi_{\mathbf{F}}(\mathbf{p}')) &= (\mathit{varset}(\mathbf{p}') \setminus \mathit{dom}(\psi_{\mathbf{F}})) \cup \mathit{range}(\psi_{\mathbf{F}}|_{\mathit{varset}(\mathbf{p}')}) \\
&= (\psi_{\mathbf{F}}: X \to \mathcal{E}^{(X)}(\mathcal{A}) \text{ total} \Longrightarrow \\
&\qquad \mathit{dom}(\psi_{\mathbf{F}}) \supseteq \mathit{varset}(\mathbf{p}')) \\
&= \emptyset \cup \mathit{range}(\psi_{\mathbf{F}}|_{\mathit{varset}(\mathbf{p}')}) \\
&\subseteq X_f(\mathbf{F}). \qquad (1)
\end{aligned}$$

From c4, (1) and lemma D3 follows that $\psi_M(\langle m, \mathbf{p}\rangle) \in \mathcal{E}^{(X_f(\mathbf{F}))}(\mathcal{A})$ for all $\langle m, \mathbf{p}\rangle \in D(M)$. Thus, $E(M) \subseteq \mathcal{E}^{(X_f(\mathbf{F}))}(\mathcal{A})$.

$E(M)$ evaluable: Consider an arbitrary compound expression $\mathbf{p} = f \bullet \langle\, \mathbf{p}_i \mid i \in I_f\,\rangle$ in $E(M)$. Let us prove that all $\mathbf{p}_i$ must be in $E(M)$ as well. There exists a $\langle m', \mathbf{p}'\rangle \in D(M)$ such that $\psi_M(\langle m', \mathbf{p}'\rangle) = \mathbf{p}$,

or, by c4, $\mathbf{p} = \psi_{\mathbf{F}}(\mathbf{p}')$. By lemma H3, $\mathbf{p}'$ is compound. Thus, $\mathbf{p}' = f \bullet \langle \mathbf{p}_i' \mid i \in I_f \rangle$ and

$$\begin{aligned} \mathbf{p} &= \psi_{\mathbf{F}}(f \bullet \langle \mathbf{p}_i' \mid i \in I_f \rangle) \\ &= f \bullet \langle \psi_{\mathbf{F}}(\mathbf{p}_i') \mid i \in I_f \rangle \\ &= f \bullet \langle \psi_M(\langle m_i, \mathbf{p}_i' \rangle) \mid i \in I_f \rangle \end{aligned}$$

for some $m_i \in M$. So for every $\mathbf{p}_i$ there is a $\langle m_i, \mathbf{p}_i' \rangle$ such that $\mathbf{p}_i = \psi_M(\langle m_i, \mathbf{p}_i' \rangle)$, which implies that $\mathbf{p}_i \in E(M)$.

$D(M)$ SoCE of $E(M)$: By the above $E(M)$ is evaluable. By definition, $\psi_M: D(M) \to E(M)$ is onto. Thus, $D(M)$ is a SoCE of $E(M)$.

$SF(M)$ family of selector functions w.r.t. $D(M)$ and ψ_M: Consider any $s_{\langle m, \mathbf{p} \rangle}$ in $SF(M)$. By c7, $\mathbf{p}$ is compound, or $\mathbf{p} = f \bullet \langle \mathbf{p}_i \mid i \in I_f \rangle$. Let us verify that $s_{\langle m, \mathbf{p} \rangle}$ really is a selector function, that is: if $\psi_M(\langle m, \mathbf{p} \rangle) = f \bullet \langle \mathbf{p}_i' \mid i \in I_f \rangle$, then for all $i \in I$

$$\psi_M(s_{\langle m, \mathbf{p} \rangle}(i)) = \mathbf{p}_i'.$$

From c4

$$\psi_M(\langle m, \mathbf{p} \rangle) = \psi_{\mathbf{F}}(f \bullet \langle \mathbf{p}_i \mid i \in I_f \rangle) = f \bullet \langle \psi_{\mathbf{F}}(\mathbf{p}_i) \mid i \in I_f \rangle,$$

so for every $i \in I_f$

$$\mathbf{p}_i' = \psi_{\mathbf{F}}(\mathbf{p}_i) = \begin{cases} \psi_{\mathbf{F}}(\mathbf{p}_i), & \mathbf{p}_i \notin X_a(\mathbf{F}) \\ \psi_{\mathbf{F}}(\mathbf{F}(\mathbf{p}_i)), & \mathbf{p}_i \in X_a(\mathbf{F}). \end{cases}$$

From c4, c6

$$\psi_M(s_{\langle m, \mathbf{p} \rangle}(i)) = \begin{cases} \psi_{\mathbf{F}}(\mathbf{p}_i), & \mathbf{p}_i \notin X_a(\mathbf{F}) \\ \psi_{\mathbf{F}}(\mathbf{F}(\mathbf{p}_i)), & \mathbf{p}_i \in X_a(\mathbf{F}) \end{cases} = \mathbf{p}_i'.$$

Thus, every $s_{\langle m, \mathbf{p} \rangle}$ in $SF(M)$ is really a selector function. By lemma H3, $\psi_M(\langle m, \mathbf{p} \rangle)$ is compound iff $\mathbf{p}$ is compound. Thus, by c7,

$$SF(M) = (\, s_{\langle m, \mathbf{p} \rangle} \mid \langle m, \mathbf{p} \rangle \in comp(D(M), \psi_M)\,).$$

So $SF(M)$ is really a family of selector functions with respect to $D(M)$ and ψ_M. ■

Lemma 6.8: *There is a binary relation $\prec_{C(M)}$ on $C(M)$ such that $\langle C(M), \prec_{C(M)} \rangle$ is a cell action structure derived from $\langle D(M), \prec_{SF(M)} \rangle$. C is a graph isomorphism $\langle M, \prec_M \rangle \to \langle C(M), \prec_{C(M)} \rangle$.*

Proof. We must do the following: 1. Prove that $C(M)$ is a partition of $D(M)$. 2. Find a relation $\prec_{C(M)}$ on $C(M)$ such that $\mathit{fi}(\langle D(M), \prec_{SF(M)}\rangle, \langle C(M), \prec_{C(M)}\rangle, \varphi_{C(M)})$. 3. Prove that C is a graph isomorphism between $\langle M, \prec_M\rangle$ and $\langle C(M), \prec_{C(M)}\rangle$.

1. *$C(M)$ partition of $D(M)$*: This follows directly, since for any $m \in M$ all elements in $C(m)$ are af the form $\langle m, \mathbf{p}\rangle$ and thus not in $C(m')$ when $m' \neq m$. By definition, $\bigcup(C(m) \mid m \in M) = D(M)$.
2. $\mathit{fi}(\langle D(M), \prec_{SF(M)}\rangle, \langle C(M), \prec_{C(M)}\rangle, \varphi_{C(M)})$: Let us define $\prec_{C(M)}$ by fi2$(\langle D(M), \prec_{SF(M)}\rangle, \langle C(M), \prec_{C(M)}\rangle, \varphi_{C(M)})$ (note that for all $\langle m, \mathbf{p}\rangle$ in $D(M)$ holds that $\varphi_{C(M)}(\langle m, \mathbf{p}\rangle) = C(m)$): for all $C(m), C(m') \in C(M)$,

$$C(m) \prec_{C(M)} C(m') \iff$$
$$C(m) \neq C(m') \wedge$$
$$\exists \langle m, \mathbf{p}\rangle, \langle m', \mathbf{p}'\rangle \in D(M)[\langle m, \mathbf{p}\rangle \prec_{SF(M)} \langle m', \mathbf{p}'\rangle].$$

We now prove that $\langle C(M), \prec_{C(M)}\rangle$ is finitely inherited from $\langle D(M), \prec_{SF(M)}\rangle$ under $\varphi_{C(M)}$.
fi1$(\varphi_{C(M)})$: By c1,

$$C(m) = \{ \langle m, \mathbf{p}\rangle \mid \exists \langle x', \mathbf{p}'\rangle \in m[\mathbf{p} \in ec(\mathbf{p}') \setminus X_a(\mathbf{F})] \}$$
$$= \bigcup(\{ \langle m, \mathbf{p}\rangle \mid \mathbf{p} \in ec(\mathbf{p}') \setminus X_a(\mathbf{F}) \} \mid \langle x', \mathbf{p}'\rangle \in m).$$

For every $\langle x', \mathbf{p}'\rangle \in m$ holds that

$$ec(\mathbf{p}') = (\text{by lemma D2}) = sp(\mathbf{p}', \prec_e^*) = sp(\mathbf{p}', \prec_e^+) \cup \{\mathbf{p}'\}$$

which by theorem 4.2 is finite. m is a finite set, therefore $C(m)$ is a union of finitely many finite set and finite itself.
fi2$(\langle D(M), \prec_{SF(M)}\rangle, \langle C(M), \prec_{C(M)}\rangle, \varphi_{C(M)})$: $\prec_{C(M)}$ is defined according to fi2.
fi3$(\prec_{SF(M)}, \prec_{C(M)})$: By theorem 5.1 holds that $\prec_{SF(M)}^+$ is finite-downward. Thus, $\prec_{C(M)}^+$ must be finite-downward too, if fi3 is to hold. Below we will prove that C is a graph isomorphism between $\langle M, \prec_M\rangle$ and $\langle C(M), \prec_{C(M)}\rangle$. This will, since $\prec_M$ is finite-downward, yield the desired result.
3. *C graph isomorphism* $\langle M, \prec_M\rangle \to \langle C, \prec_{C(M)}\rangle$: As stated in 1, $C(m)$ and $C(m')$ are disjoint when $m \neq m'$. Thus, C is 1-1. Trivially C is onto. Let us show that for all $m, m' \in M$

$$m \prec_M m' \iff C(m) \prec_{C(M)} C(m'). \quad (1)$$

For all $m, m' \in M$,

$$m \prec_M m' \iff m \neq m' \wedge dom(m) \cap range(m') \neq \emptyset \tag{2}$$

and

$$\begin{aligned} C(m) \prec_{C(M)} C(m') \iff \\ C(m) \neq C(m') \wedge \\ \exists \langle m, \mathbf{p} \rangle, \langle m', \mathbf{p}' \rangle \in D(M)[\langle m, \mathbf{p} \rangle \prec_{SF(M)} \langle m', \mathbf{p}' \rangle]. \end{aligned} \tag{3}$$

We show both directions in the equivalence (1) separately:

$\Longrightarrow$: Assume that $m \prec_M m'$. By (2), $m \neq m'$ which implies $C(m) \neq C(m')$. Furthermore, $dom(m) \cap range(m') \neq \emptyset$. Let us show that then there are $\langle m, \mathbf{p} \rangle, \langle m', \mathbf{p}' \rangle$ in $D(M)$ such that $\langle m, \mathbf{p} \rangle \prec_{SF(M)} \langle m', \mathbf{p}' \rangle$. We immediately obtain that there is an $\langle x, \mathbf{p} \rangle \in m$ and an $\langle x', \mathbf{p}'' \rangle \in m'$ such that $x \in varset(\mathbf{p}'')$. Also $x \in X_a(\mathbf{F})$. $\langle x, \mathbf{p} \rangle \in m$ implies $\langle m, \mathbf{p} \rangle \in C(m)$ and $\langle x', \mathbf{p}'' \rangle \in m'$ implies $\langle m', \mathbf{p}'' \rangle \in C(m')$. By proposition D2, $x \in varset(\mathbf{p}'')$ implies $x \in ec(\mathbf{p})$ which, by lemma D2, gives $x \prec_e^* \mathbf{p}''$. Since $\mathbf{F}$ is guarded it holds that $\mathbf{p}'' \notin X_a(\mathbf{F})$, thus $x \neq \mathbf{p}''$ which implies $x \prec_e^+ \mathbf{p}''$. Then, by lemma B2, $x \prec_e \mathbf{p}''$ or there is a $\mathbf{p}'$ such that $x \prec_e \mathbf{p}'$ and $\mathbf{p}' \prec_e^+ \mathbf{p}''$. In both cases there is a $\mathbf{p}'$ in $ec(\mathbf{p}'')$ such that $x \prec_e \mathbf{p}'$. Since x precedes $\mathbf{p}'$ it holds that $\mathbf{p}'$ must be compound, so $\mathbf{p}' \notin X_a(\mathbf{F})$. From c1 follows that $\langle m', \mathbf{p}' \rangle \in D(M)$, since $\langle x', \mathbf{p}'' \rangle \in m'$. Furthermore, we can write $\mathbf{p}' = f \bullet \langle \mathbf{p}_i \mid i \in I_f \rangle$, and $x = \mathbf{p}_i$ for some $i \in I_f$. C7 gives $s_{\langle m', \mathbf{p}' \rangle}(i) = \langle m, \mathbf{F}(x) \rangle = \langle m, \mathbf{p} \rangle$, thus $\langle m, \mathbf{p} \rangle \prec_{SF(M)} \langle m', \mathbf{p}' \rangle$ which we wanted to show.

$\Longleftarrow$: Assume that $C(m) \prec_{C(M)} C(m')$. Then, by (3), $C(m) \neq C(m')$ and there are $\langle m, \mathbf{p} \rangle, \langle m', \mathbf{p}' \rangle$ in $D(M)$ such that $\langle m, \mathbf{p} \rangle \prec_{SF(M)} \langle m', \mathbf{p}' \rangle$. $m \neq m'$ follows directly. Let us show that also $dom(m) \cap range(m') \neq \emptyset$. Since $\langle m, \mathbf{p} \rangle \prec_{SF(M)} \langle m', \mathbf{p}' \rangle$, it holds that $\mathbf{p}' = f \bullet \langle \mathbf{p}_i \mid i \in I_f \rangle$ and that $s_{\langle m', \mathbf{p}' \rangle}(i) = \langle m, \mathbf{p} \rangle$ for some $i \in I_f$. Since $m \neq m'$, definitions c1 and c7 imply that there is an $x \in X_a(\mathbf{F})$ such that $\langle x, \mathbf{p} \rangle \in m$. Furthermore, $\mathbf{p}_i = x$ and therefore $x \in varset(\mathbf{p}')$. From c1, since $\langle m', \mathbf{p}' \rangle \in D(M)$, there is an $\langle x'', \mathbf{p}'' \rangle \in m'$ such that $\mathbf{p}' \in ec(\mathbf{p}'')$ or, by lemma D2, $\mathbf{p}' \prec_e^* \mathbf{p}''$. By lemma D4, $varset(\mathbf{p}') \subseteq varset(\mathbf{p}'')$, so $x \in varset(\mathbf{p}'')$. We have thus proved the following: there is an $\langle x, \mathbf{p} \rangle \in m$ and an $\langle x'', \mathbf{p}'' \rangle \in m'$ such that $x \in varset(\mathbf{p}'')$, or equivalently $dom(m) \cap range(m') \neq \emptyset$. ■

Theorem 6.7: *For any guarded CCSA structure $\langle M, \prec_M \rangle$, derived from a recursion scheme* **F** *in X over $\mathcal{A}$, $\langle C(M), \prec_{C(M)} \rangle$ is a cell action structure. Furthermore, C is a graph isomorphism $\langle M, \prec_M \rangle \rightarrow \langle C(M), \prec_{C(M)} \rangle$ such that $o_{\mathbf{F}}(m) \subseteq o(C(m))$ for all $m \in M$.*

Proof. It remains to prove that $o_{\mathbf{F}}(m) \subseteq o(C(m))$ for all $m \in M$. Consider an arbitrary $m \in M$. C1 gives $\{\langle m, \mathbf{p} \rangle \mid \langle x, \mathbf{p} \rangle \in m\} \subseteq C(m)$, since **F** is guarded. According to definition c4,

$$\psi_{\mathbf{F}}(x) = \psi_{\mathbf{F}}(\mathbf{p})$$

for every $\langle x, \mathbf{p} \rangle \in m$. For every $\langle m, \mathbf{p} \rangle \in C(m)$ holds that

$$\psi_M(\langle m, \mathbf{p} \rangle) = \psi_{\mathbf{F}}(\mathbf{p}).$$

Thus,

$$\{\psi_{\mathbf{F}}(x) \mid \langle x, \mathbf{p} \rangle \in m\} \subseteq \{\psi_M(\langle m, \mathbf{p} \rangle) \mid \langle m, \mathbf{p} \rangle \in C(m)\}$$

which implies

$$\{\phi \circ \psi_{\mathbf{F}}(x) \mid \langle x, \mathbf{p} \rangle \in m\} \subseteq \{\phi \circ \psi_M(\langle m, \mathbf{p} \rangle) \mid \langle m, \mathbf{p} \rangle \in C(m)\}$$

or, by theorem 6.1,

$$o_{\mathbf{F}}(m) \subseteq o(C(m)).$$

■

Before proving lemma 6.9 and theorem 6.9, we prove the following lemmas about the associated expressions defined in definition 6.15.

Lemma H4: *Let $\langle C, \prec_c \rangle$ be a cell action structure derived from a computational event structure $\langle D, \prec \rangle$ of a SoE $E \subseteq \mathcal{E}^{(X)}(\mathcal{A})$ under ψ, where $D \cap X = \emptyset$. Assume that $\prec$ is given by the family of selector functions $\{s_{\mathrm{p}} \mid \mathrm{p} \in comp(D)\}$. For all $c \in C$ and for all $\mathrm{p}, \mathrm{p}' \in c \cup ice(c)$ holds that*

$$ae_c(\mathrm{p}) = ae_c(\mathrm{p}') \implies \psi(\mathrm{p}) = \psi(\mathrm{p}').$$

Proof. The lemma is shown, for any $c \in C$, by induction over $\prec|_{c \cup ice(c)}$. Let us first note that from definitions 5.4 and 6.15 follows that p is minimal with respect to $\prec|_{c \cup ice(c)}$ iff $\psi(\mathrm{p})$ is not compound or $\mathrm{p} \in ice(c)$. The inductive sentence is chosen to be

$$\forall \mathrm{p}' \in c \cup ice(c)[ae_c(\mathrm{p}) = ae_c(\mathrm{p}') \implies \psi(\mathrm{p}) = \psi(\mathrm{p}')].$$

In all cases below, we will consider an arbitrary p' such that $ae_c(p) = ae_c(p')$. For notational convenience we write "$\prec|$" for $\prec|_{c \cup ice(c)}$.

p minimal with respect to $\prec|$: then there are three cases:

$p \in ice(c)$: then $ae_c(p) = p = ae_c(p')$. Thus, $ae_c(p')$ is not compound, and since $D \cap X = \emptyset$ it follows that $p' \in ice(c)$. Then $ae_c(p') = p' = p$ which implies $\psi(p) = \psi(p')$.

$p \in c$ and $\psi(p) \in X$: then $ae_c(p) = \psi(p) = ae_c(p')$. Thus, $ae_c(p') \in X$, so it is not compound, and $D \cap X = \emptyset$ implies that $p' \notin ice(c)$. It follows that $p' \in c$ and $\psi(p') \in X$. Thus, $ae_c(p') = \psi(p')$, and $\psi(p) = \psi(p')$ follows.

$p \in c$ and $\psi(p) = f \bullet \langle\rangle$: then $ae_c(p) = f \bullet \langle\rangle = \psi(p) = ae_c(p')$. Since $ae_c(p') = f \bullet \langle\rangle$, the only possible case is that $\psi(p') = f \bullet \langle\rangle = \psi(p)$.

Inductive step, p not minimal with respect to $\prec|$: assume as induction hypothesis that for all p'' such that $p'' \prec| p$ holds that

$$\forall p''' \in c \cup ice(c)[ae_c(p'') = ae_c(p''') \implies \psi(p'') = \psi(p''')].$$

Let us show that the inductive sentence holds for p as well. When p is not minimal with respect to $\prec|$, then $\psi(p)$ must be compound, or according to definition 5.3:

$$\psi(p) = f \bullet \langle\, \psi(s_p(i)) \mid i \in I_f \,\rangle. \tag{1}$$

The computational events $s_p(i)$ are, by definition 5.4 and definition 6.15 of $ice(c)$, exactly those computational events for which $s_p(i) \prec| p$. By definition 6.15,

$$ae_c(p) = f \bullet \langle\, ae_c(s_p(i)) \mid i \in I_f \,\rangle.$$

When $ae_c(p') = ae_c(p)$ it follows that

$$ae_c(p') = f \bullet \langle\, ae_c(s_p(i)) \mid i \in I_f \,\rangle. \tag{2}$$

Since $ae_c(p')$ is compound it must hold that $\psi(p)$ is compound, or

$$\psi(p') = f \bullet \langle\, \psi(s_{p'}(i)) \mid i \in I_f \,\rangle. \tag{3}$$

Then definition 6.15 yields

$$ae_c(p') = f \bullet \langle\, ae_c(s_{p'}(i)) \mid i \in I_f \,\rangle \qquad (4).$$

(2) and (4) now implies that $ae_c(s_p(i)) = ae_c(s'_p(i))$ for all $i \in I_f$. By the induction hypothesis follows that $\psi(s_p(i)) = \psi(s'_p(i))$ for all $i \in I_f$. (1) and (3) then implies that $\psi(p) = \psi(p')$, which is the desired result. ∎

Lemma H5: *Let $\langle C, \prec_c \rangle$ be a cell action structure derived from a computational event structure $\langle D, \prec \rangle$ of a SoE $E \subseteq \mathcal{E}^{(X)}(\mathcal{A})$ under ψ, such that $D \cap X = \emptyset$. Let $\prec$ be given by the family of selector functions $\{ s_{\mathrm{p}} \mid \mathrm{p} \in comp(D) \}$. For all $c \in C$ and all* $\mathrm{p}, \mathrm{p}' \in c \cup ice(c)$ *then holds that*

$$ae_c(\mathrm{p}) \prec_e|_{c \cup ice(c)}{}^+ \, ae_c(\mathrm{p}') \implies \psi(\mathrm{p}) \prec_e^+ \psi(\mathrm{p}').$$

Proof. For notational convenience we write "$\prec_e|$" for $\prec_e|_{c \cup ice(c)}$. Let us first prove that for all $c \in C$ and all $\mathrm{p}, \mathrm{p}' \in c \cup ice(c)$ holds that

$$ae_c(\mathrm{p}) \prec_e| \, ae_c(\mathrm{p}') \implies \psi(\mathrm{p}) \prec_e \psi(\mathrm{p}').$$

For any $c \in C$ and any two $\mathrm{p}, \mathrm{p}' \in c \cup ice(c)$ holds that if $ae_c(\mathrm{p}) \prec_e| \, ae_c(\mathrm{p}')$, then $ae_c(\mathrm{p}') = f \bullet \langle ae_c(s_{\mathrm{p}}(i)) \mid i \in I_f \rangle$ and $ae_c(\mathrm{p}') = ae_c(s_{\mathrm{p}}(i))$ for some $i \in I_f$. Lemma H4 then implies that $\psi(\mathrm{p}') = \psi(s_{\mathrm{p}}(i))$. Since $ae_c(\mathrm{p})$ is compound, $\psi(\mathrm{p})$ must be compound, or $\psi(\mathrm{p}) = f \bullet \langle \psi(s_{\mathrm{p}}(i)) \mid i \in I_f \rangle$. It follows that $\psi(\mathrm{p}') \prec_e \psi(\mathrm{p})$.

Let us now consider the functions $ae_c : c \cup ice(c) \to ae_c(c \cup ice(c))$ and $\psi|_{c \cup ice(c)} : c \cup ice(c) \to \mathcal{E}^{(X)}(\mathcal{A})$. ae_c is by definition onto. Thus, lemma B4 can be applied. The result is

$$ae_c(\mathrm{p}) \prec_e|_{c \cup ice(c)}{}^+ \, ae_c(\mathrm{p}') \implies \psi(\mathrm{p}) \prec_e^+ \psi(\mathrm{p}')$$

as desired. ∎

Lemma 6.9: $\mathbf{F}_C$ *is a causal recursion scheme in $X \cup D$ over $\mathcal{A}$ and $X_f(\mathbf{F}_C) = X$. $M(C)$ is a CCSA-set derived from* $\mathbf{F}_C$.

Proof.

1. $\mathbf{F}_C$ *recursion scheme in $X \cup D$ over $\mathcal{A}$ with $X_f(\mathbf{F}_C) = X$, $M(C)$ derived from* $\mathbf{F}_C$: $D \cap X = \emptyset$ and $c \neq c'$ implies that $c \cap c' = \emptyset$. Hence ae1-ae3, m1 and m3 gives that $\mathbf{F}_C$ is a partial function $X \cup D \to \mathcal{E}^{(X \cup D)}(\mathcal{A})$. Thus, $\mathbf{F}_C$ is a recursion scheme in $X \cup D$ over $\mathcal{A}$. From m1 and m3 $\mathbf{F}_C(\mathrm{p})$ is defined for every $\mathrm{p} \in D$ and no other p, so $X_a(\mathbf{F}_C) = D$ and consequently $X_f(\mathbf{F}_C) = X$. If $\mathrm{p} \neq \mathrm{p}'$, then $\langle \mathrm{p}, \mathbf{p} \rangle \neq \langle \mathrm{p}', \mathbf{p}' \rangle$ for any $\mathbf{p}, \mathbf{p}'$. Thus, from m1, $c \neq c' \implies M(c) \cap M(c') = \emptyset$. M2 and m3 then implies that M is a set of cs-assignments derived from $\mathbf{F}_C$.
2. $\mathbf{F}_C$ *causal*: $\mathbf{F}_C$ is causal iff $\prec_{\mathbf{F}_C}$ is finite-downward. Let us first prove that $\prec_{\mathbf{F}_C}$ is well-founded. Define $f: \mathbf{F}_C \to \mathcal{E}^{(X)}(\mathcal{A})$ by $f(\langle \mathrm{p}, \mathbf{p} \rangle) =$

$\psi(p)$ for all $\langle p, \mathbf{p}\rangle \in \mathbf{F}_C$. If we can show that for all $\langle p, \mathbf{p}\rangle, \langle p', \mathbf{p}'\rangle \in \mathbf{F}_C$ holds that

$$\langle p, \mathbf{p}\rangle \prec_{\mathbf{F}_C} \langle p', \mathbf{p}'\rangle \implies f(\langle p, \mathbf{p}\rangle) \prec_e^+ f(\langle p', \mathbf{p}'\rangle),$$

then it will follow that that $\prec_{\mathbf{F}_C}$ is well-founded, since $\prec_e^+$ is well-founded. (See for instance [MaWa85].) Consider two arbitrary $\langle p, \mathbf{p}\rangle$, $\langle p', \mathbf{p}'\rangle \in \mathbf{F}_C$ for which $\langle p, \mathbf{p}\rangle \prec_{\mathbf{F}_C} \langle p', \mathbf{p}'\rangle$. Now

$$\begin{aligned} \langle p, \mathbf{p}\rangle \prec_{\mathbf{F}_C} \langle p', \mathbf{p}'\rangle &\iff p \in \mathit{varset}(\mathbf{p}') \\ &\iff \text{(by m1 and m3)} \\ &\iff p \in \mathit{varset}(ae_c(p')). \end{aligned}$$

It follows that $p \in \mathit{ice}(c)$ and thus $ae_c(p) = p$. Therefore

$$ae_c(p) \in \mathit{varset}(ae_c(p'))$$

which implies

$$ae_c(p) \prec_e|_{c \cup ice(c)}{}^+ \, ae_c(p').$$

Lemma H5 now yields

$$\psi(p) \prec_e^+ \psi(p').$$

Thus, $\prec_{\mathbf{F}_C}$ is well-founded. Lemma H1 implies that $sp(\langle p, \mathbf{p}\rangle, \prec_{\mathbf{F}_C})$ is finite for all $\langle p, \mathbf{p}\rangle \in \mathbf{F}_C$. Then it follows from lemma B3 that $\prec_{\mathbf{F}_C}$ is finite-downward. ■

Theorem 6.9: *M is a graph isomorphism $\langle C, \prec_c\rangle \to \langle M(C), \prec_{M(C)}\rangle$. For all $c \in C$ holds that $o(c) = o_{\mathbf{F}_C}(M(c))$.*

Proof.

1. *M graph isomorphism $\langle C, \prec_c\rangle \to \langle M(C), \prec_{M(C)}\rangle$*: C is a partition of D. M1, m2, m3 and the fact that there is exactly one $\langle p, \mathbf{p}\rangle \in \mathbf{F}_C$ for each $p \in D$ and vice versa, then gives that $M(C)$ is a partition of $\mathbf{F}_C$ and that $M: C \to M(C)$ is 1-1. it remains to show that for all $c, c' \in C$

$$c \prec_c c' \iff M(c) \prec_{M(C)} M(c')$$

or equivalently, by theorem 6.3,

$$c \prec_c c' \iff M(c) \neq M(c') \wedge \mathit{dom}(M(c)) \cap \mathit{range}(M(c')) \neq \emptyset.$$

The two directions are now shown separately:

$\Longrightarrow$: Assume that $c \prec_c c'$. By definition follows that $c \neq c'$ and that there are p, p' such that $p \in c$, $p' \in c'$ and $c \prec c'$. Then, by *ice*, there is a p in $c \cap ice(c')$. By m1, $p \in dom(M(c))$. By ae1, ae3 and *ice* there must be a $p' \in c'$ such that $p \in varset(ae_{c'}(p'))$. By m1, $p \in range(M(c'))$. Thus, $dom(M(c)) \cap range(M(c')) \neq \emptyset$. Finally, since m is 1-1, $c \neq c'$ implies $M(c) \neq M(c')$.

$\Longleftarrow$: Assume that $M(c) \prec_{M(C)} M(c')$, or equivalently $M(c) \neq M(c')$ and $dom(M(c)) \cap range(M(c')) \neq \emptyset$. $dom(M(c)) \cap range(M(c'))$ thus contains an element p. By m1, $dom(M(c)) = c$. From ae1-ae3 and m1 follows that $range(M(c')) = ice(c') \cup (X \cap \psi(c'))$. Apparently, since $X \cap D = \emptyset$, $p \in c \cap ice(c')$. Then, by *ice*, there is a $p' \in c'$ such that $p \prec p'$, Furthermore, $M(c) \neq M(c')$ implies that $c \neq c'$. So $c \neq c'$ and there are p, p' such that $p \in c$, $p' \in c'$ and $p \prec p'$. In other words, $c \prec_c c'$.

2. $o(c) = o_{\mathbf{F}_C}(M(c))$ *for all* $c \in C$: If we can prove for all $p \in D$ that $\psi(p) = \psi_{\mathbf{F}_C}(p)$, then the desired result will follow, since $o_\psi(p) = \phi \circ \psi(p)$ for all $p \in D$, $o_{\mathbf{F}_C}(x) = \phi \circ \psi_{\mathbf{F}_C}(x)$ for all $x \in X_a(\mathbf{F}_C)$ and for all $c \in C$ holds that $o(c) = \{ o_\psi(p) \mid p \in c \}$ and $o_{\mathbf{F}_C}(M(c)) = \{ o_{\mathbf{F}_C}(x) \mid \langle x, \mathbf{F}_C(x) \rangle \in M(c) \} = \{ o_{\mathbf{F}_C}(x) \mid x \in dom(M(c)) \} = \{ o_{\mathbf{F}_C}(p) \mid p \in c \}$. Let us prove this by induction over $\prec$. (Assume that $p \in c$. Then, by m1 and m3, $\mathbf{F}_C(p) = ae_c(p)$):
   - p minimal w.r.t. $\prec$: Then there are two possible cases:
     $\psi(p) \in X$: Then $\psi_{\mathbf{F}_C}(p) = \psi_{\mathbf{F}_C}(ae_c(p)) = \psi_{\mathbf{F}_C}(\psi(p)) = \psi(p)$.
     $\psi(p) = f \bullet \langle\rangle$: Then $\psi_{\mathbf{F}_C}(p) = \psi_{\mathbf{F}_C}(ae_c(p)) = \psi_{\mathbf{F}_C}(f \bullet \langle\rangle) = f \bullet \langle\rangle = \psi(p)$.
   - Inductive step: assume as induction hypothesis, that for all p' such that $p' \prec p$ holds that $\psi_{\mathbf{F}_C}(p) = \psi(p)$. Since p is not minimal, $\psi(p)$ must by definition 5.4 be compound, that is: $\psi(p) = f \bullet \langle \mathbf{p}_i \mid i \in I_f \rangle$. The computational events preceding p are exactly $\{ s_p(i) \mid i \in I_f \}$. On the other hand,

$$\begin{aligned}\psi_{\mathbf{F}_C}(p) &= \psi_{\mathbf{F}_C}(ae_c(p)) \\ &= \text{(by ae3)} \\ &= \psi_{\mathbf{F}_C}(f \bullet \langle ae_c(s_p(i)) \mid i \in I_f \rangle) \\ &= f \bullet \langle \psi_{\mathbf{F}_C}(ae_c(s_p(i))) \mid i \in I_f \rangle. \end{aligned} \qquad (1)$$

     For each $s_p(i)$ there are two possible cases:
     $s_p(i) \in c$: By m1, $\psi_{\mathbf{F}_C}(s_p(i)) = \psi_{\mathbf{F}_C}(ae_c(s_p(i)))$.
     $s_p(i) \notin c$: Then $s_p(i) \in ice(c)$ and $ae_c(s_p(i)) = s_p(i)$. Thus, $\psi_{\mathbf{F}_C}(s_p(i)) = \psi_{\mathbf{F}_C}(ae_c(s_p(i)))$.

In both cases $\psi_{\mathbf{F}_C}(ae_{\mathbf{c}}(\mathrm{p})) = \psi_{\mathbf{F}_C}(s_{\mathbf{p}}(i))$, so by (1)

$$\begin{aligned}\psi_{\mathbf{F}_C}(\mathrm{p}) &= f \bullet \langle\, \psi_{\mathbf{F}_C}(s_{\mathbf{p}}(i)) \mid i \in I_f \,\rangle \\ &= \text{ (by the induction hypothesis)} \\ &= f \bullet \langle\, \psi(s_{\mathbf{p}}(i)) \mid i \in I_f \,\rangle \\ &= \text{ (by definition 5.4)} \\ &= f \bullet \langle\, \mathbf{p}_i \mid i \in I_f \,\rangle \\ &= \psi(\mathrm{p}).\end{aligned}$$

■

Appendix I. Proofs of theorems of chapter 7

Proposition 7.1: $\emptyset \in \mathcal{C}(S)$. *If* $\prec_s, \prec_s' \in \mathcal{C}(S)$, *then* $\prec_s \cap \prec_s' \in \mathcal{C}(S)$.

Proof. $\emptyset$, considered as a binary relation over S, is trivially finite-downward. It also trivially fulfils the time causality criterion co1. Thus, $\emptyset \in \mathcal{C}(S)$.

Consider two arbitrary $\prec_s, \prec_s'$ in $\mathcal{C}(S)$. By co2, $\prec_s^+$ and $\prec_s'^+$ are finite-downward, so by lemma B8 follows that $(\prec_s \cap \prec_s')^+$ is finite-downward. $\prec_s \cap \prec_s' \subseteq \prec_s$, and since $\prec_s$ fulfils co1,

$$\langle t,r\rangle \prec_s \cap \prec_s' \langle t',r'\rangle \implies \langle t,r\rangle \prec_s \langle t',r'\rangle \implies t \leq t'$$

for all $\langle t,r\rangle$, $\langle t',r'\rangle$ in S. ∎

Proposition 7.2: $\prec_s \in \mathcal{C}(S) \wedge \prec \subseteq \prec_s \implies \prec \in \mathcal{C}(S)$.

Proof. Assume that $\prec_s \in \mathcal{C}(S)$ and $\prec \subseteq \prec_s$. Let us show that then $\prec \in \mathcal{C}(S)$.

co1: For all $\langle t,r\rangle, \langle t',r'\rangle \in S$, $\langle t,r\rangle \prec \langle t',r'\rangle \implies \langle t,r\rangle \prec_s \langle t',r'\rangle \implies t \leq t'$.

co2: Since $\prec_s^+$ is finite-downward, lemma B9 implies that $\prec^+$ is finite-downward. ∎

Proposition 7.4: *For any space-time* S, $\langle \mathcal{R}(S); \cup, \cap\rangle$ *is a lattice.* $\emptyset$ *is a least element.* $\prec_T$, *defined by* $\langle t,r\rangle \prec_T \langle t',r'\rangle$ *iff* $t < t'$, *belongs to* $\mathcal{R}(S)$ *and is a greatest element.*

Proof. Let us first show that $\langle \mathcal{R}(S); \cup, \cap\rangle$ is a lattice. Since $\cup$ and $\cap$ are idempotent, commutative and associative it suffices to show that $\mathcal{R}(S)$ is closed under these operations. Consider two arbitrary ripple-free relations $\prec, \prec'$:

$\mathcal{R}(S)$ *closed under* $\cup$: For any $\langle t,r\rangle, \langle t',r'\rangle \in S$ holds that

$$\langle t,r\rangle \prec \cup \prec' \langle t',r'\rangle \implies \langle t,r\rangle \prec \langle t',r'\rangle \vee \langle t,r\rangle \prec' \langle t',r'\rangle.$$

In any case, since both $\prec$ and $\prec'$ are ripple-free, follows that $t < t'$.

$\mathcal{R}(S)$ *closed under* $\cap$: For any $\langle t,r\rangle, \langle t',r'\rangle \in S$,

$$\langle t,r\rangle \ (\prec \cap \prec') \ \langle t',r'\rangle \implies \langle t,r\rangle \prec \langle t',r'\rangle \wedge \langle t,r\rangle \prec' \langle t',r'\rangle$$

which directly implies $t < t'$, since both $\prec$ and $\prec'$ are ripple-free. Let us now show that $\emptyset$ is a least and $\prec_T$ a greatest element in $\mathcal{R}(S)$:

$\emptyset$ least element: Trivially $\emptyset \in \mathcal{R}(S)$, and it directly follows that it is a least element.

$\prec_\mathsf{T}$ greatest element: By the definition of $\prec_\mathsf{T}$, $\langle t, r\rangle \prec_\mathsf{T} \langle t', r'\rangle \implies t < t'$ for all $\langle t, r\rangle, \langle t', r'\rangle \in S$, so $\prec_\mathsf{T} \in \mathcal{R}(S)$. For any $\prec \in \mathcal{R}(S)$ holds that

$$\langle t, r\rangle \prec \langle t', r'\rangle \implies t < t' \implies \langle t, r\rangle \prec_\mathsf{T} \langle t', r'\rangle,$$

so $\prec_\mathsf{T}$ is really a greatest element in $\mathcal{R}(S)$. ∎

Proposition 7.5: *Let S be a space-time. If $\prec \in \mathcal{R}(S)$, then $\prec \in \mathcal{C}(S)$ iff for all $s \in S$ $sp(s, \prec)$ is finite.*

Proof. Since $\prec$ is ripple-free, it fulfils co1. Thus, $\prec$ is a communication ordering iff $\prec^+$ is finite-downward. By lemma 7.1 $\prec$ is well-founded, so by lemma B3 $\prec^+$ is finite-downward iff $sp(s, \prec)$ is finite for all $s \in S$. ∎

Proposition 7.6: *For any space-time S, $\langle \mathcal{R}(S) \cap \mathcal{C}(S); \cup, \cap\rangle$ is a lattice with $\emptyset$ as a least element.*

Proof. Since both $\mathcal{R}(S)$ and $\mathcal{C}(S)$ are closed under $\cap$, it holds that $\mathcal{R}(S) \cap \mathcal{C}(S)$ is closed under $\cap$. In order to show that $\langle \mathcal{R}(S) \cap \mathcal{C}(S); \cup, \cap\rangle$ is a lattice, it remains to show that $\mathcal{R}(S) \cap \mathcal{C}(S)$ is closed under $\cup$. Consider two arbitrary $\prec_s, \prec'_s \in \mathcal{R}(S) \cap \mathcal{C}(S)$. Since they are in $\mathcal{R}(S)$, proposition 7.4 gives

$$\prec_s \cup \prec'_s \in \mathcal{R}(S). \tag{1}$$

$\prec_s$ and $\prec'_s$ also belong to $\mathcal{C}(S)$. Thus, they are finite-downward, so $sp(s, \prec_s)$ and $sp(s, \prec'_s)$ are finite for all $s \in S$. This implies that $sp(s, \prec_s \cup \prec'_s) = sp(s, \prec_s) \cup sp(s, \prec'_s)$ is finite. By (1) and proposition 7.5, $\prec_s \cup \prec'_s \in \mathcal{C}(S)$ which implies that $\prec_s \cup \prec'_s \in \mathcal{R}(S) \cap \mathcal{C}(S)$.

Propositions 7.1 and 7.4 gives that $\emptyset \in \mathcal{R}(S) \cap \mathcal{C}(S)$. Trivially it is a least element in $\mathcal{R}(S) \cap \mathcal{C}(S)$. ∎

Proposition 7.7: *For any space R, $\prec_\mathsf{T} \in \mathcal{R}(T \times R) \cap \mathcal{C}(T \times R)$ iff R is finite. If $\prec_\mathsf{T} \notin \mathcal{R}(T \times R) \cap \mathcal{C}(T \times R)$, then there is no greatest element in $\mathcal{R}(T \times R) \cap \mathcal{C}(T \times R)$.*

Proof.

$\implies$: Assume that $\prec_\mathsf{T} \in \mathcal{R}(T\times R) \cap \mathcal{C}(T\times R)$. Then $\prec_\mathsf{T} \in \mathcal{C}(T\times R)$, so $\prec_\mathsf{T}$ is finite-downward. From the definition of $\prec_\mathsf{T}$ follows that $\{\langle t', r'\rangle \mid t' < t\}$ is finite for all $t \in T$. $\{\langle t-1, r'\rangle \mid t' < t\} \subseteq \{\langle t', r'\rangle \mid t' < t\}$, so $\{\langle t-1, r'\rangle \mid t' < t\}$ is also finite. But when $t > 0$ there is a bijection

f from this set to R given by $f(\langle t-1, r\rangle) = r$ for all $\langle t-1, r\rangle$ in the set. Thus, R must also be finite.

$\Longleftarrow$: Assume that R is finite. Then, for all $\langle t, r\rangle \in T \times R$,

$$sp(\langle t,r\rangle, \prec_T) = \{\langle t', r'\rangle \mid t' < t\} = \{t' \mid t' < t\} \times R.$$

This set is finite, since R is finite and since there are only finitely many $t' < t$.

Let us now show that there is no greatest element in $\mathcal{R}(T \times R) \cap \mathcal{C}(T \times R)$ when $\prec_T \notin \mathcal{R}(T \times R) \cap \mathcal{C}(T \times R)$. By the above, we can equally well prove that there is no greatest element in $\mathcal{R}(T \times R) \cap \mathcal{C}(T \times R)$ when R is infinite. Assume that R is infinite. Any $\prec \in \mathcal{R}(T \times R) \cap \mathcal{C}(T \times R)$ must be finite-downward. Select a time $t > 0$ and consider the set $R_t = \{r' \mid \exists r \in R[\langle t-1, r'\rangle \prec \langle t, r\rangle]\}$. Since $\prec$ is finite-downward, there are only finitely many $\langle t-1, r'\rangle$ such that $\langle t-1, r'\rangle \prec \langle t, r\rangle$ for any $\langle t, r\rangle$. Thus, R_t must be finite. Since R is infinite, there must be an $r'' \notin R_t$. Form the relation $\prec'$ over $T \times R$ that consists of the single pair $\langle\langle t-1, r''\rangle, \langle t, r''\rangle\rangle$. Clearly $\prec'$ is ripple-free and finite-downward. Thus, $\prec' \in \mathcal{R}(T \times R) \cap \mathcal{C}(T \times R)$. But from the definition of R_t above follows that $\prec' \not\subseteq \prec$. So for any $\prec \in \mathcal{R}(T \times R) \cap \mathcal{C}(T \times R)$, we can find a $\prec' \in \mathcal{R}(T \times R) \cap \mathcal{C}(T \times R)$ which is not contained in $\prec$. Thus, there is no greatest element in $\mathcal{R}(T \times R) \cap \mathcal{C}(T \times R)$. ∎

Proposition 7.8: *For any space R, where $|R| = 1$, $\mathcal{R}(T \times R) = \mathcal{C}(T \times R)$.*

Proof. Corollary 7.1 implies $\mathcal{R}(T \times R) \subseteq \mathcal{C}(T \times R)$, since R is finite. It remains to show that $\mathcal{C}(T \times R) \subseteq \mathcal{R}(T \times R)$, that is: every communication ordering over $T \times R$ is ripple-free. R contains a single element r. $\langle t, r\rangle \prec_s \langle t', r\rangle \implies t \leq t'$ for any communication ordering $\prec_s$. If $t = t'$ it cannot be the case that $\langle t, r\rangle \prec_s \langle t', r\rangle$, since then $\prec_s$ would not be irreflexive and thus not finite-downward. It follows that $\langle t, r\rangle \prec_s \langle t', r\rangle \implies t < t'$, or in other words: $\prec_s$ is ripple-free. ∎

Before proving proposition 7.10, we show the following lemma. It tells that the condition fi3$(\prec_c, \prec_{cW})$ actually is redundant for a weakly correct mapping $W: \langle C, \prec_c\rangle \to \langle S, \prec_s\rangle$.

Lemma I1: *Let $\langle C, \prec_c\rangle$ be a cell action structure and let $\langle S, \prec_s\rangle$ be a communication structure. Let $\prec_{cw}$ be given by fi2$(\langle C, \prec_c\rangle, \langle S, \prec_{cw}\rangle, W)$. Then W is a weakly correct mapping $\langle C, \prec_c\rangle \to \langle S, \prec_s\rangle$ iff fi1(W) and $\prec_{cw} \subseteq \prec_s$.*

Proof.

$\Longrightarrow$: Follows immediately.

$\Longleftarrow$: Assume that fi2($\langle C, \prec_c\rangle, \langle S, \prec_{cw}\rangle, W$), fi1($W$) and $\prec_{cw} \subseteq \prec_s$. If fi3($\prec_c, \prec_{cw}$) holds, then the desired result follows. But $\prec_{cw} \subseteq \prec_s \subseteq \prec_s^+$. Since $\prec_s$ is a communication ordering, $\prec_s^+$ is finite-downward. By lemma B9, $\prec_{cw}^+$ is then finite-downward. fi3($\prec_c, \prec_{cw}$) follows. ∎

Proposition 7.10: *Let $\langle C, \prec_c\rangle$ be a cell action structure where C is finite and let $\langle S, \prec_s\rangle$ be a communication structure. Let W be a function $C \to S$ and let $\prec_{cw}$ be given by* fi2($\langle C, \prec_c\rangle, \langle S, \prec_{cw}\rangle, W$). Then W is a weakly correct mapping $\langle C, \prec_c\rangle \to \langle S, \prec_s\rangle$ iff $\prec_{cw} \subseteq \prec_s$.

Proof.

$\Longrightarrow$: If W is a weakly correct mapping $\langle C, \prec_c\rangle \to \langle S, \prec_s\rangle$, then $\prec_{cw} = \prec_{cW}$. $\prec_{cw} \subseteq \prec_s$ then follows directly from wcm1.

$\Longleftarrow$: Assume that $\prec_{cw} \subseteq \prec_s$, where $\prec_{cw}$ is given by fi2($\langle C, \prec_c\rangle, \langle S, \prec_{cw}\rangle$, W). Since C is finite, fi1(W) holds trivially. By lemma I1 follows that W is a weakly correct mapping $\langle C, \prec_c\rangle \to \langle S, \prec_s\rangle$. ∎

Proposition 7.11: *Let $\langle C, \prec_c\rangle$ be a cell action structure and let $\langle S, \prec_s\rangle$ be a communication structure. Let F be a function $C \to S$ that is 1-1. Then F is a correct mapping $\langle C, \prec_c\rangle \to \langle S, \prec_s\rangle$ iff $c \prec_c c' \implies F(c) \prec_s F(c')$ for all $c, c' \in C$.*

Proof. Assume that $F: C \to S$ is 1-1.

$\Longrightarrow$: Assume that F also is a correct mapping $\langle C, \prec_c\rangle \to \langle S, \prec_s\rangle$. Then $\prec_{cF}$ exists, so fi2($\langle C, \prec_c\rangle, \langle S, \prec_{cF}\rangle, F$) holds. Furthermore, $\prec_{cF} \subseteq \prec_s$. Thus, for all $s, s' \in S$,

$$s \neq s' \wedge \exists c, c' \in C[F(c) = s \wedge F(c') = s' \wedge c \prec_c c'] \implies s \prec_s s'. \quad (1)$$

Consider the premise in (1). If $c \prec_c c'$, then, since $\prec_c$ is finite-downward, $c \neq c'$. Since F is 1-1, $c \neq c'$ implies $F(c) \neq F(c')$. So if there are c, c' in C such that $F(c) = s$, $F(c') = s'$ and $c \prec_c c'$, then $s \neq s'$ which therefore is redundant in the premise in (1). (1) is thus equivalent with

$$\forall s, s' \in S[\exists c, c' \in C[F(c) = s \wedge F(c') = s' \wedge c \prec_c c'] \implies s \prec_s s']$$

which implies that for all $c'', c''' \in C$, there exists $c, c' \in C$ such that

$$F(c) = F(c'') \wedge F(c') = F(c''') \wedge c \prec_c c' \implies F(c'') \prec_s F(c'''). \quad (2)$$

$c'' \prec_c c'''$ implies $\exists c, c' \in C[F(c) = F(c'') \wedge F(c') = F(c''') \wedge c \prec_c c']$, so (2) implies

$$\forall c'', c''' \in C[c'' \prec_c c''' \implies F(c'') \prec_s F(c''')]$$

which is exactly what we wanted to show.

$\Longleftarrow$: Assume that for all $c, c' \in C$

$$c \prec_c c' \implies F(c) \prec_s F(c'). \tag{3}$$

Let us show that F is a correct mapping $\langle C, \prec_c\rangle \to \langle S, \prec_s\rangle$. Fil$(F)$ holds trivially, since F is 1-1. Define $\prec_{cf}$ by fi2$(\langle C, \prec_c\rangle, \langle S, \prec_{cf}\rangle, F)$. Then, for all $s, s' \in S$,

$$\begin{aligned}
&s \prec_{cf} s' \iff \\
&s \neq s' \wedge \exists c, c' \in C[F(c) = s \wedge F(c') = s' \wedge c \prec_c c'] \implies \\
&\text{(by the assumption (3))} \implies \\
&s \neq s' \wedge \exists c, c' \in C[F(c) = s \wedge F(c') = s' \wedge F(c) \prec_s F(c')] \implies \\
&s \neq s' \wedge s \prec_s s' \implies \\
&s \prec_s s'.
\end{aligned}$$

Thus, $\prec_{cf} \subseteq \prec_s$. By lemma I1, F is a correct mapping $\langle C, \prec_c\rangle \to \langle S, \prec_s\rangle$. ■

Before proving theorem 7.1 we show the following lemma. We consider two sets A and B, binary relations R_A on A and R_B on B, and a function $f: A \to B$.

Lemma I2: *Let π be the partition of A given by the equivalence relation induced by f. Define the binary relation R_π on π by: for all $a, a' \in A$*

$$\varphi_\pi(a) R_\pi \varphi_\pi(a') \iff f(a) R_B f(a')$$

(φ_π is the natural mapping $A \to \pi$). Then

$$fi(\langle A, R_A\rangle, \langle B, R_B\rangle, f) \implies fi(\langle A, R_A\rangle, \langle \pi, R_\pi\rangle, \varphi_\pi).$$

Proof. Assume that

$$fi(\langle A, R_A\rangle, \langle B, R_B\rangle, f). \tag{1}$$

Let us show that then $fi(\langle A, R_A\rangle, \langle \pi, R_\pi\rangle, \varphi_\pi)$.

fil(φ_π): fil(φ_π) is equivalent to fil(f).

fi2($\langle A, R_A\rangle, \langle \pi, R_\pi\rangle, \varphi_\pi$): For all elements in $f(A)$, fi2($\langle A, R_A\rangle, \langle B, R_B\rangle, f$) can be written

$$\forall a, a' \in A[f(a) R_B f(a') \iff f(a) \neq f(a') \wedge \exists a'', a''' \in A[a'' R_A a''' \wedge f(a'') = f(a) \wedge f(a''') = f(a')]].$$

Now we can use the following equivalences to replace the corresponding subclauses from the logical sentence above:

$$\begin{aligned} f(a) R_B f(a') &\iff \varphi_\pi(a) R_\pi \varphi_\pi(a') \\ f(a) \neq f(a') &\iff \varphi_\pi(a) \neq \varphi_\pi(a') \\ f(a'') = f(a) &\iff \varphi_\pi(a'') = \varphi_\pi(a) \\ f(a''') = f(a') &\iff \varphi_\pi(a''') = \varphi_\pi(a'). \end{aligned}$$

The result is exactly fi2($\langle A, R_A\rangle, \langle \pi, R_\pi\rangle, \varphi_\pi$).
fi3(R_A, R_π): Define the function $g: f(A) \to \pi$ by

$$\forall a \in A[g(f(a)) = \varphi_\pi(a)]. \tag{2}$$

g is well-defined, 1-1 and onto, since for all a, a' in A

$$f(a) = f(a') \iff \varphi_\pi(a) = \varphi_\pi(a').$$

From (2) and the definition of R_π we obtain

$$f(a) R_B f(a') \iff g(f(a)) R_\pi g(f(a'))$$

for all $a, a' \in A$ or, equivalently,

$$b R_B b' \iff g(b) R_\pi g(b')$$

for all $b, b' \in f(A)$. Thus, g is a graph isomorphism from $\langle f(A), R_B|_{f(A)}\rangle$ to $\langle \pi, R_\pi\rangle$. By lemma 6.3 it holds that $R_B|_{f(A)} = R_B$ and (1) thus implies fi3($R_A, R_B|_{f(A)}$). So when R_A^+ is finite-downward, then $R_B|_{f(A)}$ and thus R_π are finite-downward. ■

Theorem 7.1: *Let $\langle C, \prec_c \rangle$ be a cell action structure derived from the computational event structure $\langle D, \prec \rangle$. Let $\langle S, \prec_s \rangle$ be a communication structure. For every weakly correct mapping $W: \langle C, \prec_c \rangle \to \langle S, \prec_s \rangle$ it holds that $\langle C', \prec_c' \rangle$, where $C' = D/\varepsilon_{W \circ \varphi_C}$, is a cell action structure derived from D, and there exists a correct mapping $M: \langle C', \prec_c' \rangle \to \langle S, \prec_s \rangle$ such that $W \circ \varphi_C = M \circ \varphi_{C'}$. The immediate precedence relation $\prec_c'$ on C' is given by: For all* $\mathrm{p}, \mathrm{p}' \in D$,

$$\varphi_{C'}(\mathrm{p}) \prec_c' \varphi_{C'}(\mathrm{p}') \iff W \circ \varphi_C(\mathrm{p}) \prec_{cW} W \circ \varphi_C(\mathrm{p}').$$

Proof.

A fundamental theorem from function theory is that every function $f: A \to B$ can be decomposed into $f_i \circ f_o$, where f_o is onto and f_i is 1-1. f_o can always be chosen as the natural mapping $A \to A/\varepsilon_f$, and f_i as the mapping $A/\varepsilon_f \to B$ for which $f_i([a]\varepsilon_f) = f(a)$ for all $a \in A$. Thus, we can decompose the function $W \circ \varphi_C = M \circ \varphi_{C'}$ according to this, where $C' = D/\varepsilon_{W \circ \varphi_C}$, $\varphi_{C'}$ is the natural mapping $D \to C'$ and $M: C' \to S$ is 1-1. The theorem is proved by showing that $\langle C', \prec_c' \rangle$ is a cell action structure derived from $\langle D, \prec \rangle$ and that M is a correct mapping $\langle C', \prec_c' \rangle \to \langle S, \prec_s \rangle$.

$\langle C', \prec_c' \rangle$ cell action structure derived from $\langle D, \prec \rangle$: Since W is a weakly correct mapping, it holds that

$$\mathit{fi}(\langle C, \prec_c \rangle, \langle S, \prec_{cW} \rangle, W).$$

$\langle C, \prec_c \rangle$ is a cell action structure derived from $\langle D, \prec \rangle$, thus

$$\mathit{fi}(\langle D, \prec \rangle, \langle C, \prec_c \rangle, \varphi_C)$$

holds. Therefore, by the transitivity of *fi* (lemma 6.1), we can conclude that

$$\mathit{fi}(\langle D, \prec \rangle, \langle S, \prec_{cW} \rangle, W \circ \varphi_C)$$

holds. By lemma I2, this implies

$$\mathit{fi}(\langle D, \prec \rangle, \langle C', \prec_c' \rangle, \varphi_{C'}),$$

where $C' = D/\varepsilon_{W \circ \varphi_C}$ and $\prec_c'$ is defined as in the formulation of the theorem.

M correct mapping $C' \to S$: M is by definition 1-1. It remains to show that $\prec_{cM}'$ exists and that $\prec_{cM}' \subseteq \prec_s$. If $\prec_{cM}'$ exists, then it is given by

$$\mathit{fi}(\langle C', \prec_c' \rangle, \langle S, \prec_{cM}' \rangle, M).$$

We just proved that that $\langle C', \prec_c' \rangle$ is a cell action structure derived from $\langle D, \prec \rangle$, or

$$fi(\langle D, \prec \rangle, \langle C', \prec_c' \rangle, \varphi_{C'}).$$

fi-transitivity (lemma 6.1) then gives

$$fi(\langle D, \prec \rangle, \langle S, \prec_{cM}' \rangle, M \circ \varphi_{C'}) \tag{1}$$

if $\prec_{cM}'$ exists. W is a weakly correct mapping, thus

$$fi(\langle C, \prec_c \rangle, \langle S, \prec_{cW} \rangle, W)$$

and we know that $\prec_{cW}$ exists. Moreover, $\langle C, \prec_c \rangle$ is a cell action structure derived from $\langle D, \prec \rangle$, thus

$$fi(\langle D, \prec \rangle, \langle C, \prec_c \rangle, \varphi_C)$$

which by *fi*-transitivity yields

$$fi(\langle D, \prec \rangle, \langle S, \prec_{cW} \rangle, W \circ \varphi_C).$$

$M \circ \varphi_{C'} = W \circ \varphi_C$, so this can be written as

$$fi(\langle D, \prec \rangle, \langle S, \prec_{cW} \rangle, M \circ \varphi_{C'}). \tag{2}$$

From (1), (2) and *fi*-uniqueness (lemma 6.2) follows that $\prec_{cM}'$ exists and $\prec_{cM}' = \prec_{cW}$. Since W is a weakly correct mapping, $\prec_{cW} \subseteq \prec_s$. Thus, $\prec_{cM}' \subseteq \prec_s$. ■

Theorem 7.2: *For every cell action structure $\langle C, \prec_c \rangle$ and for every space-time S, there exists a correct mapping $M: \langle C, \prec_c \rangle \to \langle S, \prec_{cM} \rangle$ such that $\prec_{cM}$ is ripple-free.*

Proof. Let $\langle C, \prec_c \rangle$ be a cell action structure. Denote $sp(c, \prec_c^+) \cup \{c\} = sp(c, \prec_c^*)$ by $p(c)$, for any $c \in C$. $p(c)$ is the set of cell actions that must have been evaluated when c has been evaluated.

Consider an enumeration $c_0, c_1, \ldots$ of the elements in C. C is a partition of a SoCE which is a countable set, thus C is countable itself and an enumeration exists. Define $p_0 = p(c_0)$ and for all $n > 0$

$$p_n = p(c_n) \setminus \bigcup (p_i \mid 0 \le i < n). \tag{1}$$

Given that $c_0, \ldots, c_{n-1}$ have been evaluated, p_n is the set of cell actions that must be evaluated in addition when c_n is evaluated. $\prec_c$ is finite-downward, thus $p(c_n)$ and p_n are finite for any $n \in N$.

The usefulness of p_n depends on the following facts: for all $n \in N$

$$p(c_n) \subseteq \bigcup(\, p_i \mid 0 \leq i \leq n\,), \tag{2}$$

$$C = \bigcup(\, p_n \mid n \in N\,) \tag{3}$$

and for all $m, n \in N$

$$m \neq n \implies p_m \cap p_n = \emptyset. \tag{4}$$

($\{\, p_n \mid n \in N\,\}$ is not necessarily a partition of C, since there may be some $n \in N$ such that $p_n = \emptyset$.) From (2) follows, that for all $c_{n'}, c_n \in C$

$$c_{n'} \prec_c c_n \implies n' \leq n. \tag{5}$$

(3), (4) implies that we for some space-time S can form a function $M{:}\,C \to S$ as $M = \bigcup(\, m_n \mid n \in N\,)$, where every m_n is a function $p_n \to S$ (and $m_n = \emptyset$ if $p_n = \emptyset$). Furthermore, if we iteratively form the functions $M_n : \bigcup(\, p_i \mid 0 \leq i \leq n\,)$ by $M_0 = m_0$ and, for $n > 0$, $M_n = m_n \cup M_{n-1}$, then we can see that $M_n \subseteq M$ for all $n \in N$ and from (2) that $p(c_n)$ is totally scheduled by M_n.

(2), (3) and (4) may not be obvious. (2) follows from

$$\begin{aligned}\bigcup(\, p_i \mid 0 \leq i \leq n\,) &= p_n \cup \bigcup(\, p_i \mid 0 \leq i < n\,)\\ &= (p(c_n) \setminus \bigcup(\, p_i \mid 0 \leq i < n\,)) \cup \bigcup(\, p_i \mid 0 \leq i < n\,)\\ &= p(c_n) \cup \bigcup(\, p_i \mid 0 \leq i < n\,).\end{aligned}$$

(3) follows directly from (2) and (4) from (1).

Let us now prove that we, for any space-times S, can construct a function M according to above that is a ripple-free correct mapping. From lemma B6 follows that $\prec_c^+|_{p_n}$, the restriction of $\prec_c$ to p_n, is a s.p.o. A s.p.o. on a finite set can be embedded in a linear (total) order by a topological sort (see for instance [Kn69]), in a way preserving the s.p.o. That is: when p_n is nonempty, there is a bijection $f_n{:}\,p_n \to \{1, \ldots, |p_n|\}$ such that $c \prec_c^+|_{p_n} c' \implies f_n(c) < f_n(c')$ for all $c, c' \in p_n$.

Consider an arbitrary space-time $S = T \times R$. R is nonempty and thus it contains an element r. Define $\mu(-1) = 0$, and for every $n \in N$ $\mu(n) = \sum_{i=0}^{n} |p_i|$ (apparently $\mu(n) - \mu(n-1) = |p_n|$). Consider a family of mappings $(\, m_n \mid n \in N\,)$, where every $m_n = \langle t_n, r_n\rangle$ is a mapping $p_n \to \{\langle \mu(n-1), r\rangle, \ldots, \langle \mu(n) - 1, r\rangle\}$ if $p_n \neq \emptyset$ and $m_n = \emptyset$ otherwise. According to the topological sort reasoning above, every $m_n \neq \emptyset$ can be chosen to be a bijection, such that for all $c, c' \in p_n$

$$c \prec_c c' \implies t_n(c) < t_n(c'). \tag{6}$$

Furthermore, $n' < n$ implies that for all $c' \in p_{n'}$ and $c \in p_n$ holds that $t_{n'}(c') < t_n(c)$. Let us now form $M = \langle t, r \rangle = \bigcup(m_n \mid n \in N)$. M is 1-1, since all m_n are bijections and their domains are disjoint. Consider the ripple-free communication ordering $\prec_s$ given by: $s \prec_s s'$ iff $s = \langle t, r \rangle$, $s' = \langle t', r \rangle$ and $t < t'$. (5), (6) implies that $c \prec_c c' \implies M(c) \prec_s M(c')$ for all $c, c' \in C$. By proposition 7.11, M is then a correct mapping $\langle C, \prec_c \rangle \to \langle S, \prec_s \rangle$. Thus, $\prec_{cM}$ exists and $\prec_{cM} \subseteq \prec_s$, so $\prec_{cM} \cap \prec_s = \prec_{cM}$ which, by proposition 7.6, implies that $\prec_{cM}$ is a ripple-free communication ordering. Finally, we can note that M by proposition 7.9 also is a correct mapping $\langle C, \prec_c \rangle \to \langle S, \prec_{cM} \rangle$. ∎

Theorem 7.3: *Let $\langle S, \prec_s \rangle$ and $\langle S', \prec'_s \rangle$ be communication structures and let A be a weak space-time transformation $\langle S, \prec_s \rangle \to \langle S', \prec'_s \rangle$. Let $\langle C, \prec_c \rangle$ be a cell action structure. Then, for every weakly correct mapping $W : \langle C, \prec_c \rangle \to \langle S, \prec_s \rangle$, $A \circ W$ is a weakly correct mapping $\langle C, \prec_c \rangle \to \langle S', \prec'_s \rangle$.*

Proof. For an arbitrary cell action structure $\langle C, \prec_c \rangle$, consider any weakly correct mapping $W : \langle C, \prec_c \rangle \to \langle S, \prec_s \rangle$. Let A be a weak space-time transformation $\langle S, \prec_s \rangle \to \langle S', \prec'_s \rangle$. Since W is a weakly correct mapping $\langle C, \prec_c \rangle \to \langle S, \prec_s \rangle$, there is a $\prec_{cW}$ such that

$$fi(\langle C, \prec_c \rangle, \langle S, \prec_{cW} \rangle, W) \tag{1}$$

and

$$\prec_{cW} \subseteq \prec_s. \tag{2}$$

Since A is a weak space-time transformation $\langle S, \prec_s \rangle \to \langle S', \prec'_s \rangle$, there is a $\prec_{sA}$ such that

$$fi(\langle S, \prec_s \rangle, \langle S', \prec_{sA} \rangle, A) \tag{3}$$

and

$$\prec_{sA} \subseteq \prec'_s. \tag{4}$$

$\prec'_s$ is a communication ordering so $\prec'^{+}_s$ is finite-downward. (4) and lemma B9 then implies that $\prec^{+}_{sA}$ is finite-downward. By (2), (3) and lemma 6.4, there is a $\prec_{cWA} \subseteq \prec_{sA}$ such that

$$fi(\langle S, \prec_{cW} \rangle, \langle S', \prec_{cWA} \rangle, A) \tag{5}$$

and by (4) $\prec_{cWA} \subseteq \prec'_s$. (1), (5) and fi-transitivity (lemma 6.1) now yields

$$fi(\langle C, \prec_c \rangle, \langle S', \prec_{cWA} \rangle, A \circ W).$$

∎

Proposition 7.16: $\overset{stt}{\rightarrow}$ *is transitive and reflexive.*

Proof.

Transitivity: assume that $\langle S, \prec_s\rangle$, $\langle S', \prec_s'\rangle$ and $\langle S'', \prec_s''\rangle$ are communication structures with space-time transformations $A: \langle S, \prec_s\rangle \rightarrow \langle S', \prec_s'\rangle$ and $B: \langle S', \prec_s'\rangle \rightarrow \langle S'', \prec_s''\rangle$. $B \circ A$ is then 1-1. The proof that $B \circ A$ is a space-time transformation $\langle S, \prec_s\rangle \rightarrow \langle S'', \prec_s''\rangle$ is exactly analogous to the proof of theorem 7.3 above, where it is shown that $A \circ W$ is a weakly correct mapping $\langle C, \prec_c\rangle \rightarrow \langle S', \prec_s'\rangle$ (substitute $\langle S, \prec_s\rangle$ for $\langle C, \prec_c\rangle$, $\langle S', \prec_s'\rangle$ for $\langle S, \prec_s\rangle$, $\langle S'', \prec_s''\rangle$ for $\langle S', \prec_s'\rangle$, A for W and B for A). Therefore it is omitted here.

Reflexivity: $\overset{stt}{\rightarrow}$ is reflexive, since the identity function from a space-time S to itself always is a space-time transformation $\langle S, \prec_s\rangle \rightarrow \langle S, \prec_s\rangle$ for any communication ordering $\prec_s$ on S. ∎

In order to prove theorem 7.5, the following lemmas are needed.

Lemma I3: *For any fixed hardware structure Δ on a space R, for all r_1, r_2 in R and t_1, t_2 in T,*

$$\langle r_1, t_1\rangle \prec(\Delta) \langle r_2, t_2\rangle \iff \langle r_1, r_2, t_2 - t_1\rangle \in \Delta.$$

Proof. Trivial, directly from the definition of $\prec(\Delta)$. (Include the left-hand side of the equivalence in the quantifier range of δ and substitute t_1 for t and t_2 for $t + \delta$.) ∎

Lemma I4: Δ_0 *well-founded on* $R \implies \prec(\Delta)$ *well-founded on* $T \times R$.

Proof. Assume that Δ_0 is well-founded. Consider the relation $\prec$ on $T \times R$, defined by:

$$\langle t, r\rangle \prec \langle t', r'\rangle \iff t < t' \vee (t = t' \wedge r \Delta_0 r')$$

for all $\langle t, r\rangle, \langle t', r'\rangle \in T \times R$. $\prec$ is the lexicographical relation on $T \times R$ with respect to $<$ on N and Δ_0 on R. Since they both are well-founded, $\prec$ must also be well-founded. $\prec(\Delta)$ is easily seen to be a subrelation to $\prec$, since $\langle t, r\rangle \prec(\Delta) \langle t', r'\rangle \implies t \leq t'$ and $\langle t, r\rangle \prec(\Delta) \langle t, r'\rangle \iff r \Delta_0 r'$. Thus, $\prec(\Delta)$ must also be well-founded. ∎

Theorem 7.5: *If Δ on R has finite fanin and no infinite zero-delay paths, then $\prec(\Delta)$ is a communication ordering on $T \times R$.*

Proof. Assume that Δ has finite fanin and no infinite zero-delay paths.

co1: $\langle t', r' \rangle \prec(\Delta) \langle t, r \rangle \implies t \leq t'$ follows immediately from lemma I3 and definition 7.9.

co2: $\prec(\Delta)$ is by lemma I well-founded, since Δ has no infinite zero-delay paths. According to lemma B3, it then suffices to show that for all $\mathbf{p} \in \mathcal{E}^{(X)}(\mathcal{A})$ $sp(\mathbf{p}, \prec_e)$ is finite. By lemma I3 follows that for all $\langle t, r \rangle \in T \times R$

$$sp(\langle t, r \rangle, \prec(\Delta)) = \{ \langle t', r' \rangle \mid \langle r', r, t - t' \rangle \in \Delta \}.$$

Since Δ has finite fanin, this set is finite. ■

Proposition 7.18: *Any systolic fixed hardware structure Δ generates a ripple-free space-time ordering.*

Proof. Consider a systolic fixed hardware structure Δ over a space R. Lemma I3 implies, that for two arbitrary space-time points $\langle r_1, t_1 \rangle$, $\langle r_2, t_2 \rangle$

$$\langle r_1, t_1 \rangle \prec(\Delta) \langle r_2, t_2 \rangle \implies \langle r_1, r_2, t_2 - t_1 \rangle \in \Delta.$$

Since Δ is systolic, it follows that $t_2 - t_1 > 0$, or $t_1 < t_2$. ■

Appendix J. Proofs of theorems of chapter 8

Theorem 8.1: *For any pure CCSA-set M derived from a recursion scheme $\mathbf{F}$ in X, for any variable x in $X_a(\mathbf{F})$ and for any causal x-enumeration I in M, MIx is a pure CCSA-set derived from a recursion scheme $\mathbf{F}'$ in $(X \setminus \{x\}) \cup X_I$. For all $x' \in X \setminus \{x\}$ it holds that $o_{\mathbf{F}'}(x') = o_{\mathbf{F}}(x')$, and $o_{\mathbf{F}'}(x_i) = o_{\mathbf{F}}(x)$ for all $i \in dom(I)$. Furthermore, $\prec_{MIx} = (\prec_M \setminus \prec_{Mx}) \cup \prec_{MIxX_I}$.*

Proof. We must prove the following:

1. That there exists a causal recursion scheme $\mathbf{F}'$ in $(X \setminus \{x\}) \cup X_I$, such that MIx is derived from it.
2. That MIx is pure.
3. That $\prec_{MIx} = (\prec_M \setminus \prec_{Mx}) \cup \prec_{MIxX_I}$.
4. That $o_{\mathbf{F}'}(x') = o_{\mathbf{F}}(x')$ for all $x' \in X \setminus \{x\}$, and that $o_{\mathbf{F}'}(x_i) = o_{\mathbf{F}}(x)$ for all $i \in dom(I)$.

We now do this. For notational convenience, we denote $\{0, \ldots, n(I) - 1\}$ by $dom^-(I)$ and $\{1, \ldots, n(I)\}$ by $dom^+(I)$. (Note that if $n(I) = \infty$, then $dom^-(I) = dom(I)$.)

1. *causal recursion scheme $\mathbf{F}'$ in $(X \setminus \{x\}) \cup X_I$, MIx derived from $\mathbf{F}'$*: From definition 8.8 follows, that for all $i \in dom^+(I)$

$$range(f_{Ix}(m_i)) = (range(m_i) \setminus \{x\}) \cup \{x_{i-1}\}$$

and for all $i \in dom^-(I)$

$$dom(f_{Ix}(m_i)) = (dom(m_i) \setminus \{x\}) \cup \{x_i\}.$$

For all other $m \in M$

$$dom(f_{Ix}(m)) = dom(m) \text{ and } range(f_{Ix}(m)) = range(m).$$

We can thus conclude that

$$MIx \subseteq \mathcal{S}^{((X \setminus \{x\}) \cup X_I)}(\mathcal{A}). \tag{1}$$

Also, since M is a pure CCSA-set and since X_I is a set of fresh variables, it holds that for all $m, m' \in MIx$

$$dom(m) \cap dom(m') = \emptyset. \tag{2}$$

Since M is a CCSA-set, all $m \in M$ are finite which implies that all $m \in MIx$ are finite. Consider the relation $\prec$ on MIx, defined by

$$m \prec m' \iff dom(m) \cap range(m') \neq \emptyset$$

for all $m, m' \in MIx$. We will now show that this relation is finite-downward. By the above follows, that for any $m \in M$ not enumerated by I holds that $x \notin dom(f_{Ix}(m))$ and $x \notin range(f_{Ix}(m))$, and for all $i \in dom^{-}(I)$ holds that $x_i \notin dom(f_{Ix}(m))$ and $x \notin range(f_{Ix}(m))$. For all $m \in M$ and $x' \in X \setminus \{x\}$ holds that $x \in dom(f_{Ix}(m)) \iff x \in dom(m)$ and $x \in range(f_{Ix}(m)) \iff x \in range(m)$. So for all $m, m' \in M$ not enumerated by I and for all $i \in dom(I)$ holds, since M is pure and thus $m \prec_M m' \iff dom(m) \cap range(m') \neq \emptyset$ for all $m, m' \in M$, that

$$\begin{aligned} f_{Ix}(m) \prec f_{Ix}(m') &\iff m \prec_M m' \iff m\,(\prec_M \setminus \prec_{Mx})\, m' \\ f_{Ix}(m_i) \prec f_{Ix}(m') &\iff m_i \prec_M m' \iff m_i\,(\prec_M \setminus \prec_{Mx})\, m' \\ f_{Ix}(m) \prec f_{Ix}(m_i) &\iff m \prec_M m_i \iff m\,(\prec_M \setminus \prec_{Mx})\, m_i. \end{aligned}$$

For all $i, j \in dom(I)$ holds that

$$f_{Ix}(m_i) \prec f_{Ix}(m_j) \iff m_i\,(\prec_M \setminus \prec_{Mx})\, m_j \vee j = i + 1.$$

We can also note that

$$m \prec_{x_i} m' \iff m = m_{i-1} \wedge m' = m_i.$$

Together this implies that $\prec = (\prec_M \setminus \prec_{Mx}) \cup \prec_{MIxX_I}$. Furthermore, $\prec_M \setminus \prec_{Mx}$ is well-founded, since it is a subset of the well-founded relation $\prec_M$. $\prec_{MIxX_I}$ is a total order on $\{m_i \mid i \in dom(I)\}$ and m_0 is the least element with respect to $\prec_{MIxX_i}$. Since I is a causal x-enumeration it also holds, for all $m, m' \in MIx$, that

$$m \prec_{MIxX_i} m' \implies m \not\prec_M m' \implies \neg(m'\,(\prec_M \setminus \prec_{Mx})\, m').$$

Thus, lemma B10 can be applied and it implies that $(\prec_M \setminus \prec_{Mx}) \cup \prec_{MIxX_I} = \prec$ is well-founded. Since every $m \in MIx$ is finite, $sp(m, \prec)$ is finite for all $m \in MIx$. Lemma B3 then gives that $\prec$ is finite-downward.

(1), (2), the finiteness of each $m \in MIx$ and the fact that $\prec$ is finite-downward now through theorem 6.5 implies, that MIx is a pure CCSA-set derived from $\mathbf{F}' = \bigcup(m \mid m \in MIx)$. Since $MIx \subseteq \mathcal{S}^{((X \setminus \{x\}) \cup X_I)}(\mathcal{A})$, it follows that $\mathbf{F}'$ is a recursion scheme in $(X \setminus \{x\}) \cup X_I$. Furthermore, theorem 6.5 yields that

$$\prec_{MIx} = \prec = (\prec_M \setminus \prec_{Mx}) \cup \prec_{MIxX_I}. \tag{3}$$

2. *MIx pure*: Follows from (2).

3. $\prec_{MIx} = (\prec_M \setminus \prec_{Mx}) \cup \prec_{MIxX_I}$: Proved under 1 above and stated in (3).
4. $o_{\mathbf{F}'}(x') = o_{\mathbf{F}}(x')$, $o_{\mathbf{F}'}(x_i) = o_{\mathbf{F}}(x)$: We will prove the stronger assertion, that $\psi_{\mathbf{F}'}(x') = \psi_{\mathbf{F}}(x')$ for all $x' \in (X \setminus \{x\}) \cup X_I$. Consider $sp(f_{Ix}(m_0), \prec_{MIx})$. For no $i \in dom(I)$ it can be the case that $f_{Ix}(m_i) \prec^+_{MIx} f_{Ix}(m_0)$. So for all $f_{Ix}(m) \in sp(f_{Ix}(m_0), \prec_{MIx})$ holds that $f_{Ix}(m) = m$. A trivial inductive proof over $sp(f_{Ix}(m_0), \prec_{MIx})$, omitted here, shows that $o_{\mathbf{F}'}(x') = o_{\mathbf{F}}(x')$ for all such $f_{Ix}(m)$ and for all x' in $dom(f_{Ix}(m))$. Especially this holds for all $x' \in range(m_0)$ and thus for all $x' \in varset(\mathbf{F}'(x_0)) = varset(m_0(x))$.
   Let us now prove by induction over $dom(I)$ that $\psi_{\mathbf{F}'}(x_i) = \psi_{\mathbf{F}}(x)$ for all $i \in dom(I)$:

$i = 0$:

$$\begin{aligned}\psi_{\mathbf{F}'}(x_0) &= \psi_{\mathbf{F}'}\mathbf{F}'(x_0)\\ &= \psi_{\mathbf{F}'}|_{varset(m_0(x))}(m_0(x))\\ &= \text{(by the above)}\\ &= \psi_{\mathbf{F}}|_{varset(m_0(x))}(m_0(x))\\ &= \psi_{\mathbf{F}}(m_0(x))\\ &= \psi_{\mathbf{F}}(x).\end{aligned}$$

$i > 0$: Assume as induction hypothesis that $\psi_{\mathbf{F}'}(x_{i-1}) = \psi_{\mathbf{F}}(x)$. Then

$$\psi_{\mathbf{F}'}(x_i) = \psi_{\mathbf{F}'}(\mathbf{F}'(x_i)) = \psi_{\mathbf{F}'}(f_{Ix}(m_i)(x_i)) = \psi_{\mathbf{F}'}(x_{i-1}) = \psi_{\mathbf{F}}(x).$$

It remains to show that $\psi_{\mathbf{F}'}(x') = \psi_{\mathbf{F}}(x')$ for all $x' \in X \setminus \{x\}$. We prove this by induction over $\mathbf{F}'$, taking the inductive sentence to be $x' \in X \setminus \{x\} \implies \psi_{\mathbf{F}'}(x') = \psi_{\mathbf{F}}(x')$.

- $\langle x', \mathbf{F}'(x')\rangle$ minimal with respect to $\prec_{\mathbf{F}'}$: Then $x' \in X_a(\mathbf{F}') = X_a(\mathbf{F})$, so trivially $\psi_{\mathbf{F}'}(x') = x' = \psi_{\mathbf{F}}(x')$.
- Inductive step: Assume as induction hypothesis that for all x'', such that $\langle x'', \mathbf{F}'(x'')\rangle \prec_{\mathbf{F}'} \langle x', \mathbf{F}'(x')\rangle$, holds that if $x'' \in X \setminus \{x\}$, then $\psi_{\mathbf{F}'}(x'') = \psi_{\mathbf{F}}(x'')$. There are three cases:

a. $x' \notin X_I$ and there is no $i \in dom^+(I)$ such that $x' \in dom(f_{Ix}(m_i))$: Then $\mathbf{F}'(x') = m(x') = \mathbf{F}(x')$ and $varset(\mathbf{F}(x')) \subseteq X \setminus \{x\}$. So in this case

$$\begin{aligned}\psi_{\mathbf{F}'}(x') &= \psi_{\mathbf{F}'}(\mathbf{F}'(x'))\\ &= \psi_{\mathbf{F}'}|_{varset(\mathbf{F}(x'))}(\mathbf{F}(x'))\\ &= \text{(by the induction hypothesis)}\end{aligned}$$

$$
\begin{aligned}
&= \psi_{\mathbf{F}}|_{varset(\mathbf{F}(x'))}(\mathbf{F}(x')) \\
&= \psi_{\mathbf{F}}(\mathbf{F}(x')) \\
&= \psi_{\mathbf{F}}(x').
\end{aligned}
$$

b. $x' \notin X_I$ and $x' \in dom(f_{Ix}(m_i))$ for some $i \in dom^+(I)$: Then

$$
\begin{aligned}
\psi_{\mathbf{F}'}(x') &= \\
\psi_{\mathbf{F}'}(\mathbf{F}'(x')) &= \\
\psi_{\mathbf{F}'}(\{x \leftarrow x_{i-1}\}(m_i(x'))) &= \\
(\psi_{\mathbf{F}'}\{x \leftarrow x_{i-1}\})(m_i(x')) &= \\
(\{x \leftarrow \psi_{\mathbf{F}'}(x_{i-1})\} \cup \psi_{\mathbf{F}'}|_{dom(\psi_{\mathbf{F}'})\setminus\{x\}})(m_i(x')) &= \\
(\{x \leftarrow \mathbf{F}'(x_{i-1})\} \cup \psi_{\mathbf{F}'}|_{((X\setminus\{x\})\cup X_I)\setminus\{x\}})(m_i(x')) &= \\
\text{(by the result about } x_i \text{ earlier in this proof)} &= \\
(\{x \leftarrow \mathbf{F}(x)\} \cup \psi_{\mathbf{F}'}|_{(X\cup X_I)\setminus\{x\}})(m_i(x')) &= \\
(\{x \leftarrow \mathbf{F}(x)\} \cup \psi_{\mathbf{F}'}|_{(X\cup X_I)\setminus\{x\}})|_{varset(m_i(x'))}(m_i(x')) &= \\
(\{x \leftarrow \mathbf{F}(x)\}|_{varset(m_i(x'))} \cup \psi_{\mathbf{F}'}|_{(X\cup X_I)\setminus\{x\}}|_{varset(m_i(x'))})(m_i(x')) &= \\
\text{(since } x \in varset(m_i(x)) \text{ and}& \\
varset(m_i(x')) \cap ((X \cup X_I) \setminus \{x\}) = varset(m_i(x')) \setminus \{x\}) &= \\
(\{x \leftarrow \mathbf{F}(x)\} \cup \psi_{\mathbf{F}'}|_{varset(m_i(x'))\setminus\{x\}})(m_i(x')) &= \\
\text{(by the induction hypothesis, since } varset(m_i(x')) \subseteq X) &= \\
(\{x \leftarrow \mathbf{F}(x)\} \cup \psi_{\mathbf{F}}|_{varset(m_i(x'))\setminus\{x\}})(m_i(x')) &= \\
\psi_{\mathbf{F}}|_{varset(m_i(x'))}(m_i(x')) &= \\
\psi_{\mathbf{F}}(m_i(x')) &= \\
\psi_{\mathbf{F}}(\mathbf{F}(x')) &= \\
\psi_{\mathbf{F}}(x').&
\end{aligned}
$$

c. $x' \in X_I$: In this case the premise in the inductive sentence is always false, so the sentence is always true. ■

Theorem 8.2: *Let M be a pure CCSA-set and let W be a correct mapping $\langle M, \prec_M\rangle \rightarrow \langle S, \prec_s\rangle$. For any $m \in M$ and for any $x \in dom(m)$, if there is an enumeration $I': \{1, \ldots, n(I')\} \rightarrow \{W(m') \mid x \in range(m')\}$, where $n(I') = \left|\{W(m') \mid x \in range(m')\}\right|$, such that $W(m'_i) \prec_s W(m'_{i+1})$ for all $i \in \{1, \ldots, n(I') - 1\}$, then I, which is I' extended with $m_0 = m$, is a causal x-enumeration in M, and $W': MIx \rightarrow S$, defined by $W'(f_{Ix}(m)) = W(m)$ for all $m \in M$, is a correct mapping $\langle MIx, \prec_{MIx}\rangle \rightarrow \langle S, \prec_s\rangle$.*

Proof. We first prove that I is a causal x-enumeration.

ce1: For all m' such that $x \in range(m')$ holds that $m_0 = m \prec_M m'$. By proposition 7.11 follows that $W(m_0) \prec_s W(m')$ and especially that $W(m_0) \prec_s W(m_1)$. We can therefore conclude, that for *all* $i \in dom^-(I)$ (using the notation in the proof of theorem 8.1 above) $W(m_i) \prec_s W(m_{i+1})$. This implies that for all $i, j \in dom(I)$

$$i < j \implies W(m_i) \prec_s^+ W(m_j). \tag{1}$$

(1) implies that I is bijective. (Otherwise there would be some $i \in dom(I)$ such that $W(m_i) \prec_s W(m_i)$, which is impossible since $\prec_s$ is finite-downward and thus irreflexive.)

ce2: Follows directly from the assumptions in the theorem.

ce3: (1) also implies

$$i < j \implies W(m_j) \not\prec_s^+ W(m_i) \tag{2}$$

since $\prec_s$ is irreflexive. Since W is a correct mapping, proposition 7.11 implies that for all $i, j \in dom(I)$

$$m_j \prec_M m_i \implies W(m_j) \prec_s W(m_i)$$

which by lemma B5 implies

$$m_j \prec_M^+ m_i \implies W(m_j) \prec_s^+ W(m_i)$$

which, in turn, implies

$$W(m_j) \not\prec_s^+ W(m_i) \implies m_j \not\prec_M^+ m_i. \tag{3}$$

(2), (3) now yields

$$i < j \implies m_j \not\prec_M^+ m_i$$

for all $i, j \in dom(I)$.

We now prove that W' is a correct mapping $\langle MIx, \prec_{MIx}\rangle \rightarrow \langle S, \prec_s\rangle$. W' is 1-1. Then, according to proposition 7.11, W' is a correct mapping

$\langle MIx, \prec_{MIx}\rangle \to \langle S, \prec_s\rangle$ iff $m \prec_{MIx} m' \implies W'(m) \prec_s W'(m')$ for all $m, m' \in MIx$. According to theorem 8.1, $\prec_{MIx} = (\prec_M \setminus \prec_{Mx}) \cup \prec_{MIxX_I}$. So for all $m, m' \in MIx$

$$\begin{aligned} m \prec_{MIx} m' &\iff m\, ((\prec_M \setminus \prec_{Mx}) \cup \prec_{MIxX_I})\, m' \\ &\iff m\, (\prec_M \setminus \prec_{Mx})\, m' \vee m \prec_{MIxX_I} m'. \end{aligned}$$

There are two cases.

1. $m\, (\prec_M \setminus \prec_{Mx})\, m'$. Since $m \prec_M m' \iff f_{Ix}^{-1}(m) \prec_M f_{Ix}^{-1}(m')$ for all $m, m' \in MIx$, it follows that

$$\begin{aligned} m\, (\prec_M \setminus \prec_{Mx})\, m' &\implies f_{Ix}^{-1}(m) \prec_M f_{Ix}^{-1}(m') \\ &\implies W(f_{Ix}^{-1}(m)) \prec_s W(f_{Ix}^{-1}(m')) \\ &\implies W'(m) \prec_s W(m'). \end{aligned}$$

2. $m \prec_{MIxX_I} m'$. Then there is an $i \in dom^-(I)$ such that $m = f_{Ix}(m_i)$ and $m' = f_{Ix}(m_{i+1})$. (For no other $m \in MIx$ holds that any $x_i \in dom(m)$ or $x_i \in range(m)$.) Since for all $i \in dom^-(I)$ holds that $W(m_i) \prec_s W(m_{i+1})$, it follows that $W'(f_{Ix}(m_i)) \prec_s W'(f_{Ix}(m_{i+1}))$, or $W'(m) \prec_s W'(m')$.

So in both cases above, $W'(m) \prec_s W'(m')$ follows from $m \prec_{MIx} m'$. Then, by proposition 7.11, W' is a correct mapping $\langle MIx, \prec_{MIx}\rangle \to \langle S, \prec_s\rangle$. ■

Proposition 8.6: $\geq_\phi$ *is transitive.*

Proof. Consider three arbitrary substitutions σ_1, σ_2 and σ_3. Assume that $\sigma_2 \geq_\phi \sigma_1$ and $\sigma_3 \geq_\phi \sigma_2$. Let us show that $\sigma_3 \geq_\phi \sigma_1$ follows. By definition 8.11 of $\geq_\phi$, the assumptions give

$$\exists \sigma \in s^{(X)}(\mathcal{A}) \forall x \in dom(\sigma_1) \exists x' \in dom(\sigma_2)[\phi(\sigma\sigma_2(x')) = \phi(\sigma_1(x))]$$

and

$$\exists \sigma' \in s^{(X)}(\mathcal{A}) \forall x' \in dom(\sigma_2) \exists x'' \in dom(\sigma_3)[\phi(\sigma'\sigma_3(x'')) = \phi(\sigma_2(x'))].$$

When combined, they yield

$$\exists\sigma,\sigma' \in s^{(X)}(\mathcal{A})[\forall x \in dom(\sigma_1)\exists x' \in dom(\sigma_2)[\sigma\sigma_2(x') =_\phi \sigma_1(x)] \wedge \forall x' \in dom(\sigma_2)\exists x'' \in dom(\sigma_3)[\sigma'\sigma_3(x'') =_\phi \sigma_2(x')]]$$

$$\Longrightarrow \text{(since } \exists x[p(x)] \wedge \forall x[q(x)] \Longrightarrow \exists x[p(x) \wedge q(x)])$$

$$\Longrightarrow \exists\sigma,\sigma' \in s^{(X)}(\mathcal{A})[\forall x \in dom(\sigma_1)\exists x' \in dom(\sigma_2)\exists x'' \in dom(\sigma_3)[\sigma\sigma_2(x') =_\phi \sigma_1(x) \wedge \sigma'\sigma_3(x'') =_\phi \sigma_2(x')]]$$

$$\Longrightarrow \text{(using 5 in theorem 2.3)}$$

$$\Longrightarrow \exists\sigma,\sigma' \in s^{(X)}(\mathcal{A})[\forall x \in dom(\sigma_1)\exists x' \in dom(\sigma_2)\exists x'' \in dom(\sigma_3)[\sigma\sigma_2(x') =_\phi \sigma_1(x) \wedge \sigma\sigma'\sigma_3(x'') =_\phi \sigma\sigma_2(x')]]$$

$$\Longrightarrow \text{(using 3 in theorem 2.3)}$$

$$\Longrightarrow \exists\sigma,\sigma' \in s^{(X)}(\mathcal{A})\forall x \in dom(\sigma_1)\exists x'' \in dom(\sigma_3)[\sigma\sigma'\sigma_3(x'') =_\phi \sigma_1(x)]$$

$$\Longrightarrow (\sigma'' = \sigma\sigma')$$

$$\Longrightarrow \exists\sigma'' \in s^{(X)}(\mathcal{A})\forall x \in dom(\sigma_1)\exists x'' \in dom(\sigma_3)[\sigma''\sigma_3(x'') =_\phi \sigma_1(x)]$$

$$\Longrightarrow \sigma_3 \geq_\phi \sigma_1.$$

■

Theorem 8.3: *For any pure cs-assignment structure $\langle M, \prec_M\rangle$ and for any substitution field Φ on S holds that if W is a correct mapping with respect to Φ, then $\prec_{MW}$ exists and $\prec_{MW} \subseteq \prec_\Phi$.*

Proof. Assume that $W: M \rightarrow S$ is a correct mapping with respect to Φ. Let us prove that $\prec_{MW}$ exists and that $s \prec_{MW} s' \Longrightarrow s \prec_\Phi s'$ for all $s, s' \in S$. W is 1-1. By corollary dummy, $\prec_M$ is irreflexive. Thus, lemma 6.5 implies that $\prec_{MW}$ exists and that W is a graph isomorphism from $\langle M, \prec_M\rangle$ to $\langle W(M), \prec_{WM}\rangle$. $\prec_M$ and thus also $\prec_{MW}$ is irreflexive. Therefore, it suffices to show $s \prec_{MW} s' \Longrightarrow s \preceq_\Phi s'$. Consider two arbitrary $s, s' \in S$ such that $s \prec_{MW} s'$. By lemma 6.5 follows that there are $m, m' \in M$ such that $W(m) = s$, $W(m') = s'$ and $m \prec_M m'$. $m \prec_M m'$ implies that there exists an x in $dom(m) \cap range(m')$. For this

x holds, according to definition 8.17, that

$$\begin{aligned}
&\exists x' \in dom(\sigma^{\Phi}_{W(m)})\exists x'' \in range(\sigma^{\Phi}_{W(m')}) \\
&\qquad [\langle t_W(m') - t_W(m), r_W(m'), x''\rangle \in ot^{\Phi}_{W(m)}(x')] \\
\Longrightarrow\ &\exists x' \in dom(\sigma^{\Phi}_{W(m)})\exists x'' \in range(\sigma^{\Phi}_{W(m')}) \\
&\qquad [\langle t_W(m') - t_W(m), r_W(m'), x''\rangle \in ot^{\Phi}_{\langle t_W(m), r_W(m)\rangle}(x')] \wedge \\
&\qquad t_W(m') = t_W(m) + (t_W(m') - t_W(m)) \\
\Longrightarrow\ &\text{(according to definition 8.16 of } \preceq_{\Phi}) \\
\Longrightarrow\ &\langle t_W(m), r_W(m)\rangle \preceq_{\Phi} \langle t_W(m'), r_W(m')\rangle \\
\Longrightarrow\ &W(m) \preceq_{\Phi} W(m') \\
\Longrightarrow\ &s \preceq_{\Phi} s'.
\end{aligned}$$

■

Theorem 8.4: *For all pure sets of cs-assignments M, for all space-times S and for all functions $W: M \to S$ that are 1-1, $\prec_{MW}$ exists and $\prec_{(MW)} = \prec_{MW}$.*

Proof. Assume that W is a function from a pure set of cs-assignments M to a space-time S, that is 1-1. W is correct with respect to (MW). By theorem 8.3, then $\prec_{MW}$ exists and $\prec_{MW} \subseteq \prec_{(MW)}$. It remains to show that $\prec_{(MW)} \subseteq \prec_{MW}$.

By lemma 6.5, $\prec_{MW}$ always exists and W is a graph isomorphism from $\langle M, \prec_M\rangle$ to $\langle W(M), \prec_{MW}|_{W(M)}\rangle$. So since M is pure, it holds that for all $s, s' \in W(M)$

$$s \prec_{MW}|_{W(M)} s' \iff dom(W^{-1}(s)) \cap range(W^{-1}(s')) \neq \emptyset. \qquad (1)$$

$s \not\prec_{MW} s'$ whenever s or s' is not in $W(M)$. Thus, we can use W^- to extend (1) to all $s, s' \in S$:

$$s \prec_{MW} s' \iff dom(W^-(s)) \cap range(W^-(s')) \neq \emptyset. \qquad (2)$$

On the other hand, for all $\langle t, r\rangle, \langle t', r'\rangle \in S$ holds that

$$
\begin{aligned}
& \langle t, r\rangle \prec_{(MW)} \langle t', r'\rangle \\
\Longleftrightarrow\ & (\exists x \in dom(W^{-}(\langle t, r\rangle))\exists x' \in X[\langle t'-t, r', x'\rangle \in ot^{(MW)}_{\langle t,r\rangle}(x)]\vee \\
& \exists x' \in range(W^{-}(\langle t', r'\rangle))\exists x \in X[\langle t'-t, r, x\rangle = it^{(MW)}_{\langle t',r'\rangle}(x')]) \\
& \wedge\langle t, r\rangle \neq \langle t', r'\rangle \\
\Longleftrightarrow\ & (\exists x \in dom(W^{-}(\langle t, r\rangle))\exists x' \in X \\
& \qquad [\langle t'-t, r', x'\rangle \in \\
& \qquad \{\langle t''-t, r'', x\rangle \mid x \in range(W^{-}(\langle t'', r''\rangle))\}]\vee \\
& \exists x' \in range(W^{-}(\langle t', r'\rangle))\exists x \in X \\
& \qquad [\exists\langle t'', r''\rangle \in S[\langle t'-t, r, x\rangle = \langle t'-t'', r'', x'\rangle\wedge \\
& \qquad\qquad x' \in dom(W^{-}(\langle t'', r''\rangle))]\vee \\
& \qquad \langle t'-t, r, x\rangle = \langle 0, r', x'\rangle]) \\
& \wedge\langle t, r\rangle \neq \langle t', r'\rangle \\
\Longleftrightarrow\ & (\exists x \in dom(W^{-}(\langle t, r\rangle))\exists x' \in X\,\exists\langle t'', r''\rangle \in S \\
& \qquad [\langle t'-t, r', x'\rangle = \langle t''-t, r'', x\rangle\wedge \\
& \qquad x \in range(W^{-}(\langle t'', r''\rangle))]\vee \\
& \exists x' \in range(W^{-}(\langle t', r'\rangle))\exists x \in X \\
& \qquad [\exists\langle t'', r''\rangle[\langle t'-t, r, x\rangle = \langle t'-t'', r'', x'\rangle\wedge \\
& \qquad\qquad x' \in dom(W^{-}(\langle t'', r''\rangle))]]) \\
& \wedge\langle t, r\rangle \neq \langle t', r'\rangle \\
\Longleftrightarrow\ & (\exists x \in dom(W^{-}(\langle t, r\rangle))[x \in range(W^{-}(\langle t', r'\rangle))]\vee \\
& \exists x' \in range(W^{-}(\langle t', r'\rangle)) \\
& \qquad [\langle t'-t, r, x\rangle = \langle t'-t'', r'', x'\rangle \wedge x' \in dom(W^{-}(\langle t, r\rangle))]) \\
& \wedge\langle t, r\rangle \neq \langle t', r'\rangle \\
\Longleftrightarrow\ & dom(W^{-}(\langle t, r\rangle)) \cap range(W^{-}(\langle t', r'\rangle)) \neq \emptyset \\
& \wedge\langle t, r\rangle \neq \langle t', r'\rangle \\
\Longrightarrow\ & dom(W^{-}(\langle t, r\rangle)) \cap range(W^{-}(\langle t', r'\rangle)) \neq \emptyset.
\end{aligned}
$$

Thus, by (2), $s \prec_{(MW)} s' \Longrightarrow s \prec_{MW} s'$ for all $s, s' \in S$. ■

Proposition 8.8: *For any substitution field Φ in R, $\hat{\Phi}$ is a substitution field in $T \times R$.*

Proof. It follows immediately that every $\sigma_s^{\hat{\Phi}} \in \mathcal{S}^{(X)}(\mathcal{A})$ and that $it_s^{\hat{\Phi}}$, $ot_s^{\hat{\Phi}}$ are functions of the proper kind, for every $s \in T \times R$. It remains to show that the connection between input and output transfer functions holds for these. For all $\langle t, r\rangle \in T \times R$, for all $\langle \delta, r', x'\rangle \in Z \times R \times X$ and for all $x \in dom(\sigma^{\hat{\Phi}}_{\langle t,r\rangle})$,

$$\begin{aligned}\langle \delta, r', x'\rangle \in ot^{\hat{\Phi}}_{\langle t,r\rangle}(x) \wedge t+\delta \in T \wedge x' \in range(\sigma^{\hat{\Phi}}_{\langle t+\delta, r'\rangle}) &\Longrightarrow \\ \langle \delta, r', x'\rangle \in ot^{\Phi}_{r}(x) \wedge t+\delta \in T \wedge x' \in range(\sigma^{\Phi}_{r'}) &\Longrightarrow \\ it^{\Phi}_{r'}(x') = \langle \delta, r, x\rangle \wedge t+\delta \in T &\Longrightarrow \\ it^{\hat{\Phi}}_{\langle t+\delta, r'\rangle}(x') = \langle \delta, r, x\rangle. &\end{aligned}$$

For all $x \in range(\sigma^{\hat{\Phi}}_{\langle t,r\rangle})$

$$\begin{aligned}it^{\hat{\Phi}}_{\langle t,r\rangle}(x) = \langle \delta, r', x'\rangle \wedge t-\delta \in T \wedge x' \in dom(\sigma^{\hat{\Phi}}_{\langle t-\delta, r'\rangle}) &\Longrightarrow \\ it^{\Phi}_{r}(x) = \langle \delta, r', x'\rangle \wedge t-\delta \in T \wedge x' \in dom(\sigma^{\Phi}_{r'}) &\Longrightarrow \\ \langle \delta, r, x\rangle \in ot^{\Phi}_{r'}(x') \wedge t-\delta \in T &\Longrightarrow \\ \langle \delta, r, x\rangle \in ot^{\hat{\Phi}}_{\langle t-\delta, r'\rangle}(x'). &\end{aligned}$$

∎

Theorem 8.5: *For any substitution field Φ in the space R, $\prec_{\hat{\Phi}} = \prec(\Delta_\Phi)$.*

Proof. Consider two arbitrary $\langle t, r\rangle$, $\langle t', r'\rangle$ in $T \times R$. By definition 8.16,

$$\begin{gathered}\langle t, r\rangle \preceq_{\hat{\Phi}} \langle t', r'\rangle \\ \Longleftrightarrow \\ (\exists x \in dom(\sigma^{\hat{\Phi}}_{\langle t,r\rangle}) \exists x' \in X[\langle t'-t, r', x'\rangle \in ot^{\hat{\Phi}}_{\langle t,r\rangle}(x)] \vee \\ \exists x' \in range(\sigma^{\hat{\Phi}}_{\langle t',r'\rangle}) \exists x \in X[\langle t'-t, r, x\rangle \in it^{\hat{\Phi}}_{\langle t',r'\rangle}(x')]) \\ \wedge \langle t, r\rangle \neq \langle t', r'\rangle.\end{gathered} \tag{1}$$

By lemma I3,

$$\begin{aligned}\langle t, r\rangle \prec(\Delta_\Phi) \langle t', r'\rangle &\Longleftrightarrow \langle r, r', t'-t\rangle \in \Delta_\Phi \\ &\Longleftrightarrow \text{(by definition 8.23)}\end{aligned}$$

$$\iff (\exists x \in dom(\sigma_r^{\Phi}) \exists x' \in X[\langle t' - t, r', x' \rangle \in ot_r^{\Phi}(x)] \vee$$
$$\exists x' \in range(\sigma_{r'}^{\Phi}) \exists x \in X[\langle t' - t, r, x \rangle \in it_{r'}^{\Phi}(x')])$$
$$\wedge \neg(t - t' = 0 \wedge r = r'). \tag{2}$$

(2) is now easily seen to be equivalent with (1), since

$$\neg(t - t' = 0 \wedge r = r') \iff \langle t, r \rangle \neq \langle t', r' \rangle$$

and for all $t \in T$ and $r \in R$ holds that $\sigma_{\langle t,r \rangle}^{\hat{\Phi}} = \sigma_r^{\Phi}$, $it_{\langle t,r \rangle}^{\hat{\Phi}} = it_r^{\Phi}$ and $ot_{\langle t,r \rangle}^{\hat{\Phi}} = ot_r^{\Phi}$. ■

Appendix K. Proofs of theorems of chapter 9

Before proving theorem 9.2 we prove the following lemma:

Lemma K1: *Let A be a set, R_A a relation on A, B a subset of an index space Z^n, R_B a relation on B and f a function $A \to B$ such that $fi(\langle A, R_A\rangle, \langle B, R_B\rangle, f)$. Then, for all $d \in Z^n$,*

$$d \in \mathcal{D}(R_B) \iff \exists a, a' \in A[f(a) \neq f(a') \wedge aR_Aa' \wedge f(a') - f(a) = d].$$

Proof. From the definitions of fi and $D(R_B)$ we obtain

$$\begin{aligned} d \in \mathcal{D}(R_B) &\iff \exists b, b' \in B[bR_Bb' \wedge b' - b = d] \\ &\iff \exists b, b' \in B[b \neq b' \wedge \\ &\qquad\qquad \exists a, a' \in A[f(a) = b \wedge f(a') = b' \wedge aR_Aa'] \wedge \\ &\qquad\qquad b' - b = d] \\ &\iff \exists b, b' \in B \exists a, a' \in A[b \neq b' \wedge f(a) = b \wedge f(a') = b' \wedge \\ &\qquad\qquad\qquad aR_Aa' \wedge b' - b = d] \\ &\iff \exists a, a' \in A[f(a) \neq f(a') \wedge aR_Aa' \wedge f(a') - f(a) = d]. \end{aligned}$$

■

Theorem 9.2: *Let $\langle M, \prec_M\rangle$ be a CCSA structure, let S be a discrete space-time and let G be a weakly correct indexing function from M. If $L: G(M) \to S$ is linear and if $\prec_{L\circ G}$ exists, then*

$$\mathcal{D}(\prec_{L\circ G}) = L(\mathcal{D}(\prec_{MG})) \setminus \{\overline{0}\} = \{\, L(d) \mid d \in \mathcal{D}(\prec_{MG})\,\} \setminus \{\overline{0}\}.$$

Proof. By lemma K1 follows that $d \in D(\prec_{MG})$ iff there are m, m' in M such that $G(m) \neq G(m')$, $m \prec_M m'$ and $G(m') - G(m) = d$. Thus,

$$\begin{aligned} &\delta \in L(D(\prec_{MG})) \setminus \{\overline{0}\} \\ \iff &\exists m, m' \in M[G(m) \neq G(m') \wedge m \prec_M m' \wedge \\ &\qquad\qquad L(G(m') - G(m)) = \delta] \wedge \delta \neq \overline{0} \\ \iff &\text{(because } L \text{ is linear)} \\ \iff &\exists m, m' \in M[G(m) \neq G(m') \wedge m \prec_M m' \wedge \\ &\qquad\qquad L(G(m')) - L(G(m)) = \delta \wedge L(G(m)) \neq L(G(m'))] \\ \iff &\text{(since } L(G(m)) \neq L(G(m')) \implies G(m) \neq G(m')) \end{aligned}$$

$$\iff \exists m, m' \in M[m \prec_M m' \wedge L \circ G(m') - L \circ G(m) = \delta \wedge$$
$$L \circ G(m) \neq L \circ G(m')]$$
$$\iff \text{ (by lemma K1)}$$
$$\iff \delta \in D(\prec_{L \circ G}).$$

■

Appendix L. Proofs of theorems of chapter 10

Proposition 10.2: $r^{A'}_{A''} \circ \phi_{A'} = \phi_{A''}$.

Proof. We will prove that $r^{A'}_{A''}(\phi_{A'}(\mathbf{p})) = \phi_{A''}(\mathbf{p})$ for all $\mathbf{p} \in \mathcal{E}^{(X)}(\mathcal{A})$. The proof is by induction on $\mathbf{p}$, taking the induction hypothesis to be exactly the above:

1a. $\mathbf{p} \in X$: Then, $r^{A'}_{A''}(\phi_{A'}(\mathbf{p})) = r^{A'}_{A''}(e^X_{\mathbf{p}}|_{A'}) = e^X_{\mathbf{p}}|_{A'}|_{A''} = e^X_{\mathbf{p}}|_{A''} = \phi_{A''}(\mathbf{p})$.

1b. $\mathbf{p} = f \bullet \langle\rangle$: In this case, $r^{A'}_{A''}(\phi_{A'}(\mathbf{p})) = r^{A'}_{A''}(\phi_{A'}(f \bullet \langle\rangle)) = r^{A'}_{A''}(f \circ \langle\rangle) = f \circ \langle\rangle = \phi_{A''}(f \bullet \langle\rangle) = \phi_{A''}(\mathbf{p})$.

2. $\mathbf{p} = f \bullet \langle\, \mathbf{p}_i \mid i \in I_f \,\rangle$: Assume as induction hypothesis, that

$$r^{A'}_{A''}(\phi_{A'}(\mathbf{p}_i)) = \phi_{A''}(\mathbf{p}_i)$$

for all $i \in I_f$. Now, from the definition of $\phi_{A'}$,

$$r^{A'}_{A''}(\phi_{A'}(f \bullet \langle\, \mathbf{p}_i \mid i \in I_f \,\rangle = r^{A'}_{A''}(f \circ \langle\, \phi_{A'}(\mathbf{p}_i) \mid i \in I_f \,\rangle).$$

f is an operator in $\mathcal{A}$ and therefore independent of the domain for the polynomials, thus the above is equal to

$$f \circ \langle\, r^{A'}_{A''}(\phi_{A'}(\mathbf{p}_i)) \mid i \in I_f \,\rangle.$$

By the induction hypothesis this is equal to

$$f \circ \langle\, \phi_{A''}(\mathbf{p}_i) \mid i \in I_f \,\rangle$$

and according to the definition of $\phi_{A''}$ this equals

$$\phi_{A''}(f \bullet \langle\, \mathbf{p}_i \mid i \in I_f \,\rangle).$$

■

Vol. 324: M.P. Chytil, L. Janiga, V. Koubek (Eds.), Mathematical Foundations of Computer Science 1988. Proceedings. IX, 562 pages. 1988.

Vol. 325: G. Brassard, Modern Cryptology. VI, 107 pages. 1988.

Vol. 326: M. Gyssens, J. Paredaens, D. Van Gucht (Eds.), ICDT '88. 2nd International Conference on Database Theory. Proceedings, 1988. VI, 409 pages. 1988.

Vol. 327: G.A. Ford (Ed.), Software Engineering Education. Proceedings, 1988. V, 207 pages. 1988.

Vol. 328: R. Bloomfield, L. Marshall, R. Jones (Eds.), VDM '88. VDM – The Way Ahead. Proceedings, 1988. IX, 499 pages. 1988.

Vol. 329: E. Börger, H. Kleine Büning, M.M. Richter (Eds.), CSL '87. 1st Workshop on Computer Science Logic. Proceedings, 1987. VI, 346 pages. 1988.

Vol. 330: C.G. Günther (Ed.), Advances in Cryptology – EUROCRYPT '88. Proceedings, 1988. XI, 473 pages. 1988.

Vol. 331: M. Joseph (Ed.), Formal Techniques in Real-Time and Fault-Tolerant Systems. Proceedings, 1988. VI, 229 pages. 1988.

Vol. 332: D. Sannella, A. Tarlecki (Eds.), Recent Trends in Data Type Specification. V, 259 pages. 1988.

Vol. 333: H. Noltemeier (Ed.), Computational Geometry and its Applications. Proceedings, 1988. VI, 252 pages. 1988.

Vol. 334: K.R. Dittrich (Ed.), Advances in Object-Oriented Database Systems. Proceedings, 1988. VII, 373 pages. 1988.

Vol. 335: F.A. Vogt (Ed.), CONCURRENCY 88. Proceedings, 1988. VI, 401 pages. 1988.

Vol. 336: B.R. Donald, Error Detection and Recovery in Robotics. XXIV, 314 pages. 1989.

Vol. 337: O. Günther, Efficient Structures for Geometric Data Management. XI, 135 pages. 1988.

Vol. 338: K.V. Nori, S. Kumar (Eds.), Foundations of Software Technology and Theoretical Computer Science. Proceedings, 1988. IX, 520 pages. 1988.

Vol. 339: M. Rafanelli, J.C. Klensin, P. Svensson (Eds.), Statistical and Scientific Database Management. Proceedings, 1988. IX, 454 pages. 1989.

Vol. 340: G. Rozenberg (Ed.), Advances in Petri Nets 1988. VI, 439 pages. 1988.

Vol. 341: S. Bittanti (Ed.), Software Reliability Modelling and Identification. VII, 209 pages. 1988.

Vol. 342: G. Wolf, T. Legendi, U. Schendel (Eds.), Parcella '88. Proceedings, 1988. 380 pages. 1989.

Vol. 343: J. Grabowski, P. Lescanne, W. Wechler (Eds.), Algebraic and Logic Programming. Proceedings, 1988. 278 pages. 1988.

Vol. 344: J. van Leeuwen, Graph-Theoretic Concepts in Computer Science. Proceedings, 1988. VII, 459 pages. 1989.

Vol. 345: R.T. Nossum (Ed.), Advanced Topics in Artificial Intelligence. VII, 233 pages. 1988 (Subseries LNAI).

Vol. 346: M. Reinfrank, J. de Kleer, M.L. Ginsberg, E. Sandewall (Eds.), Non-Monotonic Reasoning. Proceedings, 1988. XIV, 237 pages. 1989 (Subseries LNAI).

Vol. 347: K. Morik (Ed.), Knowledge Representation and Organization in Machine Learning. XV, 319 pages. 1989 (Subseries LNAI).

Vol. 348: P. Deransart, B. Lorho, J. Maluszyński (Eds.), Programming Languages Implementation and Logic Programming. Proceedings, 1988. VI, 299 pages. 1989.

Vol. 349: B. Monien, R. Cori (Eds.), STACS 89. Proceedings, 1989. VIII, 544 pages. 1989.

Vol. 350: A. Törn, A. Žilinskas, Global Optimization. X, 255 pages. 1989.

Vol. 351: J. Díaz, F. Orejas (Eds.), TAPSOFT '89. Volume 1. Proceedings, 1989. X, 383 pages. 1989.

Vol. 352: J. Díaz, F. Orejas (Eds.), TAPSOFT '89. Volume 2. Proceedings, 1989. X, 389 pages. 1989.

Vol. 354: J.W. de Bakker, W.-P. de Roever, G. Rozenberg (Eds.), Linear Time, Branching Time and Partial Order in Logics and Models for Concurrency. VIII, 713 pages. 1989.

Vol. 355: N. Dershowitz (Ed.), Rewriting Techniques and Applications. Proceedings, 1989. VII, 579 pages. 1989.

Vol. 356: L. Huguet (Ed.), Applied Algebra, Algebraic Algorithms and Error-Correcting Codes. Proceedings, 1987. VI, 417 pages. 1989.

Vol. 357: T. Mora (Ed.), Applied Algebra, Algebraic Algorithms and Error-Correcting Codes. Proceedings, 1988. IX, 481 pages. 1989.

Vol. 358: P. Gianni (Ed.), Symbolic and Algebraic Computation. Proceedings, 1988. XI, 545 pages. 1989.

Vol. 359: D. Gawlick, M. Haynie, A. Reuter (Eds.), High Performance Transaction Systems. Proceedings, 1987. XII, 329 pages. 1989.

Vol. 360: H. Maurer (Ed.), Computer Assisted Learning – ICCAL '89. Proceedings, 1989. VII, 642 pages. 1989.

Vol. 361: S. Abiteboul, P.C. Fischer, H.-J. Schek (Eds.), Nested Relations and Complex Objects in Databases. VI, 323 pages. 1989.

Vol. 362: B. Lisper, Synthesizing Synchronous Systems by Static Scheduling in Space-Time. VI, 263 pages. 1989.